RURAL
LIVELIHOODS
CRISES AND RESPONSES

This book is the third in a series published by
Oxford University Press
in association with The Open University

Poverty and Development in the 1990s
edited by Tim Allen and Alan Thomas

Industrialization and Development
edited by Tom Hewitt, Hazel Johnson and David Wield

Rural Livelihoods: crises and responses
edited by Henry Bernstein, Ben Crow and Hazel Johnson

Development Policy and Public Action
edited by Marc Wuyts, Maureen Mackintosh and Tom Hewitt

The final form of the text is the joint responsibility of chapter authors,
book editors and course team commentators

The books are one component of the Open University course
U208 *Third World Development*. If you would like a copy of *Studying
with The Open University*, please write to the Central Enquiry Service,
PO Box 71, The Open University, Walton Hall, Milton Keynes,
MK7 6AG. If you have not already enrolled on the course and would
like to buy this or any other Open University material, please write to
Open University Educational Enterprises Limited, 12 Cofferidge Close,
Stony Stratford, Milton Keynes MK11 1BY

RURAL LIVELIHOODS
CRISES AND RESPONSES

Edited by
Henry Bernstein   Ben Crow
Hazel Johnson
for an Open University Course Team

OXFORD UNIVERSITY PRESS

in association with

1992

This book has been printed on paper produced from pulps
bleached without use of chlorine gases and produced in Sweden from
wood from continuously farmed forests. The paper mill concerned,
Papyrus Nymölla AB, is producing bleached pulp in which dioxin
contaminants do not occur.

Published in the United Kingdom by Oxford University Press, Oxford
in association with
The Open University, Milton Keynes

Oxford University Press, Walton Street, Oxford OX2 6DP

Oxford New York Toronto
Delhi Bombay Calcutta Madras Karachi
Petaling Jaya Singapore Hong Kong Tokyo
Nairobi Dar es Salaam Cape Town
Melbourne Auckland

and associated companies in
Berlin Ibadan

Oxford is a trade mark of Oxford University Press

The Open University, Walton Hall, Milton Keynes MK7 6AA

First published in the United Kingdom 1992

Copyright © 1992 The Open University

British Library Cataloguing in Publication Data

Data available

ISBN 0–19–877334–X
ISBN 0–19–877335–8 (Pbk)

Library of Congress Cataloging in Publication Data

Data available

ISBN 0–19–877334–X
ISBN 0–19–877335–8 (Pbk)

Edited, designed and typeset by The Open University

Printed and bound by Butler & Tanner Ltd, Frome and London

In memory of Krishna Bharadwaj
(1935–1992)
with deep appreciation

CONTENTS

THE OPEN UNIVERSITY
U208 *THIRD WORLD DEVELOPMENT*
COURSE TEAM AND AUTHORS

Tom Hewitt, Lecturer in Development Studies, The Open University (Course Team Chair 1991–92)

Ben Crow, Lecturer in Development Studies, The Open University (Course Team Chair 1989–90)

Tim Allen, Lecturer in Development Studies, The Open University

Paul Auerbach, Reader in Economics, Kingston Polytechnic (Part Assessor)

Carolyn Baxter, Course Manager, The Open University

Henry Bernstein, Senior Lecturer in Agricultural and Rural Development, Institute of Development Policy and Management, University of Manchester

Krishna Bharadwaj, Professor, Centre for Economic Studies and Development, Jawaharlal Nehru University, New Delhi, India

Suzanne Brown, Course Manager, The Open University

Janet Bujra, Lecturer in Sociology, Department of Social and Economic Studies, University of Bradford

Angus Calder, Staff Tutor in Arts, The Open University, Edinburgh

Kate Crehan, Associate Professor, New School for Social Research, New York, USA

Sue Dobson, Graphic Artist, The Open University

Harry Dodd, Print Production Controller, The Open University

Kath Doggett, Project Control, The Open University

Joshua Doriye, Professor, Institute of Finance and Management, Dar es Salaam, Tanzania

Chris Edwards, Senior Lecturer in Economics, School of Development Studies, University of East Anglia

Diane Elson, Lecturer in Development Economics, University of Manchester (Part Assessor)

Sheila Farrant, Tutor Counsellor and Assistant Staff Tutor, The Open University, Cambridge (Course Reader)

Gerry Farrell, Freelance Musician and Writer

Jo Field, Project Control, The Open University

Jayati Ghosh, Associate Professor, Centre for Economic Studies and Development, Jawaharlal Nehru University, New Delhi, India

Heather Gibson, Lecturer in Economics, University of Kent at Canterbury

Angela Grunsell, Oxfam Primary Education, London

Liz Gunner, Lecturer in Literature, Languages and Cultures of Africa, School of Oriental and African Studies, London

Garry Hammond, Senior Editor, The Open University

Barbara Harriss, Lecturer in Agricultural Economics and Governing Body Fellow, Wolfson College, University of Oxford

John Harriss, Director, Centre for Development Studies, London School of Economics (Part Assessor)

Pamela Higgins, Graphic Designer, The Open University

Lakshmi Holström, Freelance Writer

Caryl Hunter-Brown, Liaison Librarian, The Open University

Gillian Iossif, Lecturer in Statistics, The Open University

Rhys Jenkins, Reader in Economics, School of Development Studies, University of East Anglia

Hazel Johnson, Lecturer in Development Studies, The Open University

Sabrina Kassam, Research Assistant, The Open University

Andrew Kilminster, Lecturer in Economics, School of Business Studies, Oxford Polytechnic

Patti Langton, Producer, BBC

Christina Lay, Course Manager, The Open University

Anthony McGrew, Lecturer in Government, Social Science Faculty, The Open University

Maureen Mackintosh, Reader in Economics, Kingston Polytechnic

Mahmood Mamdani, Professor, Centre for Basic Research, Kampala, Uganda

Charlotte Martin, Teacher and Open University Tutor (Course Reader)

Mahmood Messkoub, Lecturer in Economics, University of Leeds

Richard Middleton, Staff Tutor in Arts, The Open University, Newcastle-on-Tyne

Alistair Morgan, Lecturer in Institute of Educational Technology, The Open University

Eleanor Morris, Producer, BBC

Ray Munns, Cartographer, The Open University

Kathy Newman, Secretary, The Open University

Debbie Payne, Secretary, The Open University

Ruth Pearson, Lecturer in Economics, School of Development Studies, University of East Anglia

Richard Pinder, Training consultant, Sheffield (Course Reader)

David Potter, Professor of Government, The Open University

Janice Robertson, Editor, The Open University

Carol Russell, Editor, The Open University

Vivian von Schelling, Lecturer in Development Studies, Polytechnic of East London

Gita Sen, Fellow (Professor), Centre for Development Studies, Kerala, India

Meg Sheffield, Senior Producer, BBC

Paul Smith, Lecturer in Environmental Studies, The Open University

Ines Smyth, Senior Lecturer, Institute of Social Studies, The Hague, Netherlands; Research Associate, Department of Applied Social Studies and Social Research, University of Oxford

Hilary Standing, Lecturer in Social Anthropology, School of African and Asian Studies, University of Sussex (Part Assessor)

John Taylor, Head of Centre for Chinese Studies, South Bank Polytechnic (Course Reader)

Alan Thomas, Senior Lecturer in Systems, The Open University

Steven Treagust, Research Assistant, The Open University

Euclid Tsakalotos, Lecturer in Economics, University of Kent at Canterbury

Gordon White, Professorial Fellow, Institute of Development Studies, University of Sussex

David Wield, Senior Lecturer in Technology Strategy and Development, The Open University

Gordon Wilson, Staff Tutor in Technology, The Open University, Leeds

Philip Woodhouse, Lecturer in Agricultural and Rural Development, Institute of Development Policy and Management, University of Manchester

Peter Worsley, Emeritus Professor, University of Manchester (External Assessor)

Marc Wuyts, Professor of Applied Quantitative Economics, Institute of Social Studies, The Hague, Netherlands

The Course Team would like to acknowledge the financial support of Oxfam and the European Community in the preparation of U208 *Third World Development*.

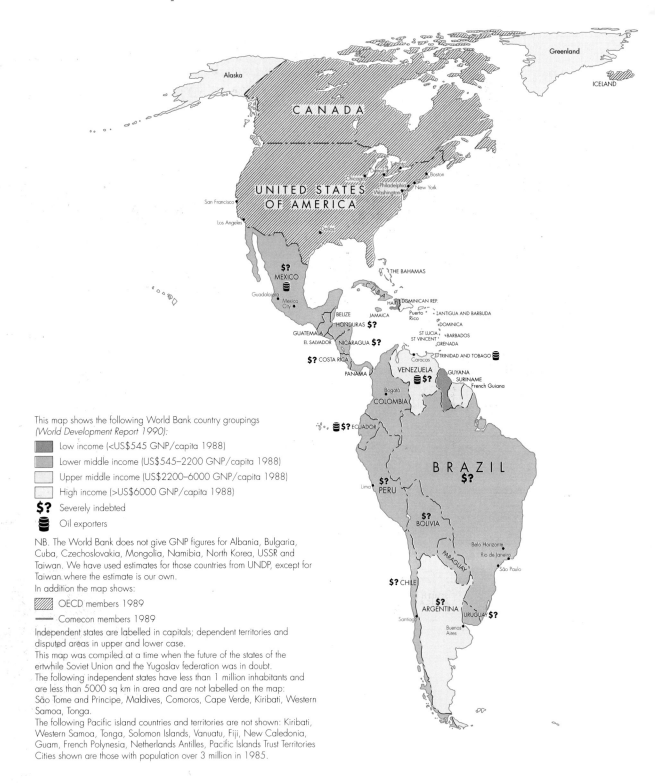

Alaska

Greenland

ICELAND

CANADA

Chicago
Detroit
Boston
Philadelphia
New York
Washington

UNITED STATES
OF AMERICA

San Francisco

Los Angeles

Dallas

$?
MEXICO

Guadalajara
Mexico City

THE BAHAMAS

CUBA

HAITI
DOMINICAN REP.
Puerto Rico
JAMAICA

ANTIGUA AND BARBUDA
DOMINICA
ST LUCIA
ST VINCENT
BARBADOS
GRENADA

BELIZE
HONDURAS **$?**
GUATEMALA
EL SALVADOR NICARAGUA **$?**
$? COSTA RICA
PANAMA

TRINIDAD AND TOBAGO

Caracas

VENEZUELA **$?**
GUYANA
SURINAME
French Guiana

Bogotá
COLOMBIA

$? ECUADOR

Lima
PERU **$?**

BRAZIL
$?

$?
BOLIVIA

Belo Horizonte
Rio de Janeiro
São Paulo

PARAGUAY

$? CHILE

$?
ARGENTINA URUGUAY **$?**

Santiago
Buenos Aires

This map shows the following World Bank country groupings
(World Development Report 1990):

Low income (<US$545 GNP/capita 1988)

Lower middle income (US$545–2200 GNP/capita 1988)

Upper middle income (US$2200–6000 GNP/capita 1988)

High income (>US$6000 GNP/capita 1988)

$? Severely indebted

Oil exporters

NB. The World Bank does not give GNP figures for Albania, Bulgaria,
Cuba, Czechoslovakia, Mongolia, Namibia, North Korea, USSR and
Taiwan. We have used estimates for those countries from UNDP, except for
Taiwan. where the estimate is our own.
In addition the map shows:

OECD members 1989

Comecon members 1989

Independent states are labelled in capitals; dependent territories and
disputed areas in upper and lower case.
This map was compiled at a time when the future of the states of the
ertwhile Soviet Union and the Yugoslav federation was in doubt.
The following independent states have less than 1 million inhabitants and
are less than 5000 sq km in area and are not labelled on the map:
São Tome and Principe, Maldives, Comoros, Cape Verde, Kiribati, Western
Samoa, Tonga.
The following Pacific island countries and territories are not shown: Kiribati,
Western Samoa, Tonga, Solomon Islands, Vanuatu, Fiji, New Caledonia,
Guam, French Polynesia, Netherlands Antilles, Pacific Islands Trust Territories
Cities shown are those with population over 3 million in 1985.

INTRODUCTION

HENRY BERNSTEIN

"People in the countryside of the Third World gain their livelihoods in a variety of ways from different types of farming and a wide range of other activities. They do so with varying degrees of success according to their access to resources and employment and how they deal with pressures arising from social, economic and environmental change. Life for many entails a daily struggle in which much energy and ingenuity is needed to secure livelihoods in the face of various crises."

These observations, from the beginning of Chapter 1, indicate the themes and concerns of this book, and immediately suggest a number of questions. For example:

- What distinguishes different types of farming? How have they evolved? How do they change?

- What sort of other activities do rural people engage in to secure livelihoods? What are the relationships between farming and other activities?

- What affects access to resources and employment for rural people?

- What types of change affect the conditions of their livelihoods? How do they deal with them?

- How do they respond to various crises? How are they helped or hindered by the actions and responses of others, including governments and non-government organizations?

These question are pursued in various ways, from a number of viewpoints, and in a number of contexts, by the authors of this book. What we have in common, however, is a commitment to asking '*which* rural people?' '*whose* livelihoods?' That is, we believe that understanding the dramas of rural people's lives in processes of development and underdevelopment – the dangers and opportunities they confront, how they cope with poverty and insecurity, their aspirations to a better life, their diverse activities and ideas, responses and initiatives – involves recognizing a rich and variegated cast of rural (and *non*-rural) actors.

How does this position relate to widely held assumptions and images that all rural people share some important, or even definitive, condition: their 'ruralness', so to speak?

What is 'rural'?

In population censuses and similar statistical exercises, 'rural' and 'urban' are defined by *residence* in settlements respectively below or above a certain size. The cut-off point varies between countries: in one it may be settlements of 5000 inhabitants, in another settlement of 10 000. This is one factor that makes precise comparisons between countries impossible. Nonetheless, the first two columns in Table 1 show trends and patterns that are acceptable as broadly accurate. One trend is for the *proportion* of rural population in total population to decline over time in virtually all countries (while the size of the rural population may continue to grow, but more slowly than that of the urban population).

Other difficulties in defining 'rural' with any great precision are more interesting for the concerns of this book, because they involve moving beyond *where* people live to ask: *how* do they live? *what* do they do? *how* do they fare?

Seeking answers to those questions immediately

Table 1   Proportions of rural population, agriculture, and agricultural employment, selected countries, 1960s–1980s

	Rural population as % of total population		% of labour force in agriculture		Agriculture as % of GNP	
	1965	*1989*	*1965*	*1989*	*1965*	*1985–88*
Low income						
Mozambique	95	74	–	64	87	85
Bangladesh	94	84	53	44	84	57
Nigeria	83	65	54	31	72	45
Uganda	93	90	52	67	91	86
India	81	73	44	30	73	63
Lower middle income						
Senegal	67	62	25	22	83	81
Peru	48	30	18	8	50	35
Mexico	45	28	14	9	50	23
Malaysia	74	58	28	–	59	42
Upper middle income						
Brazil	50	26	19	9	49	25
Rep. of Korea	68	29	38	10	55	19
Higher income			(1960)			
UK	13	11	3	2	42	
USA	28	25	3	2	73	
Japan	33	23	10	3	33	8

Column 6 shows rounded figures.

Dashes indicate no data.

Data sources:

(a)   Rural population and agricultural share of GNP: World Bank (1991) *World Development Report 1991*, Oxford University Press, Oxford and New York, table 31, pp.264–5 and table 3, pp.208–9.

(b)   Share of agriculture in employment : UNDP (1991) *Human Development Report 1991*, Oxford University Press, New York, table 16, pp.150–1, and table 35, p.183; data for UK, USA and Japan for 1960 from Crow, B. & Thomas, A. (1983) *Third World Atlas*, Open University Press, Milton Keynes, table 3, p.70.

starts to blur any clear-cut (because arbitrary) boundaries of 'rural' and 'urban' presented by census or other statistical categories. This applies even to the criterion of residence when we inquire whether all 'rural' people reside in the countryside *at all times*. As many examples in this book illustrate, great numbers of rural people (and particular kinds of people in different contexts) move between countryside and town (and between rural areas) more or less frequently, for longer or shorter periods, in regular or sporadic patterns of migration, and for various reasons: to seek work, to market their produce, to gain education, to visit relatives, to buy goods, to petition government officials, politicians or lawyers.

More generally, how rural people live, what they do, and how they fare, always involves interaction between countryside and town, whether through patterns of movement like those indicated, or for other reasons analysed in this book. If this is not recognized, and attention is focused solely on the village or locality, then key rural–urban linkages and processes will be missed, and our understanding of rural lives and livelihoods diminished or distorted accordingly. This connects with another pervasive issue: what is the relationship between rural livelihoods and agriculture?

Rural livelihoods and agriculture

This book has a lot to say about different types of agriculture (Chapters 2–5, 8–9) and how they bear on rural livelihoods. Farming is key to rural economy virtually everywhere, as a direct source of employment and income and as an indirect source through a range of activities 'upstream' of farming (supply of inputs, like tools, seeds and fertilizers, and of services, like credit, technical assistance and construction and maintenance of infrastructure such as irrigation works and roads) and 'downstream' of farming (processing, transport and marketing of crops and livestock products). In addition, incomes from farming and its linked activities in the countryside generate demand for a range of consumer goods and services.

Another equally widespread characteristic of farming, not least in the rural economies of the Third World, is its *seasonality*. This has important effects for the distribution over the year of agricultural employment and income, of other types of employment and aspects of well-being (e.g. nutrition), with pressures on the livelihoods of more vulnerable rural groups particularly marked during certain periods of the agricultural calendar.

There are often hidden assumptions that rural residence necessarily entails farming as the means of livelihood and, indeed, the basis of a distinct way of life ('ruralness'). Powerful images of the stability of rural society, and of the immobility of its inhabitants, are conveyed in notions like rural people being 'tied to the land'. However, while farming is key, it is not the whole of rural economy. First, many rural people are unable to gain adequate and secure livelihoods from farming 'on their own account' because they are landless, or because they are 'marginal' farmers. They many be marginal for a number of reasons: their landholding may be inadequate in size or quality; they may farm in difficult and unpredictable environments; or they may be subject to pressures from big farmers and merchants, from adverse market conditions and government policies.

Second, those rural people for whom poverty is defined by their landlessness or marginalization as farmers have to pursue livelihoods through a range of means, illustrated in Table 2. Note that wage employment by other (richer) farmers may be local or involve seasonal migration to other agricultural areas; likewise, non-agricultural wage employment and self-employment may be local, in other rural areas, or in nearby or distant towns.

To summarize some key propositions:

1 Different types of farming can be distinguished by their environmental and technical conditions, the types and levels of resources they require, their market linkages, and other factors, all of which affect rural livelihood opportunities *and* how they are distributed between rural people (i.e. *whose* livelihoods).

2 Agriculture and its linked activities are key to rural economy but not identical with it, nor do they necessarily generate sufficient employment and other livelihood opportunities in the countryside.

Table 2   Means of rural livelihoods other than farming own land

	Wage employment by:	Self-employment in:
Agriculture	(Richer) farmers	Share-cropping or other tenant farming
Agriculturally linked	Input suppliers, contractors, crop merchants, transporters	Artisanal production (e.g. tools, equipment), small-scale processing
Non-agricultural	Industry, trade and other services	Handicraft production, petty trade and other services

3   The fortunes of agriculture and rural economy more generally are strongly influenced, whether positively or negatively, by their links with other sectors and by government policies.

All these ideas are explored, elaborated and illustrated in some depth and detail in the chapters of this book. To get a preliminary idea of different patterns that might exist, and their complexities, it is worth looking again at Table 1 and its last two columns on the percentage of the labour force employed in agriculture. As with defining 'rural' and 'urban', considerable caution is required here too. First, national statistics might use different definitions of sectoral employment, different techniques to measure it, and apply them with widely varying degrees of accuracy. Second, we need to know how those data take account, if at all, of seasonal and other fluctuations in agricultural employment. Third, there is a well-known tendency to underestimate self-employed or 'household' labour in farming, above all undervaluing the participation of women. Fourth, this last factor can lead to serious underestimation of total agricultural labour in countries or regions where 'household' farming is combined with marked patterns of male migration, resulting in many female-headed rural households and high rates of women's participation in agricultural work.

Even with these considerable qualifications, the last two columns of Table 1 show a declining proportion of the total labour force employed in agriculture in all the countries listed. The picture is not so straightforward, however, when the sixth column is compared with the second column (percentage of rural population in 1989). For most countries the share of the labour force in agriculture is less than the share of rural population in total population, and in some instances strikingly so (Bangladesh, Nigeria). On the other hand, for Mozambique, Senegal and Peru – three very different countries – the share of the labour force in agriculture is *greater* than the share of rural population. This may reflect idiosyncrasies of definition and data collection, or substantial settlements where many farmers and agricultural workers live that are classified as 'urban'. The discrepancy in the case of Mozambique certainly reflects massive dislocation and population movement (to larger settlements, for security) to escape terrorism perpetrated by anti-government insurgents in the countryside.

Interestingly, the largest gaps between the figures in the sixth and second columns of Table 1 are for the high-income countries of the UK, USA, and Japan, where agricultural employment accounts for a very small proportion of the total labour force, especially in the UK and USA. Here, a substantial part of the gap is accounted for by urban commuters and retired people who live in the countryside. By contrast, the corresponding gap in the case of Bangladesh reflects a large proportion of rural people who are both landless and unable to find agricultural wage employment, a situation common in much of rural South Asia and Latin America.

So far, we have seen that defining the 'rural' is not straightforward, and that there is no single or simple identity between the 'rural' and agriculture as an economic sector. Underlying these observations are the points that many rural people: (1) are unable to secure adequate livelihoods from farming, whether on their own account or by working for others, and (2) have to engage in a variety of livelihood pursuits, often requiring considerable mobility over shorter or longer periods.

It has also been suggested that rural people have different opportunities to pursue various livelihoods: in access to resources for farming on their own account or in other types of self-employment; and in the availability and conditions of wage employment in various labour markets – agricultural and non-agricultural, local and distant, rural and urban.

Rural livelihoods and development

Most countries exhibit a rural–urban 'gap' on various indicators like average per capita income, levels of poverty, and access to the means of satisfying basic needs in nutrition, health and education. Several ways of measuring rural–urban gaps, and the results obtained for selected countries, are shown in Table 3. The figures confirm the prevalence of rural–urban disparities, albeit to very different degrees for different indicators in different countries. For the four countries for which there are data on total and rural poverty (columns 5 and 6), their incidence is the same in Bangladesh and not very different in India. In Malaysia the rural poverty level is much higher than the national poverty level, while in the Republic of Korea the incidence of rural poverty is lower than that of total poverty. On the other hand, the fact that child nutrition is lower in rural than in urban areas in all six countries shown (column 4) is a striking illustration that by no means all people who live from farming are able to satisfy their food needs from its products.

Why is the incidence of rural poverty relative to total poverty less dramatic in the examples shown than we might expect? One reason, applicable to Bangladesh and India, is simply a statistical effect: in countries with predominantly rural populations and widespread rural poverty, a very high proportion of total or national poverty necessarily consists of rural poverty. Two other reasons also illustrate the concerns of this book. One is that many urban people also live in conditions of desperate poverty and insecurity – indeed, 'distress' migration from the countryside may simply have the effect of 'transferring' rural poverty to the towns. The other is that income distribution, reflecting access to resources, employment, and other means of livelihoods, is likely to be very unequal within both rural and urban populations. For example, if we could disaggregate the data on child nutrition in Table 3 by income groups, we would probably find a convergence in the scores for lower income groups in both countryside and town.

How do these observations about rural (and urban) poverty relate to processes and issues of development? First, in many instances pressures on some rural people's livelihoods are intensified by particular patterns of agricultural development (economic and technical change that increases productivity) rather than by 'lack of development'. Such patterns of development may be manifested in: processes of commoditization (commercialization) of land and labour; the formation of markets with social and institutional biases and inequalities; technical change (and its environmental effects); and government policies and practices that, intentionally or unintentionally, favour some rural groups against others. In short, agricultural growth itself often intensifies existing inequalities, or creates new inequalities in the countryside that bear on the livelihood opportunities of different types of people and social groups: large farmers and the landless, merchants and small farmers, women and men. These processes of social differentiation are explored and explained in various contexts in nearly all the chapters that follow.

Of course, it may be argued that such patterns of agrarian change are part of the problem rather than the solution, and should not be considered 'development' if they increase poverty and insecurity along with the growth of production and productivity. Indeed, this resonates a debate of great topical importance concerning the

Table 3   Rural–urban gaps, selected countries, 1980s

	Rural–urban disparity[a] in access[b]				Population below poverty line	
	Health	Water	Sanitation	Child nutrition	Total (%)	Rural (%)
Low income						
Mozambique	30	39	18	–	–	–
Bangladesh	–	–	13	77	86	86
Nigeria	40	20	–	42	–	–
Uganda	63	49	94	87	–	–
India	–	66	–	–	48	51
Lower middle income						
Senegal	–	48	–	88	–	–
Peru	–	28	18	83	–	–
Mexico	–	62	12	–	–	–
Malaysia	–	14	38	–	27	38
Upper middle income						
Brazil	–	86	–	94	–	–
Rep. of Korea	89	53	100	–	16	11

[a] Figures in the four columns for rural–urban disparity are expressed in terms of the urban average, which is indexed to equal 100. Thus, the smaller the figure shown, the bigger the rural–urban disparity; the closer to 100, the smaller the disparity. Rural–urban parity would equal 100.

[b] The following definitions of access are from UNDP (1991), pp.193–6: *Health:* ability to 'reach appropriate health services on foot or by the local means of transport in not more than one hour'. *Water:* 'reasonable access to safe (i.e. uncontaminated) water supply'. *Sanitation:* 'access to sanitary means of excreta and waste disposal'. *Child nutrition:* measured by 'the median weight-for-age of the reference population'. In addition, the poverty line is defined as 'the income level below which a minimum nutritionally adequate diet plus essential non-food requirements are not affordable' – see also Chapter 1 below.

Dashes indicate no data.

Source: UNDP (1991) *Human Development Report 1991,* Oxford University Press, Oxford and New York, table 9, pp.136–7, for rural–urban disparity, 1985–88; and table 17 pp.152–3, for poverty line, 1980–88.

'appropriateness' or otherwise of conventional paths of development, and the need to find 'alternative' development strategies that do not sacrifice employment, secure livelihoods or environmental sustainability to imperatives of rapid economic growth. These issues are touched on explicitly at several points in this book; otherwise much of what is presented on the mechanisms and effects of agrarian change provides material for considering questions about 'appropriateness' and 'sustainability' that are as complicated as they are controversial.

This can be illustrated by looking at columns 3 and 4 in Table 1, which show that the contribution of agriculture to national income or GNP (Gross National Product) declined between 1965 and 1989 for all the countries shown, with the exception of Uganda (this is similar to the trend and pattern of agricultural employment). Looking at columns 3 and 4 across the income categories also shows that the contribution of agriculture to GNP declines as national income rises. Underlying these figures are several further considerations. First, there is a strong association between high-productivity agriculture and advanced industrialization (the prospects for farmers in the UK, USA, Japan and other high-income countries is currently a sensitive political issue). Second, industrialization and urbanization do not necessarily generate sufficient employment and secure livelihood opportunities to meet the needs of those displaced or marginalized as farmers and agricultural wage workers. Third, whatever possibilities there may be for agricultural development strategies that are more 'appropriate' in terms of sustainable livelihoods and environments, it is surely utopian to suppose that any future economy based solely (or even mainly) on farming could generate adequate livelihoods for all those who live in the countryside now and will be born there in the future – let alone for the great numbers of urban poor. In short, even 'alternative' forms of agricultural development have to consider what types of rural–urban interaction, of agricultural and industrial linkages, they require to be viable.

These introductory remarks have highlighted certain issues (rather than solutions), and have suggested that:

1   the boundaries of countryside and town, the 'rural' and the 'urban', are much less easy to distinguish than our assumptions, images, and concepts often suggest;

2   rural existence and struggles for livelihoods cannot simply be equated with farming, either as a means of survival or as a 'way of life';

3   'alternative' development strategies that aspire to create more adequate and secure rural livelihoods similarly cannot afford to disregard questions about the interactions of countryside and town, agriculture and industry.

The purpose of highlighting these issues is not to celebrate complexity for its own sake (regarded by many as an occupational disease of academics and experts), but to caution against simplistic solutions to problems grounded in complex realities. Any effective action to change such realities in a desired direction requires that prescription is well informed by empirical and theoretical analysis. The authors of this book have tried to provide and illustrate some important tools of analysis

relevant to rethinking the agenda of development in the late twentieth century, rather than putting forward their own prescriptions. Even if the question 'what should be done?' remains unanswered, the authors hope that by engaging with this book readers will be somewhat better informed.

Style and structure of the book

Several means are used to facilitate learning from the chapters of this book (which includes engaging and debating with them). Each chapter has one or more key questions at the beginning that the chapter addresses. Key concepts are highlighted, defined and discussed in boxes in the text. Each chapter has a short summary at the end, as an aid to readers when they want to check back on the main points of a chapter; the index is a similarly useful aid for referencing themes and issues across the various chapters. References to other works are limited to those used in quotation, or as sources of specific evidence and ideas cited. Inevitably, many of the concepts and issues presented are complex: we have tried to explain and illustrate them clearly, to make them as accessible as possible.

In the first part of the book, Chapter 1 sets the scene by reviewing some concepts, measures and meanings of poverty, in particular rural poverty, to show how the issues they involve inform the agenda of the book. This chapter was completed before the availability of an important work that constructs 'human development indicators' and argues that they should be used to complement more conventional national income measures – Table 3 in this Introduction draws on this *Human Development Report* (UNDP, 1991), which is recommended to readers interested in learning more about these indicators.

Chapters 2–4 provide an historical and comparative framework covering agrarian structures and change in Latin America, India and sub-Saharan Africa. This framework focuses on the formation, dynamics and changes of agrarian economy and society in different experiences of colonialism and capitalism. Key topics include patterns of land distribution, commoditization of farming, markets including international market linkages,

The concluding part of the book considers responses to crises in rural livelihoods 'from above' and 'from below'. Many examples throughout the book illustrate the goals, actions and strategies of governments and powerful private interests on one hand, and those of small farmers, landless workers, and of women and men within these groups, on the other hand. Chapters 11 and 12 provide a more systematic reflection on the nature and effects of such responses. Chapter 11 is concerned primarily with government policies that affect rural livelihoods directly and indirectly. It contrasts post-war land reform and a small farmer strategy that contributed significantly to industrialization in the Republic of Korea and Taiwan with the more mixed results of rural poverty alleviation programmes in India following an incomplete land reform and the Green Revolution, which made India self-sufficient in grains but failed to benefit many smaller farmers (see also Chapter 3). The limits of poverty alleviation as an approach links with the analysis in Chapter 7. Chapter 11 concludes with a consideration of the role of non-governmental organizations, and whether they are better placed than government agencies to identify and exploit 'room for manoeuvre' concerning rural livelihoods that might exist within structures of unequal property and power.

Chapter 12 examines 'responses from below', both individual and collective, to pressures and crises in rural livelihoods. It introduces and appraises the idea of 'weapons of the weak', devised to acknowledge the importance of 'everyday resistance' by the rural poor to the conditions of their exploitation and subordination, and questions whether such 'weapons' are adequate to transform those conditions. Several examples of collective organization and action by the rural poor illustrate some of the complexities, ambiguities and 'contradictions among the people' they encompass, at the same time suggesting that the energies, capacities and initiatives of poor rural people are an essential component of any movement to establish viable and secure livelihoods, the basis of a decent human existence for everyone in the countryside.

high-productivity agriculture and advanced in-dustrialization (the prospects for farmers in the UK, USA, Japan and other high-income countries is currently a sensitive political issue). Second, industrialization and urbanization do not neces-sarily generate sufficient employment and secure livelihood opportunities to meet the needs of those displaced or marginalized as farmers and agricul-tural wage workers. Third, whatever possibilities there may be for agricultural development strat-egies that are more 'appropriate' in terms of sus-tainable livelihoods and environments, it is surely utopian to suppose that any future economy based solely (or even mainly) on farming could generate adequate livelihoods for all those who live in the countryside now and will be born there in the future – let alone for the great numbers of urban poor. In short, even 'alternative' forms of agricul-tural development have to consider what types of rural–urban interaction, of agricultural and in-dustrial linkages, they require to be viable.

These introductory remarks have highlighted certain issues (rather than solutions), and have suggested that:

1   the boundaries of countryside and town, the 'rural' and the 'urban', are much less easy to distinguish than our assumptions, images, and concepts often suggest;

2   rural existence and struggles for livelihoods cannot simply be equated with farming, either as a means of survival or as a 'way of life';

3   'alternative' development strategies that as-pire to create more adequate and secure rural livelihoods similarly cannot afford to disregard questions about the interactions of countryside and town, agriculture and industry.

The purpose of highlighting these issues is not to celebrate complexity for its own sake (regarded by many as an occupational disease of academics and experts), but to caution against simplistic solutions to problems grounded in complex reali-ties. Any effective action to change such realities in a desired direction requires that prescription is well informed by empirical and theoretical analy-sis. The authors of this book have tried to provide and illustrate some important tools of analysis

relevant to rethinking the agenda of development in the late twentieth century, rather than putting forward their own prescriptions. Even if the ques-tion 'what should be done?' remains unanswered, the authors hope that by engaging with this book readers will be somewhat better informed.

Style and structure of the book

Several means are used to facilitate learning from the chapters of this book (which includes engaging and debating with them). Each chapter has one or more key questions at the beginning that the chapter addresses. Key concepts are highlighted, defined and discussed in boxes in the text. Each chapter has a short summary at the end, as an aid to readers when they want to check back on the main points of a chapter; the index is a similarly useful aid for referencing themes and issues across the various chapters. References to other works are limited to those used in quotation, or as sources of specific evidence and ideas cited. Inevi-tably, many of the concepts and issues presented are complex: we have tried to explain and illus-trate them clearly, to make them as accessible as possible.

In the first part of the book, Chapter 1 sets the scene by reviewing some concepts, measures and meanings of poverty, in particular rural poverty, to show how the issues they involve inform the agenda of the book. This chapter was completed before the availability of an important work that constructs 'human development indicators' and argues that they should be used to complement more conventional national income measures – Table 3 in this Introduction draws on this *Human Development Report* (UNDP, 1991), which is rec-ommended to readers interested in learning more about these indicators.

Chapters 2–4 provide an historical and compara-tive framework covering agrarian structures and change in Latin America, India and sub-Saharan Africa. This framework focuses on the formation, dynamics and changes of agrarian economy and society in different experiences of colonialism and capitalism. Key topics include patterns of land distribution, commoditization of farming, markets including international market linkages, and the role of colonial and independent governments. The themes underlying these topics concern relations of property and power expressed in changing forms of social (class and gender) differentiation. Each of these chapters ends with a survey of the implications for rural livelihoods today, suggesting issues taken up in greater depth and detail in the following parts of the book.

The next part 'opens up' rural households to examine what goes on inside them, and to show why their internal dynamics are important for rural lives and livelihoods. Chapters 5 and 6 demonstrate (as Chapters 2–4 showed at a more general 'structural' level) that understanding rural livelihoods is hardly a matter for economics alone: the 'economics' of households are inextricably linked with social, political and cultural processes that connect their internal dynamics with external environments. This is first investigated conceptually and illustrated with a number of examples in relation to households as units of production, co-residence and consumption in Chapter 5, and then in relation to social reproduction and differentiation within and between households in Chapter 6.

The third part comprises four rather different 'case studies' – case studies in the sense of addressing specific themes in particular settings, ranging in scale from the subcontinental (Chapters 7 and 8) to the individual village (Chapters 9 and 10). Chapter 7 enters the debate around claims that rural poverty in India has declined dramatically in recent years. It considers a number of macro-economic and regional indicators to arrive at a more disaggregated, and less optimistic, picture of trends in rural poverty. A key emphasis is on patterns of employment and unemployment, and the impact on them of government policies. The authors conclude that ensuring secure and productive rural livelihoods requires a different development strategy, with different linkages between rural and urban, agricultural and non-agricultural sectors.

Chapter 8 reviews some of the principal environmental bases, changes and problems in farming, and in rural economy more generally, in sub-Saharan Africa. Environmental issues are placed firmly in the context of social organization, and technical changes illustrated through three examples of innovation by African farmers in different ecological zones and historical conditions. The chapter finishes by considering options for change in the light of current debates about 'indigenous technology' versus 'modernization', emphasizing some issues of social organization that are central to effective environmental action.

Some of these issues carry forward to Chapter 9, a case study of a village in northern Uganda that analyses a particular pattern of class differentiation, and also suggests some of its ramifications for gender relations. While grounded in a single village, the chapter illuminates a much wider debate about the dynamics of capitalism in rural Africa, distinguishing 'accumulation from below' by richer villagers and 'accumulation from above' by those strongly connected with the state. The effects for the livelihood opportunities of different classes in the village are discussed, and the chapter also reveals some of the 'hidden' ways in which social differentiation can proceed, in this case through ostensibly egalitarian and 'traditional' forms of co-operation between village households.

Chapter 10 examines aspects of life in a village in another area of northern Uganda, reoccupied by its inhabitants after a period of violent dislocation and flight. The focus is on beliefs and practices concerning healing that distinguish but can also combine both 'modern' biomedicine and 'traditional' medicine. The chapter shows that any simple distinction or opposition between the two is misconceived, and that villagers draw on one or the other (or both) as available and appropriate. The analysis demonstrates two important points of wider significance – that much of the 'traditional' medicine applied in the village expresses an aspiration to reconstitute and heal a community after a period of intense social and individual trauma, and that 'tradition' is not a static inheritance but a repertoire of beliefs and practices that is constructed, reconstructed, modified (and sometimes even 'invented') with the needs of changing circumstances.

THE ORIGINS OF RURAL POVERTY

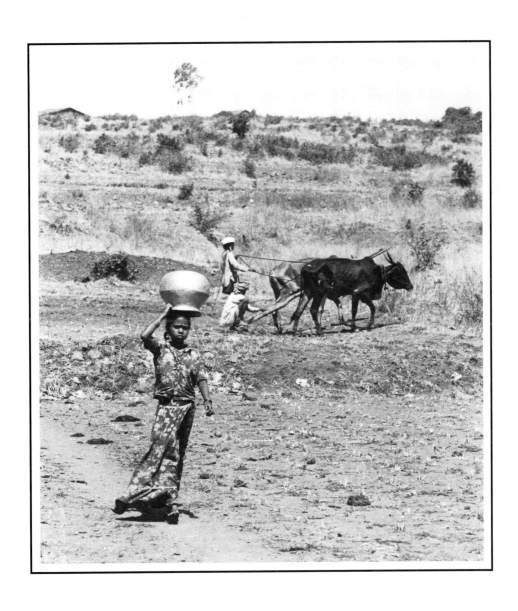

1

POVERTY AND THE POOR

HENRY BERNSTEIN

People in the countryside of the Third World gain their livelihoods in a variety of ways from different types of farming and a wide range of other activities. They do so with varying degrees of success according to their access to resources and employment and how they deal with pressures arising from social, economic and environmental change. Life for many entails a daily struggle in which much energy and ingenuity is needed to secure livelihoods in the face of various crises. Crises may be writ large, affecting whole populations or particular groups within them, or small – the loss of a cow, for example, or even a tree, can be disastrous for a poor rural household.

Q How much rural poverty is there in the Third World? What does it mean to be poor?

Because rural poverty in the Third World accounts for a large proportion of global poverty, we begin by examining some standard definitions and means of measuring poverty to see what issues they raise for the themes of this book.

1.1 Poverty: some global figures

The focus of the World Bank's annual *Development Report* in 1990 was 'Poverty'. Using a 'poverty line' of per capita annual income of US$370 for international measurement and comparison,

the Report estimated that in the mid-1980s 1116 million people in less developed countries were living in poverty – about one-third of their total population. Using a lower income measure of US$275 a year, 680 million people or 18% of the population of less developed countries were below this 'extreme' poverty line.

Table 1.1 shows the distribution of global poverty and introduces some of the indicators used to measure it. In terms of absolute *numbers*, the table shows a concentration of global poverty in three major regions: sub-Saharan Africa, East Asia and South Asia. For East and South Asia, this partly reflects the size of their total populations (and the weight of China and India within them).

The overall picture is modified by the 'headcount index' which measures the *incidence* of poverty by the percentage of the total population below the poverty line. South Asia has the highest incidence or rate of poverty, followed by sub-Saharan Africa, then the Middle East and North Africa. East Asia has the same number of extremely poor people as sub-Saharan Africa (120 million), but there they constitute 9% of total population whereas the proportion is 30% in sub-Saharan Africa. The incidence of all poverty in East Asia (20%) is less than half that in sub-Saharan Africa (47%).

Another indicator that further modifies the picture of poverty, even at this general or aggregated

Table 1.1   How much poverty is there in the developing countries? The situation in 1985

Region	Extremely poor			Poor (including extremely poor)			Social indicators		
	Number (millions)	Headcount index (%)	Poverty gap	Number (millions)	Headcount index (%)	Poverty gap	Under-5 mortality (per thousand)	Life expectancy (years)	Net primary enrolment rate (%)
Sub-Saharan Africa	120	30	4	180	47	11	196	50	56
East Asia	120	9	0.4	280	20	1	96	67	96
China	80	8	1	210	20	3	58	69	93
South Asia	300	29	3	520	51	10	172	56	74
India	250	33	4	420	55	12	199	57	81
Eastern Europe	3	4	0.2	6	8	0.5	23	71	90
Middle East and North Africa	40	21	1	60	31	2	148	61	75
Latin America and the Caribbean	50	12	1	70	19	1	75	66	92
All developing countries	633	18	1	1116	33	3	121	62	83

The poverty line in 1985 dollars is US$275 per capita a year for the extremely poor and US$370 per capita a year for the poor.
The headcount index is defined as the percentage of the population below the poverty line.
The poverty gap is defined as the aggregate income shortfall of the poor as a percentage of aggregate consumption (see text).
Under-5 mortality rates are for 1980–85, except for China and South Asia, where the period is 1975–80.
Source: World Bank (1990) *World Development Report 1990*, Oxford University Press, New York, p.29.

level, is the **poverty gap**, which concerns the extent to which the poor fall below the poverty line. Poverty gap indicators try to capture the *severity* of poverty, and by implication the magnitude of efforts required to overcome poverty. Table 1.1 shows that sub-Saharan Africa and South Asia have the most severe as well as the most widespread poverty.

The figures on the right of the table for the three social indicators of child mortality, life expectancy and net primary school enrolment correspond approximately to the size of the poverty gap, with one significant exception – that of China. China has more positive scores on these social indicators than three regions with a smaller poverty gap:

the Middle East and North Africa, East Asia as a whole, and Latin America and the Caribbean. This finding is even more striking if we compare per capita **gross national product** (GNP) and selected social indicators for China and a number of other countries (Table 1.2).

> **Poverty gap:** An indicator of poverty measured as the aggregate or total income shortfall of the poor as a percentage of total (national) consumption – in other words, the resources that would need to be transferred to bring the incomes of all poor people up to the poverty line.

Table 1.2 GNP per capita and selected social indicators

	GNP per capita (US$, 1988)	Infant mortality rate (1988) [a]	Life expectancy at birth (years, 1988)	Daily calorie supply (per capita, 1986) [b]
China	330	31	70	2630
Other low income				
Bangladesh	170	144	51	1927
India	340	97	58	2238
Pakistan	350	107	55	2315
Indonesia	440	68	61	2579
Middle income				
Egypt	660	83	63	3342
Guatemala	900	57	62	2307
Thailand	1000	30	65	2331
Turkey	1280	75	64	3229
Peru	1300	86	62	2246
Brazil	2160	61	65	2656

[a] Infant mortality rate: number of infants who die before reaching one year of age, per thousand live births in a given year (hence a different measure from that used in Table 1.1).

[b] Daily calorie supply per capita: calorie equivalent of net food supplies divided by the population. Note that this is not necessarily what people have access to or eat.

Source: Data from World Bank (1990) *World Development Report 1990*, Oxford University Press, New York, pp.178–9, 232–5, 256.

Gross national product (GNP): The World Bank defines GNP as 'the total domestic and foreign output claimed by residents of a country' in one year. What they 'claim' is also their *income*; thus GNP is a measure of national income available for private and public spending, and GNP per capita is a measure of the average income of each member of the population, including what they may earn or receive from abroad.

In broad terms, rising GNP per capita is associated with declining poverty, as indicated by the regional data in Table 1.1. What conclusions can be drawn from Table 1.2? Remember that GNP per capita and the social indicators used in Tables 1.1 and 1.2 are *averages*, which tend to conceal (significant) *differences within populations* in the distribution of income and satisfaction of basic needs (access to adequate food, shelter, health care).

This, of course, is where the 'exception' of China comes in. First, it suggests that income distribution is an important variable intervening between GNP per capita and levels of satisfaction of basic needs. Second, given very low average income in China, it further suggests that more public effort has been directed to meeting basic needs, and with greater effectiveness, than in countries with a similarly low, or indeed substantially higher, GNP per capita. China is also an interesting example because it appears that this impressive achievement is being undermined by the economic 're-forms' from the late 1970s towards the individualization and privatization of the rural economy. There is evidence of growing inequality in rural income distribution (Martin, 1990), of falling average life expectancy and rising infant mortality, especially of female children, in recent years (Drèze & Sen, 1989, pp.215–21).

1.2 What is poverty?

Asking 'What is poverty?' in relation to the kinds of indicators introduced above is to inquire, first, *what* they measure and *how effectively* they do it.

The 1990 *World Development Report* defines poverty as 'the inability to attain a minimal standard of living' measured in terms of basic *consumption* needs or *income* required to satisfy them: the poverty line is based on 'the expenditure necessary to buy a minimum standard of nutrition and other necessities'. This expenditure varies between countries, therefore country-specific poverty lines have to be constructed. In fact, the lower poverty line of US$275 per capita annual income in the Report's international comparison (Table 1.1) is also that used by the Government of India (see Chapter 7).

Consumption-based poverty lines are thus directed above all to *physical* measures of relative well-being. The inability to attain minimal standards of consumption to satisfy basic physiological criteria is often termed *absolute* poverty or deprivation. It is most directly expressed in not having enough to eat (hunger or malnutrition) to which other indicators can be added, as in Tables 1.1 and 1.2.

How effectively poverty can be measured in these terms is subject to some widely noted difficulties, including where to draw the poverty line. Evidently, more inclusive or generous definitions of basic needs will raise the level of income/expenditure at which the poverty line is drawn, thereby also increasing the headcount index or incidence of poverty. In addition to possible controversy about how and where to draw the poverty line, there are problems of how to apply it in practice.

One problem concerns knowing and valuing with reasonable accuracy the quantities of goods and services produced for *self-consumption* (or subsistence) rather than for market exchange. Another problem is that the prices of basic commodities like food staples often *vary* within countries, both spatially (between different rural areas, between countryside and town) and with time (for example, seasonally each year, or between good and bad harvest years). A third problem is that even when poverty indicators take distribution into account, they tend to measure distribution of income or consumption between *households*, which vary widely in their size and composition and in the distribution of income and consumption *within* them. Social relations that generate unequal distribution of income and consumption within households are discussed further in Chapters 5 and 6.

These problems of effective measurement are exacerbated when applied on an increasing scale of aggregation, or when trying to discern trends in poverty and projecting future trends. Indeed, the two are often linked. Consider, for example, distribution close to the poverty line: considerable numbers of people may cross the poverty line in either direction, temporarily or permanently, over quite short periods of time. Who is more likely to do so, and why, raises issues about *causes* and *patterns* of poverty that we will come back to.

First, though, it is worth noting another element in the *World Development Report's* discussion of the poverty line: income beyond the satisfaction of basic physical needs that is required to participate 'in the everyday life of society'. This implicitly acknowledges the notion of *relative poverty* (or deprivation) investigated in a famous work on poverty in Britain by Peter Townsend:

> "Individuals, families and groups in the population can be said to be in poverty when they lack the resources to obtain the types of diets, participate in the activities and have the living conditions and amenities which are customary, or at least widely encouraged and approved, in the societies to which they belong. Their resources are so seriously below those commanded by the average individual or family that they are, in effect, excluded from ordinary living patterns, customs and activities."
>
> (Townsend, 1979, p.31)

This broader view of poverty or deprivation moves beyond the criterion of individual physical survival to satisfying the material conditions of a full *social* existence, the ability to participate in 'ordinary living patterns, customs and activities'. However, while it is relatively straightforward to measure the former, it is far more difficult (or impossible) to measure the latter, especially in terms of international comparisons, because definitions of what is necessary to participate in the everyday life of society are 'subjective' (World Bank, 1990, p.27).

Or, to put the point more usefully, we can say that what constitutes a minimal decent standard of living and way of life is socially and culturally constructed (hence is also subject to change).

These observations suggest that we have reached the point of moving beyond our initial question – what conventional poverty line indicators measure and how effectively – to ask what *at best* they can and cannot tell us. (This is explored in greater depth and detail in Chapter 7 on interpreting poverty trends in India; the discussion of intra-household relations in Chapters 5 and 6 also indicates that poverty may be experienced differently by individual household members, illustrated in Section 1.3 below.) At issue here is the distinction between *technical* issues of accurate measurement and *conceptual* issues of understanding poverty. On one hand, statistical measures of poverty are useful only so far as their conceptual basis is sound. On the other hand, however sophisticated and rigorous statistics are, there are limits to what they can tell us.

This critical point has been developed by Robert Chambers, who cautions against the 'bias to the measurable' of professional researchers and statisticians and the governments and development agencies that employ them. The significance of this bias is not only that there are limits to what can be measured meaningfully, but also that the obsessive pursuit of the measurable (physical and monetary indicators) can actively distract attention from 'many aspects of want and disadvantage' (Chambers, 1988a, p.1), which include vulnerability, powerlessness, and the struggle for dignity.

Chambers (1988a, pp.8–9) distinguishes five dimensions or conditions of poverty/deprivation (which can be combined in various ways):

1 *'poverty proper'* in the sense used by the World Bank and others; that is, lack of adequate income or assets to generate income;

2 *physical weakness* due to undernutrition, sickness or disability;

3 *isolation* physically and/or socially due to peripheral location, lack of access to goods and services, ignorance, illiteracy;

4 *vulnerability* to any kind of emergency and contingency, and the risk of becoming even poorer;

5 *powerlessness* within existing social, economic, political and cultural structures.

How do these dimensions of poverty relate to the points made about the 'bias to the measurable' and, at the same time, the limits of the measurable? The first three conform to conventional models of poverty expressed in measurable physical indicators, while the latter two are simultaneously less amenable to measurement *and* may raise awkward issues about social inequality and the distribution of property and power. Not surprisingly, then, the first three dimensions are more acceptable to dominant groups for defining both the condition of poverty and what constitutes legitimate interventions by governments and others to alleviate poverty.

Chambers then raises the question: what are the priorities of the poor? Needless to say, far less attention has been devoted to this question than to measuring the physical manifestations of poverty. Drawing on relevant research in India, but which has general implications, Chambers suggests that the priorities of the poor are:

1 adequate incomes and consumption;

2 security;

3 independence and self-respect.

The first of these converges, of course, with the mainstream definitions of poverty that we have looked at and that inform official statements of the problem of poverty and policies to overcome it. However, the virtually exclusive concentration of governments and development agencies on consumption and income-based definitions of poverty means that they exaggerate this aspect at the expense of others. For poor people, achieving security and recognition of their dignity as human beings may be just as important as, or inseparable from, improving their incomes and standards of consumption. Indeed, they may be prepared in certain circumstances to 'trade off' possible gains in income against gains in their security and self-respect (see also Chapter 12).

Chambers (1988a, pp.23–4) cites an interesting case that highlights this point. Research in Rajasthan in India by S. N. Jodha traced the fortunes of a number of village households over 20 years or so. Thirty-five households surveyed experienced a decline in real income of more than 5% during this period, but considered themselves better off. The reason was that they had become far less dependent on the 'support, mercy or patronage' of wealthier villagers, in particular through practices of bonded labour and borrowing, leading to debt (see below). Their increased freedom from 'dependence and humiliation' counted for more, in this instance, than reduced income, which on conventional criteria would simply record them as worse off.

Chambers' multidimensional approach to poverty, and the issues it suggests, brings us to the next important question: *who* are the rural poor? This takes us a major step further in several ways. First, it focuses attention on those whose lives, problems, *and differences* are reduced to, and hence concealed by, uniform measures of poverty. Second, to ask who are the rural poor is to begin to enquire about different causes and patterns of poverty. In short, in Robert Chambers' words, it is to move beyond 'simple numbers' to investigate 'complex realities'.

1.3 Who are the rural poor?

One way of moving from statistics and abstract generalizations to complex realities is to see how people experience or perceive their poverty. The following accounts (Boxes 1.1 to 1.4) by Sharifa and her husband Abu, who lived with their six children in a village in Bangladesh in the mid-1970s, are recorded in a remarkable book by Betsy Hartmann and James K. Boyce called *A Quiet Violence: view from a Bangladesh village*.

The extracts portray key dimensions of rural poverty which are worth noting, and are addressed in various ways in different chapters of this book:

> '*structural*' factors concerning access to land and farming implements, to credit, and other resources;

- structural factors involving *social relations* with others who are richer and more powerful;

- access to resources and employment which define the conditions of the family's *livelihood strategies*;

- *existential* dimensions of poverty – its frustrations, insecurity and oppression – and how these are expressed and explained;

- social relations and interaction, of tension and solidarity, *within* a poor family or household.

Box 1.1 The path to poverty

'At least we had enough to eat at first,' Sharifa says. 'But then one by one the other brothers married, and we spent money for their weddings. Soon everyone wanted separate kitchens – it's always like that. So the land was divided four ways. My husband should have got a larger share because he supported his mother, but he was too timid to stand up to his brothers.

'I was young then and worked very hard. I used to husk rice to make money. I made four hundred *taka* that way, all by myself! That was enough to buy two *duns* of land in those days. Children were coming and I knew we would have to feed them. My dream was to buy two *duns* so we could support our family. With my husband's land and the half *dun* I brought as my dowry, we would have had four *duns* of land!

'My husband didn't think like me – he didn't see ahead to the future or appreciate my intelligence. His mother was old and dying, and he wanted to spend my money on medicines for her. He told me, "If you want to keep your money separately, I'll divorce you. I'm your husband and what's yours is mine." '

Sharifa casts an accusing glance at her husband, who slowly lifts his eyes to look at her. The bed creaks as their daughter rolls to the other side.

'I didn't want to be divorced, so I gave in. All my money was wasted. How many injections and medicines the doctor gave his mother – and of course she died. We were left with only two *duns* of land. When my father died, I inherited a cow which we sold to buy more land. But we couldn't manage. Each year another child came and our situation grew worse.

'We had to borrow money to eat. Sometimes neighbours would lend us money without interest, but we often had to sell our rice before the harvest. Moneylenders would pay us in advance, and take our rice at half the market price. No matter how hard we worked, we never had enough cash. We started selling things – our wooden bed, our cow, our plough. Then we began to sell our land bit by bit. Now we have less than one *dun* left, and most of that is mortgaged to Mahmud Haji.'

(Hartmann & Boyce, 1983, pp.162–3)

What do you get from this extract about the strengths (and tensions) of kinship relations?

What does it suggest about gender relations?

How is the path to poverty of Sharifa and Abu charted in terms of declining assets? increasing insecurity? indebtedness?

Box 1.2 Trying to earn a living

At the mention of Mahmud Haji, Abu's back straightens and his cheeks flush with anger. His voice loses the weak, passive tone brought on by the fever. 'I sharecrop three *duns* of land from Mahmud Haji and work for him for wages, but still I can't earn enough money to pay back the mortgage. I don't even earn enough to feed my family! Most people pay their labourers a kilo of rice, one *taka* and a meal before the day's work. But Mahmud Haji is so stingy that he gives an extra quarter kilo of rice instead of the meal, even though a man eats twice that much. So I work for him on an empty stomach.

'Sharecropping is not much better. I do all the work, and then at harvest time Mahmud Haji takes half the crop. When I work for wages, at least I bring home rice every night, even if it's not enough. But when I work on my sharecropped land, I have to wait until the harvest. In the meantime I have no rice in the house, so what are we to eat? Since I have no cow or plough, I have to rent them from a neighbour. The price is high – I plough his land for two days in return for one day's use of his cattle. In this country, a man's labour is worth half as much as the labour of a pair of cows! Each time I need to plough, it means three days with no income.'

(ibid., p.163)

How do Abu's attempts to farm (through sharecropping) involve him in dependence on others?

Who sets the terms of his wage work?

Box 1.3 Poverty and hunger

Sharifa laughs, exposing her red-stained gums. 'Without betel nut I wouldn't survive. Whenever I feel hunger, I chew it and it helps the pain in my stomach. I can go for days without eating – it's only worrying about my children that makes me thin.' She looks at her daughter asleep on the bed. 'Do you know what it's like when your children are hungry? They cry because you can't feed them. When my oldest daughter comes to visit from her husband's house, I want to feed her chicken curry and sweet cakes, but I can't even give her rice. I tell you it's not easy being a mother.

'If I'd been a different woman, I would have left my husband and children and fled to a different place. But I stay here with him – it's my fate and I accept it.'

She brushes a strand of hair from her forehead. 'Why do you sit here listening to our troubles?' she asks. 'You should hear happier stories. When people in this country are happy and their bellies are full, they won't listen to stories of sorrow. They say, "Why are you telling me this? I don't want to hear."'

Abu nods, and adds, 'Allah says a rich man should care for a poor man. He should ask him if he has eaten. But in this country a rich man won't even look at a poor man.'

(ibid.)

How does their poverty affect Sharifa and Abu's ability to participate in the everyday life of society?

In what ways does Sharifa experience poverty as a woman?

Box 1.4 records a conversation while Abu was chopping down a tree he had planted for its fruit but which he now had to sell for firewood to pay off some debts and buy a few days' supply of rice.

Figure 1.1   Landowner and labourers in north India.

Box 1.4   The jackfruit tree

'These days I have no work,' [Sharifa] complains. 'If we had land, I would always be busy – husking rice, grinding lentils, cooking three times a day. You've seen how hard Jolil's wife works, haven't you? I have nothing to do, so I watch the children and worry. What kind of life is that?

'I wonder if things will ever change. People say that in four or five months there will be another war in this country. I say let it be. What difference does it make to a poor person like me? The opposition will shoot the rich people and leave people like us alone. These days the rich are getting richer and the poor are getting poorer. How long can it go on?'

After striking a root with his hoe, Abu motions for Nozi to move away and pushes the jackfruit tree. As it totters to the ground, his eyes fill, but he checks the tears, staring numbly at his wife and son. Nozi moves closer to inspect the tree, tearing the leaves from its slender branches.

As Abu takes the axe to chop off a root, the blank expression on his face slowly gives way to indignation. 'People get rich in this country by taking interest,' he says sharply. 'They have no fear of Allah – they care only about this life. When they

buy our rice before the harvest at half its value, they say they're not taking interest, but they are.'

Abu chops off another root, and continues, 'There is no rice in my household and I have six children to feed. In June I cut down my mango tree, and now I am chopping up my jackfruit tree. My children will never eat fruit – how can I afford to buy it in the bazaar? Rich people in this country don't understand how my stomach burns.

'Yesterday I went to Mahmud Haji's house and asked him to advance me some mustard seed. The ground is ready for planting, but I have no cash to buy seed. He told me, 'Buy it yourself. My sharecroppers have to provide their own seed.' He has bags of mustard seed in his house. How can a man be so mean?'

Abu arranges the cut roots into a neat pile. 'I'll sell the roots as firewood too,' he says. 'Tomorrow I'll carry the wood to town.'

'Tomorrow night someone will cook rice over a fire of our wood,' Sharifa adds, laughing despondently. 'They say Allah makes men rich and poor, but sometimes I wonder – is it Allah's work or is it the work of men?'

(ibid.)

How fatalistic are Sharifa and Abu about their poverty?

What do they see as the causes of their poverty?

This last question is quite complex because even these brief extracts expose a number of reasons and feelings, some of which seem to be in tension with each other. In one sense it is not surprising, because complex and contradictory realities are experienced, perceived and acted on in complex and contradictory ways, as later chapters will amplify.

The most important point about these extracts is their powerful reminder that behind the statistics of poverty are the daily realities of poor people's lives and struggles, of which Sharifa and Abu's accounts give us just one glimpse in one particular setting. To move to a more general consideration of rural livelihoods, crises and responses involves examining different patterns and causes of rural poverty. Asking 'who are the rural poor?' takes us in this direction, as well as warning against viewing poor people as all the same, as one stereotyped 'mass' of helpless victims.

Robert Chambers (1988a, pp.6–8) outlines several ways of disaggregating or classifying poverty. Two points are important for this book. One, just noted, is to counter the dehumanizing effect of poverty-line statistics that aggregate the poor as a uniform 'mass' (represented in 'simple numbers'). The other is to start to connect the (differentiated) 'who' of the rural poor with the (differential) 'why' of their poverty.

The first distinction Chambers notes is between the poor and the 'ultra-poor'. This alerts us to the distribution of poverty *below* the poverty line: that some poor people may be much poorer than others. This observation can be linked with the 'poverty gap', and for purposes of statistical measurement could be accommodated by drawing 'upper' and 'lower' poverty lines, as in Table 1.1. In addition, it has been suggested that the rural poor are like other rural people only poorer, while the ultra-poor experience different patterns of poverty and behave differently. For example, Michael Lipton (1984) hypothesizes that female participation in income-generating activity increases among the poor but declines among the ultra-poor, recalling Sharifa's words in Box 1.4: 'I have nothing to do, so I watch the children and worry. What kind of life is that?' She contrasts her enforced inactivity and frustrated energies with the work of a woman neighbour (Jolil's wife), and by implication with her earlier self in days of better fortune (see Box 1.1).

A second issue is *'ascribed deprivation'* in Chambers' term, meaning that certain categories of people are more likely to be poor (and poorer) according to ascribed characteristics such as their gender (usually women), membership of oppressed ethnic or minority groups ('Indians' in Latin America (Figure 1.2); low castes, 'untouchables' or 'tribals' in India), or age (children, the elderly).

Figure 1.2 Who are the rural poor? 'Ascribed deprivation' in Peru.

Figure 1.3   Diversifying rural livelihoods: making pots in a Bangladesh village.

This also supplies clues about the distribution and patterns of poverty among the poor. For example, in some rural societies women and female children have distinctly lower levels of consumption and higher levels of insecurity, morbidity and mortality than men and male children in the same households.

A third specification of the poor is by their *spatial location* – particular regions, rural localities, or types of villages are poorer than others. The 1990 *World Development Report* notes that:

"Increasing numbers of poor people live in areas that have little agroclimatic potential and are environmentally fragile. Examples include the Loess Plateau in China, the highlands of Bolivia and Nepal, the desertic African Sahel, and much of the humid tropics. Population pressure in those areas has decreased the productivity of the land and increased its vulnerability to flooding and soil erosion. This raises the question of the links between poverty and environmental degradation...

The causes of these growing pressures on natural resources are complex and interconnected. In many countries poor farmers are being marginalized and pushed to frontier areas. In addition, population growth and the commercialization of agriculture have forced farmers who once relied on environmentally sustainable forms of agriculture to use their land more intensively."

(World Bank, 1990, pp.71–2)

Where such marginal farming areas are also physically isolated, their inhabitants lack access to goods and services (including markets for their crops and other products) because of lack of roads and other means of communication, as well as because of their poverty.

Finally, Chambers distinguishes the poor by their *livelihood strategies*, using the metaphor of 'foxes' and 'hedgehogs' from the Greek proverb that 'the fox knows many things, but the hedgehog knows one big thing'. The 'foxes' among the rural poor build a 'repertoire of different petty enterprises and activities': *diversification* is the key motif of their livelihood strategies (Figure 1.3). 'Hedgehogs', on the other hand, are locked into one predominant source of livelihood: for example, as bonded or attached labourers or sharecroppers.

Which elements of Sharifa and Abu's livelihood strategy (and aspirations) display 'fox' or 'hedgehog' characteristics? On one hand, Abu is constrained by 'hedgehog'-like ties of personal dependence on the landlord Mahmud Haji to whom his remaining land is mortgaged, from whom he sharecrops land, and for whom he works as a wage labourer, all of which suggest a kind of 'debt bondage' common in much of rural South Asia. On the other hand, Sharifa and Abu continue to struggle to secure some assets (hence 'petty enterprises') they can control and derive income from. The account in *A Quiet Violence* also notes Sharifa's intention to claim the rest of her inheritance, a quarter *dun* of land (appropriated by a brother), and that she is 'sharecropping' a calf from a friendly neighbour: 'We'll raise her and when she bears a calf of her own, we'll return her, keeping the calf for ourselves. Then, if it pleases Allah, we will have a cow of our own'; the ill-fated outcome of Abu's attempt to grow and harvest fruit trees was described in Box 1.4.

1.4 Rural poverty and agrarian structures

Everything we have said so far leads to a fundamental issue about rural poverty and how we understand it in relation to questions of development: namely, whether we conceive poverty as *residual* or *relational*.

The residual approach views poverty as a consequence of being 'left out' of processes of development, on the assumption that development brings economic growth which, sooner or later, raises everybody's income. This is termed the 'trickle down' effect: that the benefits of growth trickle down even to the poorest groups in society in the form of increased opportunities to earn (more) income. The implication for development policy is to target the rural poor in order to integrate them into processes of development they have been excluded from. In practice, this typically means

integrating them more deeply into markets and devoting more of their resources and energies to producing goods for sale (commodity production). This was the conception, for example, of the World Bank's strategy for 'reaching the poorest' through rural development programmes in the 1970s.

In contrast, relational approaches investigate the causes of rural poverty in terms of *social relations* of production and reproduction, of property and power, that characterize certain kinds of development, and especially those associated with the spread and growth of capitalism. A relational approach thus asks rather different questions: are some poor *because* others are rich (and vice versa)? What are the mechanisms that generate both wealth and poverty as two sides of the same coin of (capitalist) development?

Much of the material in this book pursues these relational questions and their ramifications in a number of contexts, through the analysis of agrarian structures and processes. For the moment, we can note that *structure* is a shorthand term for the relations that characterize the conditions of production and reproduction for different groups in the countryside, hence their life chances, experiences and struggles. The most important of these social relations, according to our perspective, are those of class and gender.

This initial and extremely general definition of agrarian structure can be made more concrete in terms of three central themes of the book: the three 'Ls' of *Land, Labour and Livelihoods*. The different ways in which land and labour are combined in farming is fundamental to understanding rural livelihoods, and *whose* livelihoods. An immediate and indispensable element in grappling with agrarian structures is to consider different *forms of production*, or the different ways in which land and labour are combined – how farming is 'structured' by relations of class and gender. This involves posing questions like:

1   Who owns what? (or has access to what?)

2   Who does what?

3   Who gets what?

4   What do they do with it?

These are questions about the social relations governing the distribution of:

1 *property* in land and other means of production necessary to farming (or other economic activities);

2 *work* in different social divisions of labour;

3 *income*, whether in kind, in money or both;

4 income between *consumption* and *accumulation* (savings for investment).

While terms associated with economic life (property, work, income, consumption, accumulation) are highlighted, this does not mean that we can analyse rural livelihoods in narrow economic terms. This is why the importance of *social relations* is also stressed (rather than, say, abstract market forces or 'laws' of demand and supply). Considering social relations always involves asking 'whose livelihoods?', and reminds us that economic inequalities (of property, work, income, consumption, accumulation) typically incorporate and express social, cultural, political and institutional inequalities as well – for example, Robert Chambers' point above about ascribed deprivation.

You may have noticed that the four questions listed above can be applied to a range of contexts or levels of economic and social activity, from that of the individual farm (be it a large commercial enterprise like a plantation, or one of the many varieties of 'peasant' farms), to that of a particular village or rural locality, the agricultural sector as a whole, the national economy, or the international economy.

This further suggests that while analysing forms of production in farming is an indispensable element in considering land, labour and livelihoods, and the connections between them, it is only *one* element in a larger picture. The larger picture includes the class and gender relations, divisions of labour, markets and linkages of specific agrarian structures or particular national economies, as well as the international economy.

In short, analysing agrarian structures and how they change can help connect and flesh out the two very different ways of constructing and communicating facts about rural poverty presented earlier: statistical *measures* of the incidence of rural poverty and its trends over time, and statements of what it *means* to be poor, given voice by poor rural people and expressing their experiences and everyday struggles (see also Chapter 12). Both kinds of evidence have their limits as well as their uses.

Statistical measurement and analysis reveal effects rather than causes. At best, they may highlight something that needs to be explained (e.g. by suggesting a strong correlation between landlessness and rural poverty, typical of South Asia, for example), but cannot tell us why in some countries land is distributed unequally. At worst, the 'bias to the measurable' gives a misleading picture of rural poverty, and of the identities of the rural poor.

On the other hand, while the voices and actions of the rural poor give unique and powerful expression to the *existential* realities of their lives (their problems, pains, aspirations, and attempted solutions) they do not provide a complete or sometimes even useful guide to the *reasons* for their poverty. For example, misfortune may be attributed to 'the will of God', or to the personal characteristics of an extortionate landlord or merchant rather than to the structures that give landlords and merchants their power. It is possible to contest stereotypes of the rural poor as passive victims of circumstance, and to recognize that they are social 'actors' struggling with the conditions of their daily existence, without believing that the actors' perspective reveals all that can be known. It is as well to bear in mind the limits of 'actors' as a sociological metaphor. As the historian Eric Hobsbawm noted in a different context of social action: 'the evident importance of the actors in the drama does not mean that they are also dramatist, producer, and stage-designer' (quoted in Skocpol, 1979, p.18).

Summary

1   Conventional poverty indicators focus on measures of (a) physical deprivation, (b) monetary income necessary to satisfy basic needs.

2   The latter is used to construct a 'poverty line'. The 'headcount index' measures the proportion of a population below the poverty line (i.e. the incidence or rate of poverty), the 'poverty gap' measures how far below the poverty line the poor are (i.e. the severity of poverty).

3   Conventional indicators of poverty and its trends over time present the poor as a uniform 'mass' – a statistical entity. Such indicators obscure the varying identities of the poor, and the different reasons they are poor.

4   Listening to the voices of poor people helps us understand who they are, their problems and struggles, and their aspirations to security, dignity and an effective role in social life.

5   To analyse the causes of rural poverty we also need to examine agrarian and other 'structures' that affect the distribution of property (especially but not exclusively land), work, income, consumption and accumulation in the countryside and elsewhere.

6   Analysing rural poverty is not just about economics but also about social relations. Some questions should always be asked. Whose livelihoods? What are the social relations (and cultural, political and institutional factors) that affect: who owns what? who does what? who gets what? what they do with it?

2

AGRARIAN STRUCTURES AND CHANGE: LATIN AMERICA

HENRY BERNSTEIN

Chapters 2, 3 and 4 provide a broad overview of agrarian structures and rural livelihoods in Latin America, India and sub-Saharan Africa (excluding South Africa). Although only one country, India provides a useful basis of comparison with the other two regions because it is subcontinental in its size, diversity and population: in mid-1988 India had nearly as many people (816 million) as Latin America and the Caribbean (414 million) and sub-Saharan Africa (464 million) combined. Also, as we saw in Chapter 1, India is characterized by extensive and severe rural poverty.

If India contains great diversity in its ecological and climatic zones, farming systems, rural social structures, and cultures, what of the diversity between *and* within the many countries of Latin America and sub-Saharan Africa? Evidently, these three chapters cannot do justice to the richness and variety of the historical experiences or contemporary realities of these regions and countries. What they try to do is to outline, with the broadest of brushstrokes, some central themes in the formation and workings of their agrarian structures, and how they affect rural livelihoods.

Q How has the development of capitalism in the Third World affected the livelihoods of rural people?

The themes above bear directly on the key questions of the relational approach indicated in Chapter 1: who owns what? who does what? who gets what? what do they do with it? This chapter and Chapters 3 and 4 suggest answers to these questions by charting some of the ways that colonialism and the development of capitalism have affected the following:

1   how land and other resources are distributed and their uses determined;

2   how labour is organized in farming and other activities;

3   how agriculture is commoditized (or commercialized) and integrated with domestic and international markets;

4   how these processes affect the livelihoods of different social classes and groups, and are affected by the social contradictions and struggles they generate.

Various kinds of evidence, both facts and interpretations, are used to illustrate these processes and their dynamics. Again, it should be stressed that the evidence selected can only be illustrative rather than comprehensive, suggestive rather than conclusive. It includes some further examples of statistics relevant to agrarian structure and rural livelihoods, and how they can be interpreted.

As capitalism is so central to the accounts in these three chapters, and throughout much of the rest of the book, it is worth outlining what we understand by capitalism before we turn to Latin America.

2.1 Capitalism

Most generally, and abstractly, **capitalism** is understood here as a 'mode of production' or socio-economic system of *generalized commodity production* (production of goods and services for market exchange) based on a social relation between (wage) *labour* and *capital*. This social relation mutually defines the two essential classes of capitalism: the working class (or proletariat) and the capitalist class (or bourgeoisie).

> **Capitalism:** A mode of production or socio-economic system of generalized commodity production (production for the market for profit) based on an essential class relationship between capitalists (owners of the means of production) and workers (sellers of labour power).

Lacking alternative means of securing a livelihood, workers have to sell their labour power or ability to work (their physical and mental capacities) to capitalists who own means of production (machines, raw materials, land) and require labour to work them to produce goods and services to make profits. This is the sense in which the social relation of labour and capital is mutually definitive: workers sell their labour (power) for wages in order to subsist, capitalists buy labour (power) in order to make profits and accumulate capital to make more profits.

This sounds like a stark and simple (or simplistic) definition, but it is not a definition in the conventional sense of the term (like looking up 'table' or 'chair' in a dictionary). Rather it expresses, albeit with extreme brevity, a particular *approach* to understanding capitalism which involves a great deal of social theory and investigation (and controversy). A number of chapters in this book illustrate the elaboration of this approach and its application to a variety of historical and contemporary processes.

One crucial aspect of the approach is that it is *historical*. The development of capitalism is a historical process that we can study to understand how the world we live in was created: how, when, and where did such a mode of production come into existence? How are classes of capitalists and workers formed? How do the forms of capitalism change as it develops? Evidently this is very different from viewing capitalism as the expression of some universal, hence timeless, element of human nature such as a desire to maximize individual income, gain or possessions. It is also different from reducing the meaning of capitalism to simply an economic system, characterized, for example, by the predominance of the 'free' market.

Again schematically, the broad historical framework underlying Chapters 2, 3 and 4 is as follows (*Allen & Thomas, 1992*, pp.169–70).

1 North-western Europe, led by Britain, underwent a transition from feudal to capitalist society over a long historical period, mainly from the sixteenth to the nineteenth centuries, when the industrial revolution took off. This period of transition was one of continuous (albeit uneven) expansion of commodity production and exchange, facilitated by a range of social, political and cultural changes. It saw the formation of the two basic classes of capitalism: the capitalist class and the working class.

2 'Primitive' or primary accumulation – the formation of the bourgeoisie (owners of productive capital) and of the working class (labour power) – went hand in hand with the destruction of feudal society, a process that included the commercialization of landed property and of agricultural production on a capitalist basis, and the dispossession of the peasantry and many other small producers who became 'free' to sell their labour power.

3 Primitive accumulation in the transition to capitalism was helped by the 'expansion of Europe' in the same period, drawing in vast amounts of wealth from the plunder, conquest and colonization of many of the pre-capitalist societies of Latin America, Asia and Africa. In itself, this flow of wealth was not different in character from the riches amassed through other great imperial ventures in history, such as those of the Ottomans, the Moguls and the Manchus. It would not have contributed to the rise of capitalism if it had not fed into changes in the social basis of production already taking place in Europe. For example, much of the treasure extracted from their colonies by Spain and Portugal went to buy commodities from north-western Europe where the transition to capitalist production in manufacturing as well as in agriculture was taking place. The relatively slow transformation of feudal relations in Iberian society resulted in the declining wealth and power of Spain and Portugal compared with those countries that were pioneering capitalism.

4   The development of capitalism had a global dimension from the beginning, which was experienced by the pre-capitalist societies of the Third World through their incorporation in an emerging world market and an international division of labour, typically initiated during a period of European colonial rule. In this sense, capitalism came to these societies from the outside rather than resulting from their internal dynamics.

5   These broad statements give some indication of the timescales that are appropriate for examining such fundamental social changes, and of the uneven character of capitalist development in different countries and areas of the world. On the first point, one is struck by the long period – about three centuries – between the beginning of the breakdown of feudal society and the onset of the industrial revolution, which provided the emerging capitalist society of Britain with its distinctive type of production process – large-scale machine production. Once capitalist industrial production was firmly established and had begun to develop elsewhere in Europe, the USA and Japan, the striking feature by contrast was the 'acceleration of history', caused by the tendency of capitalism constantly to revolutionize technology and methods of production and to accumulate capital on an ever larger scale.

6   This framework suggests key connections in the relationship between capitalism and colonialism and also significant variations in the colonial experience arising from:

- *different stages in the emergence of capitalism*, and its *uneven development* between colonizing powers and within the areas they colonized;

- *different types of colonial state* and the interests they represented;

- the *diversity of the pre-colonial societies* on which European domination was imposed.

You may think we have moved very abruptly from an initial definition to a schema imposed on the last five centuries or so of global history! Needless to say, the history is much more complex than the schema suggests, for several reasons. One, simply, is that history is *always* more complicated than the definitions and conventions we use to try to summarize it. One kind of convention is that of chronology; another is describing broad structural changes (the rise of capitalism, the impact of European colonialism) as if they were somehow self-generated and separate from the actions, ideas and conflicts of different social groups and individuals of which they are, in fact, the outcomes.

Having said that, there are other (and less obvious) reasons for historical complexity that have to do with the nature of capitalism itself. The first is that capitalism is characterized by *'uneven and combined' development*: 'uneven' because the ways in which, and pace at which, it transforms production and social life vary a great deal between one part of the capitalist system and another; 'combined' because these variations are linked to each other, as Box 2.1 illustrates.

Box 2.1   Uneven and combined development

Consider the following 'snapshot' from, say, the 1870s.

Britain has undergone a massive 'industrial revolution' (the world's first) over the previous century or so. Its leading sectors are textile manufacture and engineering. Engineering industries provide the machines for the textile factories in Britain, and built the railways in India that carry cotton grown there to the ports for shipment to the mills of Lancashire. The mills typify the new dynamic civilization of industrial capitalism, while the raw materials they process are produced by the labour of millions of Indian peasant farmers subject to a combination of feudal-like and colonial compulsion (Chapter 3).

Brief as it is, this snapshot suggests several dimensions of uneven development concerning:

- links between economic sectors, technology, and social forms of production – factory industry/peasant farming;
- a spatial (in this case international) as well as social division of labour;
- the relationship between an imperial power and a colony.

The key force in combining this uneven development is British capital, which benefits from the labour of both the Indian peasants who grow cotton and the Lancashire mill workers who produce textiles from it.

(There are similar examples of uneven and combined development today, easily demonstrated by a visit to a local supermarket or high street clothes shop.)

This is not to say that such divisions are always peculiar to capitalism. In fact, gender relations are the most widespread type of **social differentiation** (and source of inequality) in human history. However, the development of capitalism reshapes pre-existing social divisions as well as creating new ones, in both cases linked to its processes of class formation. Even when there is continuity of social and cultural *forms* (often misleadingly perceived as a static 'tradition'), typically they acquire a new and different social *content* from the ways they are integrated within capitalism (see Chapter 9 on 'communal' forms of labour in Uganda).

> **Social differentiation:** Social relations of systematic inequality along lines of class, gender and other divisions (e.g. ethnic or cultural), and the processes through which these social divisions and relations are created.

A second reason for historical complexity is that while capitalism is founded on the essential social relation of (wage) labour and capital (that is, it is a particular type of class society), its development generates a number of other social differences and divisions, including those of gender, mental and manual labour, town and countryside, ethnicity (intimately connected with the imperial project and its legacy), nationality (intimately connected with the formation of modern states), and so on. Moreover, these kinds of divisions and the identities associated with them (female/male, white collar/blue collar, urban/rural, black/white, citizen/'foreigner') combine with those of class in various ways (e.g. female black agricultural worker). According to circumstances and perception, they also may be experienced more intensely than those of class.

2.2 Capitalism, peasants, petty commodity production

The last observation is particularly relevant to those often termed '**peasants**', whose conditions of life and struggles are a central focus of this book (Box 2.2).

> **'Peasants':** Small family farmers. What is distinctive about contemporary peasants within capitalism is that they are petty commodity producers subject to processes of class and other social differentiation, which can be charted through pressures on simple reproduction on the one hand, and opportunities for accumulation on the other.

Box 2.2   What and who are 'peasants'?

'As a first approximation we can distinguish peasants as small agricultural producers who, with the help of simple equipment and the labour of their families, produce mostly for their own consumption, direct or indirect, and for the fulfilment of obligations to holders of political and economic power' (Shanin, 1987, p.3).

This is a view of peasants widely held by social scientists. It is worth 'interrogating' each of its elements in turn.

1 Peasants are small farmers: the size of their farms is limited by the use of simple technology and family labour.

- What are simple technologies? The implication is use of hand labour and animal traction (e.g. ox ploughing) versus mechanical technologies such as tractors. What about technologies that are both small scale and mechanical, e.g. electrical pumpsets for irrigation? (See Chapter 3.)

- What is the basis of distinction between peasants and 'their families'? Does it mean that *men* are the principal landholders? farmers? heads of families? Are there women peasants in their own right?

2 It is also implied that peasants *remain* small farmers because they 'produce mostly for their own consumption'. They may do this by producing food and other goods that they consume themselves (termed **'use values'** in political economy), and/or indirectly by producing commodities for sale (**'exchange values'**) to earn money to buy goods and services that they need but do not produce themselves (means of production like tools, seeds, fertilizers; means of consumption like clothing, cooking pots). In either case, this suggests that the primary concern of peasant households is **simple reproduction**: that is, meeting their needs, rather than making profits or accumulating by enlarging the scale of their farming (**'expanded reproduction'**).

- What of peasants unable to satisfy their needs from their own farming? Or peasants who strive to accumulate?

3 Peasants also have to produce a surplus beyond their own needs to fulfil their 'obligations to holders of political and economic power' (in the form of rents, taxes, etc.). This further implies that peasants are also a *class* in relation to other classes that exploit them economically and dominate them politically.

- Are there cases in which peasants (or some peasants) themselves exercise a measure of economic or political power?

- What of relations of exploitation and domination within peasantries? Are there different classes of peasants?

Another important concern is: is this view of 'peasants' restricted to those who secure their livelihoods primarily through small-scale farming? In much of the contemporary Third World, small-scale farming is combined with a range of other activities such as wage labour, trade, and craft production, in the pursuit of livelihoods (and sometimes of accumulation). In short, 'pure' or exclusive family farming is exceptional rather than typical for the majority of rural people still termed 'peasants' today.

Where does this leave us? The view of peasants quoted above may *describe* (aspects of) different rural societies more or less adequately: say, communities of serfs in medieval England, the rural majority of Tsarist Russia or Manchu China, the peasants of colonial Spanish America (see below) or of contemporary India (Chapter 3) or Africa (Chapter 4). Assuming there might be a descriptive 'fit' that suggests some apparent *similarities* across such great variation of place and time, we still face the problem of explaining *differences* between (and within) such societies and historical periods.

Use values: Goods and services produced for direct consumption.

Exchange values: Goods and services produced as commodities for sale (hence consumption by those who did not produce them).

Simple reproduction: Reproduction of producers and their means of production on the same scale of productive assets and income.

Expanded reproduction: Enlargement of the scale of production, output and income, whether of individual enterprises or larger aggregates, though productive investment or *accumulation*. The accumulation of capital (from profits) and expanded reproduction (to make more profits) are the driving force of capitalism.

The view of peasants cited at the beginning of Box 2.2 is derived historically from the great agrarian societies of the pre-capitalist and pre-industrial era, based in classes of landowners and peasants. It is also thus a view of *continuity*, of the 'persistence' of the peasantry into the world of the twentieth (and soon the twenty-first) century. We then need to ask: how deep or real are such apparent continuities, often expressed in terms like 'persistence', 'tradition', 'custom', and so on?

An alternative view is that to understand the various livelihoods, crises and responses of 'peasants' in the Third World *today*, we need an *analytical* concept that can relate small-scale production to the distinctive social relations and processes of capitalism. The concept suggested is that of **petty commodity production** ('petty' meaning small-scale). This signifies that, as well as being manifested in distinct classes of workers and capitalists, labour and capital can also be combined in small-scale (usually 'household' or 'family') enterprises which produce for the market. What is distinctive about petty commodity producers is that they are capitalists and workers at the same time: capitalists because they own or

have access to means of production (unlike landless or otherwise propertyless workers), and workers because they use their own labour (unlike capitalists who employ the labour of others). In short, they are capitalists who employ (hence exploit) themselves (Figure 2.1).

> **Petty commodity production:** A distinctive form of production in capitalism that combines the class positions of capital and labour within small, typically 'household' or 'family', enterprises.

How can this alternative conceptualization be 'put to work' to throw light on the livelihoods, crises and responses of contemporary 'peasants'? To begin with, they have to reproduce their means of production (as capitalists) and their labour (as workers) in circumstances of greater or lesser risk or opportunity, whether exerted by:

- conditions of access to key resources (land, markets, credit) and relations with powerful

Figure 2.1   Indian farmer as petty commodity producer: combining his own labour with his own capital (land, plough, draught animals).

groups and individuals (landowners, merchants, agrarian and industrial capitalists);

- nature (climatic uncertainty, ecological degradation on the one hand, the availability of land and yield enhancing technical innovations on the other);

- markets (the relative prices, or terms of trade, of what they need to buy and what they need to sell to purchase necessities);

- government policies (affecting their economic conditions and access to social goods such as health care, clean water supply, and education).

The pressures of reproduction of both capital and labour, and the degree to which different peasants succeed in dealing with them, bear directly on questions of livelihoods, and can be charted in the social or class differentiation of peasants as petty commodity producers:

1 *Middle peasants* are those able to meet the demands of *'simple' reproduction*; that is, maintaining their means of production and raising the next generation of family labour to work them.

2 *Poor peasants* are subject to a *simple reproduction 'squeeze'* on their capital or labour or both. Often their poverty and extremely depressed levels of consumption (reproduction of labour) is an effect of intense struggles to try to maintain their means of production (reproduction of capital). If the latter is a losing battle, poor peasants become 'semi-proletarians' (see further below) or full proletarians whose livelihoods have to be pursued through working for others.

3 *Rich peasants* are able to engage in *'expanded' reproduction*, that is, to expand the land and/or other means of production at their disposal beyond the capacity of family labour. They then start to employ the labour of others. In short, they may undergo a transition from better-off petty commodity producers to capitalist farmers (Chapter 9 contains a more detailed analysis of such peasant class differentiation in Uganda).

Two further points are important to understanding and applying this conception.

First, the process of *commoditization* or commercialization (that is, the increasing production of goods and services for the market) is a fundamental part of the growth and spread of capitalism. It is also the process by which peasants are integrated within capitalism as petty commodity producers. Commoditization is both uneven and historically variable. Some elements of production and social life may be commoditized before others. For example, markets for crops and basic consumption goods often develop before markets in land or labour (Bharadwaj, 1985). In specific cases, then, we need to distinguish between different forms, degrees and phases of the *intensification* of petty commodity production. A good example of intensification is provided by the Green Revolution in India (Chapter 3), which increased opportunities for some peasants and risks for many others by introducing new inputs that have to be purchased as commodities (seeds, fertilizer, irrigation water).

Second, and most critically, the model of petty commodity production does not mean that a peasant household or family represents a homogeneous or egalitarian combination of 'collective' capitalist and worker. Often these class positions are distributed unequally *within* peasant households and families, above all through structures and practices of gender relations. *Sexual divisions of labour* in agriculture (who does what in terms of 'male' and 'female' tasks) vary a great deal between different rural societies, and between different classes and groups within them. On the other hand, there seems to be considerable convergence among different peasantries concerning male control over land (who owns what) and household income (who gets what), and the almost universal responsibilities of women for the sphere of reproductive or domestic work (again, who does what). Chapters 5 and 6 look inside rural households to examine these issues in greater analytical depth.

This brief introduction to some of the ideas and preoccupations underlying the accounts of agrarian structures and rural livelihoods that follow should allow critical assessment of their relevance and plausibility. It is worth noting where the material presented addresses the four key questions of a relational approach brought forward from Chapter 1 and how answers to these questions are contained in substantive ideas about processes of capitalist development and their historical variations. The answers are not necessarily complete – or may not even seem convincing – so it is always useful to consider what one would need to know to offer better or alternative explanations.

2.3 Latin America

Latin America was the first region of the Third World colonized by Europeans. Here, as elsewhere, European expansion was spearheaded by adventurers pursuing riches and glory. Their first major impact was on the densely populated areas of complex civilizations in Central America (the Aztecs and Mayas of present-day Mexico and Guatemala) and the Andes (the Incas of Peru).

These areas of initial conquest also experienced (subsequent to the usual plunder of existing treasure) the first concentrated economic activity of Spanish colonialism: the mining of silver, which reached a peak between the mid-sixteenth and early seventeenth centuries. Labour for the mines was secured through an annual labour tribute (contribution of unpaid labour) imposed on 'Indian' communities by the colonial administration, and allocated by it to mining (and other) enterprises.

Another form of forced labour was used to establish the sugar plantations of Portuguese Brazil: slavery using African slaves. Brazil was the principal destination of the transatlantic slave trade to the mid-seventeenth century, and slavery remained important to its agricultural export economy until late in the nineteenth century (Figure 2.2). Plantations appear and reappear in Latin America's agrarian history, the booms and slumps of their production of sugar, sisal, rubber, cocoa and coffee following the cycles of demand for these tropical crops in the different phases of development of the world market (*Hewitt et al., 1992*, ch.3).

2.4 The *hacienda*

In most of Spanish America, another form of agrarian property, the *hacienda*, or landed estate, dominated the economic (and, indeed, social, political and cultural) life of the countryside from the late seventeenth to the late nineteenth and even twentieth centuries. The *hacienda* was associated with various types of farming and labour use in different places at different times, but a constant motif in its emergence and evolution was the need of landowners to secure labour in conditions of shortage.

Spanish colonial conquest ushered in a demographic catastrophe. The immediate cause was the fatal epidemics of diseases it introduced to the New World – smallpox, typhoid, measles, malaria, yellow fever – but pressures on 'Indian' cultivation and food resources were also important, generated by the reallocation of labour (e.g. to mining), land (e.g. from cultivation to cattle ranching), and (irrigation) water in the early colonial economy (Wolf, 1959, ch.9). This suggests an observation of general relevance – that demographic trends alone cannot account for:

1 the 'problem' of matching the demand for labour and its supply in agrarian systems;

2 whose 'problem' it is (those who demand and those who supply labour);

3 how the 'problem' is resolved in different social conditions and processes of change.

A system of landed property like the *hacienda* is often marked by struggles between landlords and peasants over the control and uses of land and labour, as we shall see.

The formation of the *hacienda* brought together two crucial features of early Spanish colonization in Latin America: the granting to colonists of rights to levy tribute on Indian communities in the form of goods or labour services (the *encomienda*), and rights to land (*mercedes de tierras*) originally granted for military service to the Spanish Crown. Combining labour and land in the *hacienda* created a type of landed estate structurally very similar to the manor of European feudalism.

Figure 2.2   Brazil in 1835: (top) sugar mill; (above) slaves harvesting coffee.

Kay (1974) distinguishes two basic types of *hacienda* (again comparable to the manorial system), shown in Box 2.3.

In type A, most of the land is allocated to peasants to farm (and to graze their animals on), for which they have to pay the *hacienda* owner a rent, whether in a stipulated amount of produce (rent in kind), a stipulated proportion of their crop (i.e. a sharecropping arrangement), or in money. In type B, the landlord increasingly allocates land to production on his own account, worked with the unpaid labour of the *hacienda* tenants. Labour rent thus replaces forms of rent from the produce, or income, of the tenants' plots.

It must be emphasized that terms like 'tenant' and 'rent' can easily mislead if we associate them with, say, the kind of rental contract between 'improving landlords' and commercial tenant farmers in England's eighteenth-century capitalist agricultural 'revolution'. The *hacienda* in colonial (and subsequently independent) Spanish America typically involved forms of 'tenancy' much closer to *serfdom*.

'Enserfment' was pursued by appropriating land farmed by Indian peasant communities. This was done more to populate the *hacienda*, to secure a residential and dependent labour force, than to extend land under cultivation (although cattle raising, an important activity for many *haciendas*, was done through extensive ranching). The extension of the *hacienda* system involved protracted struggles, especially in areas of higher people/land ratios with strong peasant community organization, such as parts of Central America and the Andean highlands. In more sparsely populated regions like the plains of Chile, Argentina and Uruguay, the twin processes of *hacienda* formation and enserfment proceeded more rapidly.

A key factor, then, in understanding variations in the history of the *hacienda* concerns important differences in the social conditions of different areas. Another key factor concerns time as well as place. Struggles between landlords and peasants were also affected by historical patterns of the commercialization of agriculture, and their fluctuations. Generally, the forms of land use, labour regime and rent of type A were more prevalent at times of low or declining commercialization, whereas those of type B were more characteristic of periods of accelerated commercialization. As the potential profitability of farming increased with rising market demand, landlords sought to enlarge their own farms and to work them with the labour of the *hacienda* tenants by converting rent in kind or money to labour rent.

This adds another kind of struggle between landowners and peasants to that over the encroachment of *haciendas* on Indian land, namely *within haciendas* over the forms and intensity of labour exploitation. Labour service on the landlord's estate was generally experienced as more oppressive than peasants' full-time work on their own plots and payment of rent in money or kind,

---

Box 2.3   Two types of *hacienda*

Land use	Labour regime	Form of surplus
A Multi-farm estate (principally peasant farms)	Peasant cultivation of land allocated to them, and control of the labour process	Rent in kind, rent in money, crop shares
B Landlord estate (landlord's farm plus peasant farms)	Peasants work increasingly on landlord's (enlarged) farm, while maintaining their subsistence plots	Labour rent (= unpaid labour on landlord's farm)

especially when landlords attempted to increase the demands of labour service.

The outcomes of both kinds of struggles depended to a large extent on the 'balance of forces' between landlords and peasants in particular places at particular times. Where landlords confronted a shortage of labour for their own commercial farming *and* lacked the ability to resolve this through coercive means, they might begin to pay tenants for their work, at least in part. This, then, suggests the beginning of a possible transition towards wages rather than labour rent as the means of securing labour for commercial farming.

In fact, the *hacienda* system created two types of peasantry, whose relative sizes varied with local conditions: an '*internal*' resident peasantry, and an '*external*' peasantry which escaped incorporation in the *hacienda* but whose greater degree of independence was usually won at the cost of inhabiting marginal and limited land, farmed to produce a subsistence livelihood. If this became more difficult due to pressures exerted by population growth and land fragmentation in Indian communities, and/or by commercialization and monetization of the rural economy, then a precarious grasp on subsistence gave way to 'sub-subsistence' farming. The term *minifundio* in Spanish America expresses the notion of peasant subsistence landholdings, by contrast with the extensive landholding of the *latifundio* (in effect another name for the *hacienda*). Pressures on livelihoods in *minifundio* agriculture, then, necessitated the search for other supplementary sources of income, including daily or seasonal wage employment on landlord farms as commercialization led to an increase in farm size and demand for labour.

2.5 Phases of commercial expansion

The colonial economies of Latin America declined in the seventeenth century following the end of the silver boom. This decline was linked to an economic downturn in Europe, to which economic activity in the colonies was so closely connected from the beginning. Colonial wealth accumulated from mining and commerce diversified into land, contributing to the formation of the *hacienda* from the late seventeenth century. Export production increased, if unevenly, in the eighteenth century with economic revival in Europe, although this was supplied more from slave plantations than *haciendas*. The first two decades of the nineteenth century were marked by the widespread disruption of the wars against Spanish rule, followed by a period of restoring social and political order in the fragile new states created by independence. From the mid-nineteenth century, and especially from 1870, Latin America experienced an expansion of export agriculture, not least on the *hacienda*, that was qualitatively different from, as well as much greater than, earlier phases of commercialization.

First, this was because of the scale of international demand for primary products, both agricultural and mineral, and because of the more powerful and systematic links generated in this period between Latin American economies and the centres of global economic power. Particularly important were the investment of British capital in railway construction, mining, banking and trade in the Andean region and southern cone (Argentina, Chile), and of American capital in similar activities and in plantation agriculture, notably in Central America.

Second, a commercially and sometimes 'modernization' minded landowning class became the dominant force in the state. This class had the same centrality in national economic and political life in Latin America as the new industrial bourgeoisie did in Western Europe, leading to 'a virtually unique combination in the nineteenth century of political independence and primary commodity led incorporation into the international capitalist economy' (Archetti *et al.*, 1987, p.xvi). It was 'virtually unique' because other major areas incorporated into the world economy through the export of primary commodities (agricultural and mineral raw materials) were under *colonial* rule (see Chapters 3 and 4).

What were the effects for the *hacienda*, and other types of agrarian enterprise? First, there was a new wave of expansion of *haciendas* onto the lands of 'external' peasant communities, supported by the national and local power of the landowning class: 'in Mexico beginning in the 1870s, in Guatemala where the reduction of Indian lands was accompanied by anti-vagrancy laws, in Bolivia where two-thirds of the rural population became dependent on *haciendas*, and in fact throughout the Andean spine, the resources and means of independent livelihood of a great many rural people were reduced' (Bauer, 1979, p.52).

Second, together with land grabbing and a consequent growth in the size of the 'internal' peasantry, *haciendas* secured and retained a resident and dependent labour service peasantry through indebtedness. Advances in money or goods (sometimes imposed on peasants) bound labour to landlords' estates by creating debt bondage (*Allen & Thomas, 1992*, p.190), which might be perpetuated over generations of peasant families subject to the power of landlords to enforce and reproduce it. Where conditions allowed, plantation owners also resorted to directly coercive means of securing labour. For example, in the lowlands of southern Mexico the 'combination of strong markets for tropical exports (sisal, rubber, sugar), a labour shortage, geographical isolation, and a progressive (*sic*) state willing to support the planters with force explains the virtual enslavement of masses of Mayas and Yaquis' (Bauer, 1979, p.37).

Even with these various devices, the existing supply of labour in Latin America was insufficient to meet the increased demand from typically labour-intensive export crops. As happened earlier through slavery, the mass immigration of new workers was increasingly resorted to in the nineteenth century. Some came as indentured labourers (*Allen & Thomas, 1992*, p.190) from the great agrarian societies of Asia, now subjected to the effects of a world economy being reshaped by the needs of Western industrialization. Indentured workers from India filled the gap left by the abolition of slavery in the plantation labour force of Guyana; between 1847 and 1874 over a quarter of a million Chinese worked in the plantations of Cuba and coastal Peru.

Immigrants from Europe, with or without labour contracts, came to Latin America in even larger numbers. As coffee production increased in Brazil from the mid-nineteenth century, landowners experimented with importing Swiss and German sharecroppers, while still depending heavily on slave labour (an estimated 50 000 slaves worked in coffee estates in São Paulo as late as 1885). In the early 1880s the more forward-looking coffee planters, recognizing the inevitable abolition of slavery and the difficulties of attracting immigrants from Europe who had to repay the costs of their passage, were able to pass legislation committing the Brazilian state to subsidizing the costs of mass immigration. From 1884 to 1914 'some 900 000 (European) immigrants arrived in São Paulo, mostly as cheap labour for the coffee plantations' (Stolcke & Hall, 1983, p.182).

The great primary commodity boom in Latin America from 1870 to 1930 was thus a formative phase in many important respects. Industrial crops like cotton, sisal and rubber, wheat and livestock products from the *pampas* or prairies of Argentina and Uruguay, tropical foodstuffs and beverages like sugar, bananas and coffee, were produced in massive quantities for the rapidly growing industrial economies and urban populations of Europe and North America. A commercialization of agriculture unprecedented in its comprehensiveness, scale and intensity led to both renewed land grabbing and coercion of labour in some instances, and the development, however unevenly, of a rural labour market in many other instances. These labour markets were supplied in various ways: a transition from labour service to wage work in some *hacienda* regions; the increasing reliance on at least seasonal and supplementary wage labour of many 'external' peasant households; and by massive immigration of indentured and other contract workers.

Weakening international trade links and the collapse of export agriculture in the global depression of the 1930s and the years of the Second World War stimulated a period of more 'inner directed' economic growth in Latin America, including a marked acceleration of import-substituting industrialization (*Hewitt et al., 1992*). Significantly, this happened more in those countries like Brazil where a substantial part of the profits generated by export crops had been retained by local landowners and capitalists, than in countries where foreign capital had entrenched itself more deeply in export agriculture and its linkages (finance, trade, processing and transport).

After the Second World War and especially in the 1960s and 1970s, until the global recession of the late 1970s and 1980s (and Latin America's debt crisis), some countries, notably Brazil and Mexico, underwent considerable further industrialization. But this was now in very different international conditions, including the key role of the USA in reshaping the post-war world economy and the arrival on the scene, and increasing power, of multinational corporations in industrial manufacturing. Latin America's export agriculture revived and grew in ways strongly conditioned by this new context.

2.6 Paths of development of capitalist farming

The transformation of the once feudal-like *hacienda* to large-scale capitalist farm was, to different degrees in different places, a result of the pressures of competitiveness in international markets as well as more directly social and political pressures. Both occurred in a period of rapid population growth which for the first time provided a sufficiently large supply of labour at sufficiently low wages to commercial agriculture: 'In most countries [of Latin America] a crucial moment was reached in the third or fourth decade of this century when the supply of free available labour caught up with and overtook the demand, and the [labour] market replaced coercion' (Pearse, 1975, quoted in de Janvry, 1981, p.82). In short, any remaining benefits from trying to squeeze labour from a resistant 'internal' peasantry were superseded when new labour markets, crops and farming technologies made it unnecessary.

The final demise of the *hacienda* and its colonial legacy was also stimulated by political pressures of different kinds. Armed peasant movements demanding land redistribution were important in Mexico in 1917 and in Bolivia in 1952. In Chile between 1970 and 1973 land redistribution occurred in the unique circumstances (for Latin America) of an elected socialist government (Figure 2.3). More commonly, land reform was launched from above by 'modernizing' coalitions of agrarian capitalists, industrial and urban interests, and state technocrats, seeking to hasten

Figure 2.3 Campesino (peasant) demonstration in Concepción, Chile in 1973.

the conversion of 'traditional' *haciendas* with their underutilized land and low labour productivity to more profitable uses. In the 1960s the USA threw its considerable weight behind land reform from above as part of the Alliance for Progress, a programme of civilian and military aid intended simultaneously to modernize Latin American economies and to deflect threats of 'subversion' in the wake of the Cuban revolution.

De Janvry (1981, ch.6) concludes that most peasants failed to benefit from land reforms or were even harmed by them. The final transition of *haciendas* to modern capitalist farms, whether undertaken by their owners (under threat of expropriation), or by making their lands available to commercial investors, often resulted in the eviction of former labour tenants and sharecroppers. Even when land redistribution was more extensive as a result of popular movements and peasant uprisings, peasants rarely received enough good land and other resources (improved technology, access to irrigation, to credit and other institutional support) necessary to generate a viable livelihood from household farming. In two instances, popular land reforms were followed by counter-reforms or the restoration of redistributed land to its former owners: in Guatemala after 1954 and in Chile after 1973 following the bloody overthrow, supported by the USA, of the governments of Arbenz and Allende respectively. In general, the *minifundio* persists with varying capacities to reproduce itself under capitalism, and the rural population with no or insufficient land continues to grow (see Section 2.8 below).

The transformation of the quasi-feudal *hacienda* to large-scale modern farming, investing in new technology and employing wage labour, is the most widely discussed path of development of capitalist agriculture in Latin America, but several others can be noted briefly. These include the formation of modern plantations (whether owned by corporate or individual capital, national or foreign) and of commercial farms of varying type and size established by the investment of capital originating outside landed property, and likewise employing salaried managers and wage workers.

The most interesting and contentious of the other paths of development of capitalist farming is that emerging from the class differentiation of the peasantry, a process in which some peasant households are able to acquire sufficient resources in land and finance to expand and capitalize their production, often at the expense of their poorer neighbours who may be marginalized as farmers, and possibly lose their access to land altogether (see Sections 2.2 and 2.8). Richer peasants able to travel this path of development may become 'capitalized family farmers' of the kind common in North America and Western Europe, or (small-scale) capitalist farmers employing the labour of others.

Both Kay (1974) and de Janvry (1981) are sceptical about the extent to which differentiation of the peasantry has generated capitalist farming in Latin America relative to other paths of its development, because of the limits on peasant landholding and accumulation exerted by the *hacienda* system, and its legacy of social and political subordination, bitterly contested though this has been by a long history of peasant movements and uprisings. On the other hand, it is likely that a small minority of peasants, benefiting from land reforms and/or other measures such as integrated rural development programmes and land colonization schemes that provided them with sufficient resources, succeeded in transforming themselves into 'capitalized family farmers' or indeed capitalist farmers employing labour.

2.7 Agriculture and industry

These different 'roads' to capitalist farming – of varying weight in different countries, and in different areas within them – should be placed in the broader context of the international linkages and specific patterns of industrialization of post-war economic development in Latin America. Although the extent and types of industrialization have also varied widely between different countries, de Janvry (1987, pp.392–3) noted some common characteristics:

1   While industrial growth has often been rapid, it has also been unstable and has failed to generate sufficient new employment even during boom periods.

2 Real wages of unskilled workers have not risen significantly over time, hence industrialization has not modified patterns of highly unequal income distribution.

3 The fastest growing sectors have been capital goods and luxury consumption goods rather than wage goods (commodities that people need for basic consumption).

In effect, industrialization has been fuelled by cheap labour, manifested in the downward pressure of those seeking work on (unskilled) wage rates, and persistent or increasingly unequal income distribution. Cheap labour is also the key to the average annual growth rate of agricultural GDP (see below) in Latin America of 3.3% between 1950 and 1980, and the ways in which agriculture has contributed to industrial growth.

The first way, de Janvry argues, is that the terms of trade between agriculture and industry (that is, the relative prices of agricultural and industrial commodities) have systematically undervalued the former, especially food crops, in relation to the latter. At the same time, the profitability of capitalist farming has been sustained partly by the depression of agricultural wages, and partly by 'institutional rents' (Box 2.4). Institutional rents refer to policy instruments such as subsidized credit, mechanization, public extension services, infrastructure projects, and differential pricing, all of which can be targeted to particular crops, areas, and agricultural subsectors in which capitalist farming is dominant.

Secondly, the major growth of capitalist farming has been concentrated in industrial crops and luxury commodities, especially for export. This is shown by the differential growth rates of output of major crops between 1950 and 1976 (Table 2.1).

Table 2.1 Growth rates of output of major crops in Latin America between 1950 and 1976 (%)

Industrial and export		Food	
Cotton	1942	Beans	72
Sugar cane	573	Maize	62
Soya	447	Cassava	12
Tobacco	300	Wheat	−43

Source: Kandiyoti, D. (1985) *Women in Rural Production Systems*, UNESCO, Paris, p.62.

Industrial crops provide raw materials for processing and manufacturing. Exports earn the foreign exchange needed to buy imports. In the particularly dynamic growth of Brazilian soybean products, exports increased in value from US$14.6 million in 1965 to US$1.3 billion in 1975 and US$1.6 billion in 1979, representing 0.9%, 15% and 11% of the value of Brazil's total exports in those years (Goodman & Redclift, 1981, p.170).

Thirdly, industrialization based on low wages is dependent on a supply of cheap food, given that food is the essential wage good (that is, essential to the daily reproduction of workers) (Figure 2.4). Cheap food has been supplied to some extent by

Box 2.4 The concept of 'rent'

The term 'rent' has a number of meanings in economics that differ from its colloquial usage. The most relevant meaning here is: a payment or gain, deriving from a position of special privilege, that is additional to what would be expected in a competitive market.

A common example of this is monopoly – when buyers or suppliers are able to act as 'price-makers' because they are the sole dealers in particular commodities (see Chapter 4). De Janvry argues that capitalist farmers in Latin America enjoy privileged access to goods and services supplied through government policies and institutions, because of their political influence. As a result, they acquire inputs, credit and other services on preferential terms below a competitive market price, and benefit from (institutional) 'rents' denied to peasant farmers.

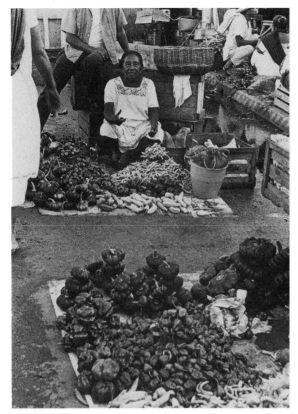

Figure 2.4   Industrialization based on low wages is dependent on a supply of cheap food: urban food market in Mexico.

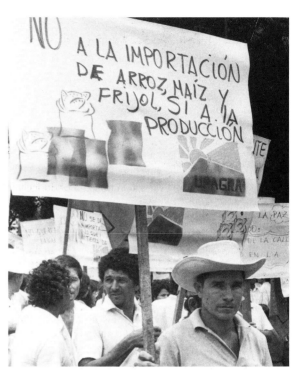

Figure 2.5   Costa Rican farmers protest against food imports: 'No to imports of rice, maize and beans. Yes to production.'

those peasant farmers with sufficient resources who are prepared to produce food for the market at prices that capitalist farmers would not accept. This source of cheap food thus rests on the implicit 'cheapening' – or undervaluing – of peasant labour that grows and sells it at prevailing prices: a particular form of peasant 'self-exploitation' (see Section 2.2).

It is perhaps not surprising that most Latin American countries have become increasingly dependent on imported grain, a process also stimulated by United States food aid in the 1950s, which paved the way for subsequent commercial imports (Figure 2.5). In the 1970s, increasing concern about the growing food import dependence of Latin America led to incentives to capitalist farmers to invest in grain production. They have done so to some extent, particularly in wheat and rice, but with rather patchy results (de Janvry, 1981, pp.69–81, ch.4).

2.8   Rural livelihoods

Now we turn to the majority of the rural population, those without land and those who have been otherwise marginalized as farmers – in short, those people whose livelihoods are likely to be most precarious. The tables below present data for three groups of countries, with three examples in each group, that together help give an idea of the wide variation in national conditions in Latin America. Latin American countries in the three groups (with those shown in the tables italicized) are as follows:

Group 1:   Bolivia, Dominican Republic, *El Salvador, Guatemala*, Haiti, Honduras, Nicaragua, *Paraguay*.

Group 2:   *Brazil, Colombia*, Costa Rica, Ecuador, Mexico, *Peru*, Panama.

Group 3:   *Argentina*, Chile, *Uruguay, Venezuela*.

To begin with, Table 2.2 gives some aggregate indicators for the three groups.

Table 2.2   Indicators for three groups of countries, 1980

	Group 1	Group 2	Group 3
GDP per capita (US$)	354	829	1075
Agricultural GDP as % of GDP	23.0	12.9	8.4
Rural population as % of total population	59.5	34.4	17.6
Agricultural EAP as % of total EAP	56.3	31.9	14.7
Ratio of share of rural population to share of agricultural EAP	1.06	1.08	1.20

GDP, gross domestic product.

EAP, economically active population.

Source: de Janvry, A., Sadoulet, E. & Wilcox Young, E. (1989) 'Land and labour in Latin American agriculture from the 1950s to the 1980s', *Journal of Peasant Studies*, 16(3), p.99.

These indicators show a correlation between rising **gross domestic product** (GDP) per capita in the national economy and declining shares of agricultural GDP, rural population, and the economically active population (EAP, or population available for employment) employed in agriculture. Interestingly, the last line of the table suggests that in the wealthier economies, rural *non*-agricultural employment is more plentiful.

Turning to Table 2.3, the third and fourth columns indicate the massive urbanization of Latin America

Gross domestic product (GDP): The World Bank defines GDP as the 'total final output of goods and services produced by an economy'. Thus GDP measures the size of the economy.

from 1960 to 1980 (Figure 2.6), when the share of rural population declined in all the countries shown, from about one-half to one-third overall. A

Figure 2.6   The urbanization of Latin America: Rio de Janeiro, Brazil.

Table 2.3 Trends in rural and agricultural population, 1960–80

	Average annual growth rates of population		Rural population as % of total population		Agricultural EAP as % of total EAP		EAP in agriculture as % change, 1960–80	
	Total	Rural	1960	1980	1960	1980	Peasant	Modern
Group 1								
El Salvador	2.9	2.7	61.7	58.9	61.6	50.5	128.4	10.0
Guatemala	3.1	2.6	67.0	61.1	66.7	57.7	17.1	24.6
Paraguay	2.6	2.3	64.4	60.6	53.6	44.9	–	–
Group 2								
Brazil	2.5	–0.1	53.9	32.4	52.1	30.4	57.4	25.3
Colombia	2.5	0.7	51.8	36.3	51.4	25.8	28.8	–1.6
Peru	2.7	0.5	53.7	34.8	52.1	39.8	15.3	–26.1
Group 3								
Argentina	1.5	–0.5	26.4	17.6	20.2	13.1	4.8	–20.0
Uruguay	0.7	–0.4	19.9	16.0	19.7	10.8	67.7	–20.1
Venezuela	3.4	–0.1	33.4	16.7	33.7	18.0	32.8	–36.6
Latin America	2.6	0.6	50.2	34.3	48.7	31.7	41.3	16.3

EAP, economically active population.

Source: Adapted from de Janvry, A., Sadoulet, E. & Wilcox Young, E. (1989) 'Land and labour in Latin American agriculture from the 1950s to the 1980s', *Journal of Peasant Studies*, 16(3), tables 1, 3, pp.398, 401.

striking feature of this process is the predominance of female urban migration from rural areas, especially of young women but also of widowed, abandoned and older unmarried women (Kandiyoti, 1985, pp.59, 65–6; see further below).

As this rural exodus exceeded the employment created in new industries, it helped keep urban (unskilled) wages low, and represents a transfer of poverty from countryside to city. De Janvry *et al.* (1989, p.402) illustrate this by using 'marginal' sector employment as a proxy for low incomes and insecure livelihoods. They define 'marginal' sector employment as the sum of the EAP in both 'traditional' agriculture (peasants) and 'traditional' urban activity (petty production and services), and point out that it declined very little between 1950 and 1980, from 47% to 42% of total EAP in Latin America, despite the high rates of economic growth and industrialization that occurred.

Despite massive urban migration, the absolute size of the rural population (second column in Table 2.3) increased in Latin America as a whole and in five of the countries shown, with only a slight decline in the other four countries. Likewise, while the share of agriculture in EAP declined in all the countries, and for Latin America from almost one half to less than one third (fifth and sixth columns), the absolute numbers of 'peasants' increased in all eight countries (there being no data for Paraguay), whereas the numbers of 'modern' sector workers (wage workers on capitalist farms) increased in El Salvador, Guatemala and above all Brazil.

In Table 2.4, the third column shows that the number of small farms increased between the dates shown in all countries except Argentina and Venezuela, and almost doubled for Latin America as a whole from 1950 to 1980. The fourth and fifth columns show the extremely unequal distribution of agricultural land in all Latin American countries, and the sixth column indicates a trend of declining size of small farms for most countries and for the continent as a whole.

As landholdings of the average size indicated in the last column of Table 2.4 tend to have poorer soils and physical access, and to lack irrigation

and other productive investments, the numbers of such farms (column 3) serve as a proxy for the numbers of rural households unable to secure adequate livelihoods from their own farming. This is supported by the data from more detailed rural surveys presented in Table 2.5.

We can now see why de Janvry argues that the great majority of rural people in Latin America who have retained access to some land have been 'semi-proletarianized' (Box 2.5), and struggle to secure their livelihoods from combining sub-subsistence farming with agricultural or urban wage work, and other activities such as artisanal production and petty trade. Even Garcia Rovira in the highlands of north-west Colombia, where smallholders produce maize and beans for their own consumption and tobacco for sale, and where wages contribute a relatively smaller share of household income (see Table 2.5), is an area of temporary labour migration and permanent out-migration to other parts of Colombia and to neighbouring Venezuela.

Box 2.5   Semi-proletarianization

Many political economists argue that the extent and continuity of semi-proletarianization expresses a specific feature of capitalism in Third World countries. That is, the scarcity of secure employment combined with low wages means that many households can gain their livelihoods only by combining wage work with various types of petty commodity production, whether rural or urban. This not only inhibits full proletarianization, but suggests that semi-proletarianization is a structural rather than transitional or temporary condition.

It seems misguided to try to decide whether rural semi-proletarians in general are really farmers (or 'peasants') who also work for others, or really workers who supplement insecure and inadequate wages with farming and other 'own account' activities. According to their circumstances, different groups of rural people at different times might be closer to the former or the latter.

Table 2.4   Number and average size of small farms over time

	Years	Maximum farm size (hectares)	Number of farms (thousands)	% of farms	% of area	Average farm size (hectares)
Group 1						
El Salvador	1950	5	140	80.6	12.4	1.4
	1971		237	86.9	19.7	1.2
Guatemala	1950	7	308	88.4	14.4	1.7
	1979		548	90.0	16.0	1.8
Paraguay	1943	5	45	48.1	8.0	2.7
	1956		69	45.9	1.0	2.4
Group 2						
Brazil	1950	5	459	22.2	0.5	2.6
	1980		1888	36.6	1.1	2.1
Colombia	1954	10	648	71.0	6.9	2.9
	1970		860	73.0	7.2	2.6
Peru	1961	5	699	82.9	5.2	1.3
	1972		1084	77.9	6.6	1.4
Group 3						
Argentina	1952	25	235	41.8	1.1	9.2
	1969		226	42.0	0.9	8.9
Uruguay	1951	20	36	42.0	1.8	8.3
	1961		40	45.8	1.9	8.0
Venezuela	1950	5	126	53.7	1.2	2.1
	1971		122	42.3	1.0	2.2
Latin America	1950	small farms	4134	–	–	2.4
	1980		7949	–	–	2.1

Source: Adapted from de Janvry, A., Sadoulet, E. & Wilcox Young, E. (1989) 'Land and labour in Latin American agriculture from the 1950s to the 1980s', *Journal of Peasant Studies*, 16(3), table 7, p.407.

The two principal sources of livelihood of peasants who are semi-proletarians are farming and wage work, both of which are subject to their own pressures especially in periods of economic recession like the 1980s (Figure 2.7). Pressures on farming arise from trying to maintain the output of smallholdings, in which the intensification of household labour – and especially that of women (see Box 2.6 below) – has to substitute for lack of irrigation, improved seed varieties, fertilizers, and other yield-raising inputs.

Pressures on wage income arise from (a) the patterns and fluctuations of demand in rural and urban labour markets, affecting how much employment is available, when, and where, to those seeking it, and (b) trends in real wages, affecting how much people are paid when they do get work. Clearly, the mechanization of farming is likely to depress demand for labour, other things being equal. While agricultural wages in Latin America as a whole increased in real terms in the 1970s at an average annual rate of 2.4%, with the gathering

Table 2.5 Sources of income of small farm households, 1970s, various years

Region	Maximum farm size (hectares)	% share of income from: Farm	Wages	Other	Sum of wages and other
Cajamarca (Peru)	3.5	23	50	27	
Puebla (Mexico)	4	32	58	11	
Garcia Rovira (Colombia)	4	79	16	5	
South Bolivia	5	38			62
Region IV (Chile)	5	47	40	13	
Vertentes (Brazil)	10	–	56	–	
North-west Altiplano (Guatemala)	3.5	29	59	12	
El Salvador	2	64	27	9	
Sierra (Ecuador)	5	37	44	19	
Coast (Ecuador)	5	48	41	11	
Chamula (Mexico)	–	11			89

–, no data available.

Source: de Janvry, A., Sadoulet, E. & Wilcox Young, E. (1989) 'Land and labour in Latin American agriculture from the 1950s to the 1980s', *Journal of Peasant Studies*, 16(3), p.410.

Figure 2.7   Wage work: cutting sugar cane in Brazil.

crisis in the early 1980s they fell at an average annual rate of 4.4% (de Janvry *et al.*, 1989, p.414).

In addition, there has been a marked trend from permanent to seasonal wage employment in agriculture. This predated accelerated mechanization from the 1960s, reflecting the desire of capitalist farmers to avoid labour legislation which regulated the conditions of employment of permanent workers. With full mechanization of farming operations in some crops and regions, wage employment tends to stabilize, albeit at low levels, whereas partial mechanization bunches demand for labour around certain periods, e.g. at harvest time, which creates a marked seasonality in employment opportunities (Figure 2.8).

These pressures on the livelihoods of the rural poor, and their responses to them, have marked gender characteristics in Latin America as elsewhere. However, the general rate of female participation in agricultural work tends to be much lower than in Asia and Africa (even though census data are notorious for underestimating women's participation both as family workers on peasant farms and as agricultural wage workers). This is partly because of the dominance of large-scale capitalist farming in Latin America with its high levels of mechanization and associated patterns of employment, and partly because the land available to poor peasant households is insufficient to utilize family labour. This necessitates the

Figure 2.8 Seasonal work: migrant labour for the coffee harvest in Guatemala.

pursuit of wage employment, including gender-specific patterns of labour migration like the urban migration of young women noted earlier.

Of course, this overall picture is subject to much specific variation. For example, some labour-intensive branches of capitalist farming and agro-industry favour employing women for the same reasons as manufacturing sectors like clothing and electronics, justified by the same ideologies of female 'traditional skills' and 'nimble fingers': for example, milking (Peru), cultivating and picking flowers (Colombia), and harvesting strawberries and other fruit (Mexico, Chile). Box 2.6 summarizes some of the factors underlying the gender characteristics, and variations, of rural livelihood strategies in Latin America.

Box 2.6   Gender and rural livelihoods in Latin America

Intimate links can be detected between different paths of rural transformation and the need for female labour in agriculture. In those cases where the peasant family has to produce for the market, or when men migrate for work, leaving women in charge of their *minifundios*, there is an intensification of their contribution as unremunerated family labourers.

When, on the other hand, peasant households lose their access to productive resources and simultaneously come to rely increasingly on additional cash incomes, their labour redeployment strategies frequently involve sending out daughters, chiefly to service jobs in cities, domestic service figuring prominently among these. There is a clear connection between the prevalence of female out-migration and women's dispensability as an agricultural labour force.

There is also a significant process of female proletarianization as increasing numbers of women join the salaried workforce. However, women enter wage-work laden with the generic limitations of their primarily domestic role definitions. Recruitment and wage policies of agro-industries employing rural women capitalize upon and reinforce this conception of women's place by favouring a young, unmarried workforce, able to settle for low wages, intermittent work and aiming to withdraw from work upon marriage.

However, low and precarious incomes contribute to create a simultaneity of conditions...whereby women may be involved in wage-work alongside unremunerated agricultural tasks and their habitual domestic chores. This multiple workload is exacerbated by the fact that there is no evidence of a more equitable redistribution of household tasks between the sexes. The effects of poverty and of the agrarian crisis are clearly compounded with women's subordination as a gender.

(Kandiyoti, 1985, pp.75, 77)

Summary

1   The dominant form of landed property in colonial and independent Spanish America to the early twentieth century was the *hacienda*, a type of feudal estate whose formation and expansion involved the dispossession, incorporation or marginalization of Indian farming communities.

2   The export agriculture boom from about 1870 to 1930 witnessed a major commercialization of *hacienda* production (as well as plantation expansion) at the expense of the Indian peasantry.

3   The *hacienda* underwent a transition to capitalist farming during the twentieth century, replacing a labour service peasantry with wage workers; this transition was partly stimulated by pressures for land reform from below and from above.

4   Industrialization and urbanization since the 1950s were made possible by cheap labour among other factors, but failed to generate sufficient employment for those seeking jobs.

5   Urban migration, partly spurred by highly unequal patterns of land distribution, thus represents a transfer of poverty from countryside to town.

6   Rural semi-proletarianization has intensified – the majority of 'peasants' depend on a combination of sub-subsistence farming and wage work for survival. Gender divisions in access to land and different types of wage employment mean that landless and semi-proletarian women are particularly subject to acute pressures and deprivation.

3

AGRARIAN STRUCTURES AND CHANGE: INDIA

HENRY BERNSTEIN

Latin America was the prize of Spanish and Portuguese attempts to find a westward route from Europe to the fabled treasures of the Indies, whence derives the name 'West Indies' for the islands of the Caribbean, and the bizarre if now habitual anomaly of the term 'Indians' for the indigenous peoples of North, Central, and South America. European explorers and adventurers also sought the Indies by an eastern maritime route involving the lengthy task of first circumnavigating Africa, in the course of which they established bases that helped supply the transatlantic slave trade from the sixteenth to nineteenth centuries.

From initial trading posts on the coasts of South Asia, British expansion into the interior eventually created a massive and exceptional colonial possession. 'British India', the 'jewel in the crown' of its colonial empire, encompassed the present-day countries of India, Pakistan, Bangladesh, Burma, and Sri Lanka (then Ceylon). Starting in Bengal (north-east India and contemporary Bangladesh), the soldiers and officials of the British East India Company enjoyed a bonanza in the second half of the eighteenth century as their conquests yielded vast sources of plunder. As the Company extended and consolidated its rule between the late eighteenth and mid-nineteenth centuries, plunder gave way to considerations of more systematic sources of income: a transition from piracy to bureaucracy, as Barrington Moore

(1966, p.342) characterized it in relation to the land revenue systems of colonial India. The first system stemmed from the 'Permanent Settlement' in Bengal and its adjacent areas from 1793, in which the *zamindars* (the tax-farmers and revenue collectors of the previous Mogul state) were converted into quasi-landlords, with certain private property rights in land contingent on collecting revenue from the peasantry. The Bengal land settlement (as well as the eventual incorporation of 600 or so princely states in the British *raj*) was also a means of trying to secure indigenous political allies in establishing and administering these vast colonial domains (Figure 3.1).

Figure 3.1   The administration of colonial India: the Rajah of Pudikkotai with the Assistant Political Agent, Mr Tagg.

The other major land 'settlement' was the *ryotwari* system (after *ryot* or 'peasant') introduced in large parts of Bombay and Madras and later in areas of north-west and north-east India. This confirmed property rights in land on those already cultivating it (at least in principle), subject to the annual payment of a money tax. The association of rent or tax with individual property rights in land and monetization, together with the increasing commercialization of the countryside in the nineteenth century, has had a persisting resonance for land, labour and rural livelihoods, as we shall see.

3.1 The colonial economy

In the latter half of the nineteenth century the economy of colonial India was comprehensively integrated into the burgeoning world market, then dominated by British industrial and banking capital. Of course, trade had also been an objective of the East India Company, not least the notorious opium trade with China. Bengali peasants were made to cultivate opium and the Chinese were forced to import this in exchange for tea, which was carried by the Company's ships to Britain.

From the 1860s, the pace of commercialization accelerated. British cloth manufacturers targeted India (and Egypt) as their principal sources of supply of raw cotton to avoid dependence on the USA, where cotton was 'king' in the plantation south (and where many former slaves became cotton sharecroppers after the Civil War). Jute was another important industrial crop grown by peasants, while plantations for tea and other products expanded massively in the late nineteenth century, particularly in Assam and Ceylon and also in parts of south India (Figure 3.2).

The commercialization of agriculture was made easier by the growth of transport – notably the

Figure 3.2   British commercial expansion: tea factories were well established in Ceylon by the time of this photograph in 1923.

expansion after 1858 of the railway network, which carried export crops to the ports and distributed manufactured imports to all parts of the country, thereby completing the decline of many of India's numerous rural and urban artisans. Before 1914, 97% of British investment in India was in government bonds, transport, plantations and finance – 'precisely those economic sectors tributary to the commercial penetration of India, enlarging the market for Indian raw materials and British manufactured goods' (Clairmonte, 1960, p.126).

It is difficult to overstate how strategic India was to Britain's economy, both during the crucial period of its industrial 'take-off' between 1770 and 1820 and subsequently, especially when it was subjected to increasing competition from other rapidly industrializing powers like Germany and the USA from the 1870s (Hobsbawm, 1969, pp.148–9). In 1910 India had the largest merchandise export surplus in the world after the USA (Patnaik, 1990b, p.7). But America's exports reflected rapid and massive economic development, whereas India's exports of raw materials reflected its subordination as a 'captive' source of supply within Britain's imperial trade system. Before the First World War, 'the key to Britain's whole payments pattern lay in India, financing as she probably did more than two-fifths of Britain's total deficits' (Saul, 1960, p.62). These dimensions of India's colonial economy provide a necessary context for considering the relations and dynamics of its agrarian structure, and their effects for rural livelihoods, that were inherited with independence in 1947.

Land and rent

To understand why colonialism left India (and South Asia as a whole) with 'perhaps the world's most refractory land problem' (Thorner & Thorner, 1962, p.57), we need to look more closely at how the land systems, outlined earlier, worked in practice. Their overall effect is stated succinctly by Barrington Moore:

> "the [land] settlements were the starting point of a whole process of rural change whereby the imposition of law and order and associated rights of property greatly intensified the problem of parasitic landlordism. More significant still, they formed the basis of a political and economic system in which the foreigner, the landlord, and the moneylender took the economic surplus away from the peasantry, failed to invest it in industrial growth and thus ruled out the possibility of repeating Japan's way of entering the modern era."
>
> (Moore, 1966, p.344)

'Parasitic landlordism' was the key to agrarian structure, because the land systems generally made it much more profitable to control land as a source of extracting *rent* from the peasants who worked it, than as an arena of investment to raise agricultural production and productivity. The somewhat flexible Mogul system of taxes in kind was replaced by the tyranny of annual fixed rents and taxes payable in money. This squeezed peasants in times of bad harvests or declining crop prices, and also smaller or profligate *zamindars* responsible for passing on land revenue to the colonial state. Many *zamindars* in Bengal sold out to merchants and speculators from Calcutta who invested in land for the rents, and came to form an urban *rentier* or absentee landlord class.

Similar oppressive effects for most peasants also occurred in *ryot* areas. The settlement of property rights followed pre-existing inequalities in

> **Parasitic landlordism:** A system characterized by exploitation by a landholding class of those who work the land though the extraction of rent. It is 'parasitic' because the income extracted is not reinvested to maintain or enhance the productivity of farming. An example was in Chapter 2: many *haciendas* in Latin America were seen as similarly parasitic, both by the peasants who worked them, and by the modernizers who wanted to convert them to more productive capitalist farming.
>
> ***Rentier:*** A French term for someone who lives from rents.

landholding, and was also subject to a great deal of chicanery, allowing the emergence of landlords, if on a more modest scale than the *zamindars* of the north. Using local influence and political connections, often directly with revenue officials, such landlords were able to consolidate their hold on land and to pass on its tax burdens to their tenants by the threat or use of force.

The **'forcible commercialization'** (Bharadwaj, 1985) of peasant farming in such conditions, initially to earn the money required for rents and taxes, succeeded in pumping out cash crops from the countryside through intense exploitation, especially of those peasants with the weakest rights of access to land (known as tenants-at-will, meaning at the will of the landlord). Their numbers were swollen by those who had lost their own land through indebtedness, or their livelihoods as artisans through competition from factory-produced imports, or had no land to start with. The growth of peasant cash crops (cotton, jute, indigo, groundnuts, sugar cane, food grains) took place without any significant investment in the quality of the soil or in improving yields, apart from some limited public investment in irrigation works. Landlords, merchants and moneylenders (often the same people) were more interested in incomes from rent, trade and interest than in productive investment. Most peasants, struggling with insecurity of tenure, debt, and fragmentation of their holdings, lacked the means to improve the quality of their farming.

> **Forcible commercialization:** The process whereby small farmers are forced into producing for the market by various means, both direct and indirect. This process is characteristic of the transition to capitalism in different parts of the world at different times. It is further illustrated in colonial Africa in Chapter 4.

These conditions generated widespread poverty and intense insecurity. As well as supplying tenants-at-will, sharecroppers, and bonded labourers or servants to landlords, rural destitution

provided a stream of indentured labourers to many parts of the British Empire, to the sugar plantations of the Caribbean (Guyana and Trinidad), the Pacific (Fiji), the Indian Ocean (Mauritius), and South Africa (Natal), and to build the railways in British East Africa. The growth of cash cropping diverted resources from peasant production of food staples for their own consumption. Major famines occurred at times when food was being exported from India: for example, in 1876–77 (Clairmonte, 1960, p.102) and 1943–44 when four million people died in Bengal (Sen, 1981).

This last point merits some comment. The colonial history of India provides a fascinating record of the uses of economic ideology, much of which is relevant to understanding (or 'decoding') current debates about Third World development (and, for that matter, about economic policy in the industrial capitalist countries and in the aftermath of state socialism in eastern Europe). On one hand, much of the formative framework of colonial rule in India was proposed by the utilitarians and free market liberals of late eighteenth and nineteenth century Britain. They hoped that the *zamindari* settlement in northern India would create a class 'akin to the great Whig landed aristocracy (of England): a class of improving landlords, under whose aegis a prosperous capitalist agriculture would emerge' (Byres, 1991, p.60). In fact, the development of parasitic landlordism and money lending, or usury, made a perverse reflection of this rosy vision, a shadow cast by the suffering of the Indian peasantry. The same applies to much of south India where Sir Thomas Munro, the architect of the *ryotwari* system in Madras, fantasized about implanting another English 'model', the sturdy yeoman farmer, thus: 'a crowd of men of small, but of independent, property who, when they are certain that they will themselves enjoy the benefits of every extraordinary exertion of labour, work with a spirit of activity which would in vain be expected from the tenants and servants of great land-holders' (quoted in Low, 1973, p.46).

On the other hand, as against such robust promises of progress under its benign paternalism, the colonial state in India was strongly bureaucratic and interventionist in promoting and protecting the interests of British capital. Its applications of free market liberalism were highly selective in practice. While 'Famine Codes' to alleviate distress were introduced after 1880, the 'British Indian administration considered governmental involvement in food trade or distribution as sacrilegious' (Drèze & Sen, 1989, p.124), even when many were dying of hunger, which recalls exactly the stance of another British colonial state during the great famine (or, as it is more tellingly known in Ireland, the 'great starvation') of 1848–49 (Gibbon, 1975).

3.2 Independence and land reform

Given the picture outlined, it is not surprising that agrarian unrest was an important component of the anti-colonial movement nor that land reform under the slogan of 'land to the tiller' (that is, to those who work it) was high on the agenda of the Congress party which formed the government of independent India in 1947.

Redistributing land, or confirming rights to occupation and ownership of those who farmed it, proved to be very difficult, however. While some of the grandest *zamindar* landed properties and the aristocratic domains of the princely states (both strongly associated with the colonial order) were more accessible targets for reform and redistribution, elsewhere the land problem remained 'refractory' for two reasons.

One was the prevalence of smaller scale landlordism – the 'squirearchy' owning 25 to 40 hectares (Warriner, 1969, p.139). Its close interconnections with crop trading, money lending, and political and cultural power at local and village level bound together rural society in a dense web of class and caste relations.

The second reason is that like all anti-colonial movements of a broad 'national' type, Congress brought together a wide range of class and other social groupings and interests, of which landlords were well represented at the level of state (i.e. regional) government within the federal structure of independent India. The decision to leave land

reform legislation and implementation to the jurisdiction of state governments, where landlord influence was strong (if sometimes bitterly contested by peasant movements in the late 1940s), therefore obstructed any radical and comprehensive distribution of land rights 'to the tiller'. Most landowners (and certainly those within the 'middling' to small range) were able to deflect any more thoroughgoing land reform, frequently with official connivance at state and local levels, where reform legislation, compromised as its final drafts usually were, turned into an administrative nightmare. In some cases, resident tenants and sharecroppers were evicted (as happened in many Latin American land reforms) or turned into bonded labourers, so that landlords could claim that they themselves were farming the land.

Thorner (1956, ch.1) characterized the agrarian structure following land reform in terms of three main groups.

1 *Maliks* Landowners living off rents, or farming on their own account, or some combination of the two. Those with larger holdings in more than one village are more likely to be absentees, employing managers to collect rents and supervise cultivation by tenants or labourers. Smaller landowners are more likely to be resident and to exercise some management themselves, whether of tenants' and sharecroppers' cultivation or of their own farms. Both types of landowners have common economic interests as the chief receivers of rent and employers of agricultural labour.

2 *Kisans* 'Working peasants', i.e. those with sufficient land to be viable family farmers with little involvement in renting land or hiring labour – in short, the owner-occupiers and 'independent' farmers that the advocates of land reform in the Congress national leadership aspired to establish and generalize (in an echo of Sir Thomas Munro's vision of the yeoman farmer, quoted above). To what extent peasants were, or are, as homogeneous as this description suggests, is a matter of some controversy in India as elsewhere (see further below). For example, Byres & Crow (1988, p.166) suggest that land reform and agricultural growth in the 1950s tended to accelerate the differentiation of the peasantry and in particular to benefit richer peasants who were 'stabilized as independent proprietors...on the way to becoming, in many areas of India, the new dominant class in the emerging agrarian structure'.

3 *Mazdurs* Agricultural labourers, recruited from the landless (estimated to be about 15% of the rural population in 1951) or from poor peasants with insufficient land to produce viable livelihoods. They provided tenants and share-croppers to the *maliks*, and labour to both *maliks* and richer peasants (whose demands for labour exceeded the supply of household labour available to them).

The relative size and strength of these three main groups varied greatly in different parts of India; moreover, trends in their growth or decline over time in different regions are both complex and contentious. Warriner (1969, pp.144–5) suggested that the poor (landless households and those cultivating less than 1 hectare) comprised 70–75% of the rural population in the 1960s.

Did this minimal land reform mean that the colonial agrarian structure remained largely unchanged? The answer must be at least a qualified 'no'. First, independence ended the long period of India's colonial 'tribute', opening up certain opportunities at the macro-economic level for the first time. Second, the Congress government was strongly committed to national economic development based on industrialization within a framework of state planning and investment. Third, these conditions and the buoyancy they created in the 1950s provided a new stimulus to agricultural growth, supported by the replacement of colonial land revenue by other forms of land taxation (never strongly enforced in practice), and the strengthening of the rich peasantry in many areas as a result of land reform. Rich peasants, as well as those former *zamindars* and other landowners who saw a new role for themselves as commercial farmers, began to invest in increasing production and productivity.

This is not to say that landlordism and oppressive rental, sharecropping, and moneylending arrangements disappeared; indeed, they still exist in many rural areas of India and are sometimes perceived as 'semi-feudal' in character (Bhaduri,

1983). Nor does it mean that the agrarian structure became significantly more egalitarian, or that chances improved for a majority of rural people. What did happen was that, in some key areas at least, productive investment in commercial farming started to provide an alternative (or, for some landlords, a complement) to the activities of renting out land, trade and moneylending. New energies were committed to farming, and there were new incentives for those who commanded the resources to respond to them. In the 1950s and until the disastrous harvests of 1965–66 and 1966–67, agricultural output grew at an average annual compound rate of just under 3%, faster than in the last 50 years of the colonial period.

The mid-1950s to the mid-1960s was also the first and most successful decade of state planned and financed industrialization based on import-substitution, and reserving strategic economic sectors (capital goods, major infrastructure, energy) for public ownership. There was substantial public investment in these sectors, with the initial support of Indian private capital 'which shrewdly recognized this as opening up future possibilities which they could not on their own steam have generated' (Saith, 1990, p.214). Agriculture had a residual role in this conception: while industrial growth was 'taking off' (with anticipated subsequent benefits to agriculture), a combination of land reform, community development, and support for employment-generating activities like hand-loom weaving (which also satisfied the ideology of 'Gandhian economics') was supposed to provide a cushion for rural livelihoods.

As it turned out, one of the factors causing public-sector-led industrialization to founder in the early 1960s was rising food prices (Saith, 1990, p.215), leading to a 'new strategy' for agriculture widely known as the Green Revolution.

3.3 The era of the Green Revolution

The 'Green Revolution' was seen by its international (and Indian) architects, and gained its name, as an alternative to 'Red' revolution in Asia, by making available a package of biochemical inputs (high-yielding variety, or HYV, seeds (Figure 3.3), water and fertilizer) which is 'scale neutral' – in other words, a technology that could substantially raise the yields and increase the incomes of *all* farmers, regardless of their scale of operation. **Scale neutrality** promised equal access to the benefits of the new technology to compensate for persisting inequities in land distribution: a 'dream scenario' for policy-makers seeking rapid agricultural growth without increasing rural inequality and its potential for social unrest.

Scale neutrality: A term for technologies that do *not* have marked 'economies of scale' (which means becoming economically viable only at a certain scale of operation, e.g. farms or enterprises of a certain size). Scale-neutral technologies in agriculture are bio-chemical (like HYV seed 'packages') rather than mechanical (tractors, combine harvesters, some types of irrigation).

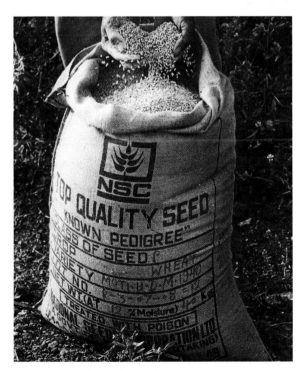

Figure 3.3 Certified wheat seeds for India's 'Green Revolution'.

The most (and probably only!) unambiguous observation about the Green Revolution in India is that it has enabled the achievement of national self-sufficiency in food grains. From being a net importer of grains from 1951 to 1975, in the latter year India achieved self-sufficiency. In 1984–85 there were nearly 30 million tonnes of grain in the government's reserve stocks. After this, the complexities and contradictions begin. Following Utsa Patnaik (1990a), we can summarize them in the extremely uneven impact of the Green Revolution on *regions, crops*, and *social classes* (hence rural livelihoods).

In regional terms, the success story of rapid output growth of HYV food grains is concentrated above all in northern India in the states of Punjab, Haryana and (the western part of) Uttar Pradesh, with some effect in parts of southern India, less in the east, and hardly any over much of the central plateau of the Deccan. The three northern states produced 36% of national grain output in 1985–86, while per capita grain production remained constant or fell in 11 of the 15 major states of India between 1960 and 1985 (Patnaik, 1990b). This stark spatial pattern is closely linked to the water requirements of HYVs, hence the distribution of irrigation.

In terms of crops, the great success has been with high-yielding varieties of wheat, while those of rice have been less effective, yet rice is the historic subsistence staple of the most densely populated rural areas of India: the 'rice belt' of the Gangetic plain and coastal and river deltas with high rainfall and canal irrigation. *Buying* wheat or rice is a luxury for many of the poor in India who rely more on the 'coarse' grains of millet, sorghum and maize, and for whom pulses are a major source of protein. With the particular crop bias of the 'new strategy', these other staples have been neglected and even displaced, and 'the nutritional balance of the average Indian's food intake has worsened, with a fall in the share of pulses relative to cereals' (Patnaik, 1990a, p.83).

These brief observations have started to link regional disparities with variations in crops, and the latter with the income (and implicitly class) distribution of consumption. Knowing what foods are grown, and how much, does not in itself tell us *who* can afford to eat them (and how much). In fact, the class and gender characteristics of a changing agrarian structure, and their social and political ramifications, are the principal key to the unevenness of the Green Revolution. The main assumption behind the new technology – 'scale neutrality' – is not the same as '**resource neutrality**', as Box 3.1 shows.

Resource neutrality: A term indicating that all farmers, irrespective of their scale of operation and assets, should be able to adopt profitably a scale-neutral technology or other appropriate innovation. However, as the experience of the Green Revolution in India shows, resource neutrality (a socio-economic assumption) is not the same as scale neutrality (a technical assumption).

Box 3.1   Risks and resources

Perhaps the crucial point here is that those disposing of more resources are in a much better position to cope with the risks associated with this higher cash-intensive technology. A small producer who has to borrow extensively in order to meet the cash costs of HYV cultivation runs the risk (in the event of failure because of miscalculation or unfavourable agro-climatic conditions) of becoming deeply indebted and thereby of placing the household in more precarious circumstances than before. Failure is a serious matter for producers with more resources, too, but they at least have a buffer which is not there for the poor small producer.

It has commonly been found, and for reasons which are clear, that those with more resources, whether economic or political, have better access to the controlled irrigation necessary for HYV cultivation, perhaps because they are able to influence the distribution of water in a bureaucratically controlled canal irrigation system.

Those with more resources in the first place are more likely to obtain them, or they may be able to obtain credit for the purchase of inputs at low rates of interest. If inputs like fertilizer are in short supply then they are more likely to be able to obtain them, or they may be able to obtain them at a lower price. Those with more resources are more likely to enjoy easy access to extension and other services supplied by the state.

Those well endowed with resources are also more likely to be able to command sufficient labour at the most appropriate times for successful cultivation; and where the hiring of labour is necessary (as it is for almost all producers at some stages in the cultivation of wet rice), the big people may well be able to pay lower wages because some labourers are indebted to them over long periods.

These are some of the reasons why the HYV package is not 'resource neutral'. They often mean not only that those with more resources are able to make better use of the package than the small producers but also that conditions are thereby created which are even more unfavourable to small producers (because the success of the relatively well off makes it more difficult for poorer cultivators to obtain labour, credit or fertilizers).

Further, the success of the few better endowed producers, and the advantages to them in the mechanization of operations (especially when the increased labour demands of the HYVs have tightened labour markets, pushed up wage rates and encouraged the political mobilization of labour) can encourage displacement of labour, eviction of tenants and the physical expansion of already large farms at the expense of the smaller.

(Harriss, 1987, pp.231–2)

The Green Revolution tends to accelerate the differentiation of the peasantry, contributing to the formation of a class of rich peasants, some of whom, depending on circumstances, may become fully-fledged capitalist farmers like those landowners who turned successfully to commercial farming, a class depicted vividly by Elizabeth Whitcombe:

> "In the new capital-intensive agricultural strategy, introduced into the provinces in the late 1960s, the Congress government had the means to realize the imperial dream: progressive farming amongst the gentry. Within a year or two of the programme's inception, virtually every district could field a fine crop of demonstration ex-*zamindars* – the Rai Sahibs with their 30-, 40-, 50-, 100-acre holdings, their multiplication farms of the latest Mexican wheat and Philippines paddy, their tube-wells gushing out 16 000 gallons an hour, much of it on highly profitable hire, their tractors, their godowns stacked with fertilizer, their cold-stores...; in short, a tenth of the *zamindari*, but ten times the income."

> (Whitcombe, 1980, p.179)

You may be struck by the fact that such 'Rai Sahibs' can prosper on the basis of farms as small (by comparative standards) as 30 acres (12 ha). In India the development of commercial farming through capital investment to intensify production and increase profits has not required, nor generated, the massive concentration of land and giant farms characteristic of some 'roads' to capitalist agriculture elsewhere (for example, in much of Latin America). To a large extent this reflects the pressure of population density on land in rural India, and the tenacity peasants display in holding on to their farms, however small and uneconomic, as an element of security. The development of a capitalist agriculture has proceeded furthest in Punjab and other parts of the north, as the earlier sketch of regional disparities would suggest. In these areas, the biochemical 'package' of HYV technology was soon associated with mechanization (not at all 'scale neutral') as well: from tubewell/pumpset irrigation to mechanical

ploughing, harvesting and threshing. Nonetheless, it is also true that many smaller farmers have adopted Green Revolution technology, especially in certain areas of the 'rice belt'.

Harriss (1987) describes one such area in North Arcot District of the state of Tamil Nadu in the south-east of India, where peasant households farm an average of 1.2 ha of paddy (rice) and groundnuts. There is inequality in Randam, the village studied by Harriss, but it is apparently stable rather than increasing, due to obstacles to richer peasants acquiring more land in this densely populated area, and the practice of land fragmentation on inheritance. Richer peasants have diversified into rice trading as an activity that is both more feasible and profitable than trying to expand the scale of their farming.

The contrast between Punjab and North Arcot shows that rural class formation is complex, and the commercialization of farming variable in its mechanisms and outcomes. We should also note that there are important players in the drama of agrarian change other than those who farm, whether peasants (rich, middle, poor) or capitalists. There are classes of rural labourers; there are crop traders (Figure 3.4) and input merchants, and those who supply credit and consumption goods; there are also the effects of different kinds of linkages between agriculture and industry (see Chapter 7 below), and of political structures and processes, formal and informal, at national and local levels.

Figure 3.4 Weighing grain for sale in a north-west Indian market.

Moreover, the purpose of studying differentiation is not only to identify whether and how richer peasants and capitalist farmers emerge. It is just as important to investigate class formation and its effects at the other end of the agrarian structure. The relatively stable peasantry discussed by Harriss comprises only half of rural households: those whose principal income comes from their own farming. More than one-third of the households in Randam are landless, deriving their income from a combination of farm labouring, wage work in nearby or distant towns (from where individual labour migrants remit part of their earnings to support their families in the countryside), and hand-loom weaving (Harriss, 1987, p.239).

3.4   Rural livelihoods

Evidently life is often very hard and precarious for those small and medium peasants whose principal livelihood is farming. The prospects of enhanced yields and incomes offered by the Green Revolution, and intensified commercialization more generally, also entail higher production costs and their associated risks (which are not distributed in a 'resource neutral' fashion). However, livelihoods are even more constrained and precarious for those who are landless or have marginal holdings (like Sharifa and Abu in Chapter 1). Estimates of the size of these rural classes (proletarians and semi-proletarians) in India are much contested, as is the measurement of rural poverty and its trends, but the figures in Table 3.1 give at least an approximation.

A first point is that these data underestimate the extent of inequality of farming because irrigated land, especially in Green Revolution areas, allows for two or even three harvests a year (multi-cropping). Even so, the figures show an unequal distribution of holdings within a comparatively narrow range of farm sizes: 74.5% of farms (marginal and small) occupy just over a quarter of cultivated land, only slightly more than the area that 2.4% of large farms occupy. Assuming that the landless are included in the marginal category, the 56% of rural households placed here are those where poverty is most concentrated. In addition, many holdings in the small category do not provide a basis for viable livelihoods from farming. While the average holding of 1.4 ha in this category may be sufficient with HYV rice and multiple cropping (recall the average of 1.2 ha in

Table 3.1   Distribution of operational farm holdings in India, 1980–81

Size category	Holdings			Area operated			Average size (ha)
	Number (millions)	%	(Cumulative %)	Hectares (million)	%	(Cumulative %)	
Marginal: below 1 ha	50.5	56.5	(56.5)	19.8	12.2	(12.2)	0.4
Small: 1–2 ha	16.1	18.0	(74.5)	23.0	14.1	(26.3)	1.4
Semi-medium 2–4 ha	12.5	14.0	(88.5)	34.6	21.2	(47.5)	1.8
Medium: 4–10 ha	8.1	9.1	(97.6)	48.3	29.7	(77.2)	6.0
Large: 10 ha and above	2.1	2.4	(100.0)	37.1	22.8	(100.0)	17.3

Source: Calculated from data in National Agricultural Census of 1980–81.

North Arcot), this is not the case in the many areas of dryland farming.

Despite considerable regional, class and cultural variation, rural households (especially in northern India as well as in Bangladesh and Pakistan) tend to exemplify a 'classic patriarchy', which 'implies the control of younger males by the old and the "shelter" of women in a highly hierarchical domestic realm. It also implies the control by men of some form of viable joint patrimony in land, animals or commercial capital' (Kandiyoti, 1985, p.58). Such 'sheltering' is a cultural ideal, most formalized in the Islamic institution of *purdah*. In households that can afford it, women do not do agricultural field work, although they carry out many other activities in seclusion: threshing and winnowing grain and husking rice, earning money for their husbands or fathers through 'domestic industries' like lace-making, and the usual wide range of reproductive activities (including preparing meals for labourers in those households that employ them). In fact, women's domestic workloads can actually increase by their being withdrawn from the fields and can result in a change in status since the work they do is no longer visible.

But what happens when 'viable joint patrimony' breaks down for many rural households like the landless and marginal farmers? Analyses of recent data and trends (Agarwal, 1986, 1990, on India; Kabeer, 1990, 1991, on Bangladesh), confirm the conclusions of Kandiyoti that because intensified commoditization

> "produces processes of proletarianization, semi-proletarianization or sheer pauperization, the material bases of classic patriarchy are eroded. Crisis for women sets in when this system, with its implicit promise of protection for weaker, particularly female, household members, breaks down under the strain of processes of rural transformation, especially if they lead to increasing landlessness and poverty... as an increasing number of women have to contribute actively to household income and as security through male protection becomes an ideal difficult to sustain, their identity as producers may become more visible, both to themselves and to others, as indeed seems to be the case among poorer and lower caste women."
>
> (Kandiyoti, 1985, p.58)

How have the technical changes of the Green Revolution package affected livelihoods of women and men among these proletarian and semi-proletarian households? Agarwal (1985, pp.93–102) suggests the following trends.

Certain tasks in the technical package (transplanting, weeding and harvesting where not mechanized) require more labour, and the numbers of casual workers, especially females, employed in these tasks have increased (Figure 3.5).

Figure 3.5 Certain tasks require more labour after technical change: transplanting rice in south India.

Figure 3.6   The gendered effects of mechanization: (above) milling by hand; (right) milling by machine.

The mechanization often accompanying the bio-chemical package has had different effects on rural employment. While mechanized ploughing (a male task) is usually done by household members or permanent employees, the mechanization of harvesting, threshing and shelling displaced both male and female workers. Most critical for the poorest women is the mechanization of rice milling (now a male task) because dehusking by hand used to be an important source of their income (Figure 3.6).

Another issue is whether increases in employment in certain tasks have been accompanied by increases in real wages. Agarwal finds that only one state, Uttar Pradesh, has experienced annual increases in both days worked *and* real wages. Overall, the most likely trend is an increase in work by women of proletarian and semi-proletarian households (except for women displaced by milling) without a concomitant increase in income. Furthermore, although women's identities as visible (as opposed to invisible) producers may be

established or reinforced by these processes, there is no automatic improvement in their levels of consumption and welfare (Agarwal, 1986). (This may also be the case in different circumstances for medium peasants who have adopted new technologies: cash crops have often displaced subsistence crops but increased money income is not necessarily translated into improved household food consumption.)

Reviewing the overall evidence on rural poverty during the first phase of the Green Revolution, from the mid-1960s to mid-1970s, Bina Agarwal (1986, pp.191–201) concludes that agricultural growth (again averaging about 3% a year after the bad harvests of 1965–66, 1966–67) probably had no significant impact in reducing poverty, although it may have prevented it increasing, at least in the highest growth Green Revolution areas. More recently, Government of India figures, in the Seventh Five Year Plan documents, indicate a notable decline in the overall incidence of rural poverty – from over 50% in 1980 to under 40% in 1984–85 (see Chapter 7 for a discussion of the evidence).

Ashwani Saith (1990, pp.177–8) is sceptical about the certainty of this finding, and the optimistic gloss put on it. Apart from the customary difficulties of statistical methodology, he points to massive government spending on rural poverty alleviation in the 1980s which, despite considerable 'leakages' in its implementation (i.e. diversion of funds from those on whom they are 'targeted'), probably provided sufficient resources to some poor rural households to lift them above the official poverty line (at least for the time being). However, this is unlikely to resolve the structural problems of Indian agriculture in terms of generating either sustainable rural livelihoods or long-term economic development (see also Chapters 7 and 11). Saith suggests another 'special effect': rural incomes in some areas (e.g. the state of Kerala) have been raised by remittances from labour migration to the Gulf states. His warning that this provides no lasting or structural

solution to rural poverty was borne out only too sharply by events after his article was published – the Gulf crisis of 1990–91 and the expulsion of millions of South Asian (and Arab) migrant workers from Kuwait and Iraq. Patnaik (1990a, p.87) also suggests that the poor are getting poorer: 'the real incomes of the rural poor are declining alarmingly in the low- and negative-growth regions.'

However elusive statistical precision (and agreement) may be, the stark fact remains that at least 43 million rural households in India (on recent government figures) live in often desperate conditions, and many may be experiencing increasing poverty and insecurity. For the great majority of the landless and marginal peasants, selling their labour power is their exclusive or major source of income. As in Latin America, and despite the great differences in patterns of agrarian structure and change, the question of rural livelihoods is thus intimately linked to questions about *labour*: who demands it, how much, when, where, on what conditions? Is there 'surplus labour' in the countryside (Figure 3.7)? What is the balance of forces between those seeking workers at minimal cost, and those seeking employment to satisfy their minimal needs?

Resistance by agricultural workers to their exploitation and oppression may stimulate those farmers who can afford it to mechanize their operations (further), as in the Punjab. If this happens on any appreciable scale in rice-growing areas, the effect on agricultural employment, especially for the large numbers of landless women wage workers, will be extremely serious (Agarwal, 1986, p.207). While continuing commoditization of the countryside and country towns generates some additional employment in crafts and small-scale industry, trade and other services, it is unlikely to be enough to meet the demand for work (Harriss & Harriss, 1984). Likewise, the sluggish growth of manufacturing industry (from 14% of GDP in 1960 to 17% in 1985) does not provide an adequate alternative source of employment and livelihoods (Saith, 1990).

Figure 3.7   An irrigation programme in Uttar Pradesh provides work for labourers at slack times of the year.

Summary

1   The land systems of colonial India generated both parasitic landlordism and forcible commercialization, to the benefit of Indian landlords and British capitalism and at the expense of the peasantry.

2   A limited land reform following independence, together with new opportunities for commercial farming, consolidated the power of smaller landowners and facilitated the development of a rich peasantry, with little gain and in some cases considerable loss to weaker classes in the countryside.

3   These pressures were further boosted by the Green Revolution from the late 1960s. Its promise of benefits to all farmers through a productive scale-neutral technology was constrained in practice by the mistaken assumption of resource neutrality. The gains of the Green Revolution are distributed unequally across crops, regions, and social classes (in terms of production *and* consumption of food grains).

4   Despite an initially dynamic period of public-sector-led industrialization and the Green Revolution, India still contains massive poverty, both rural and urban. Rural poverty and struggles for livelihoods are experienced most sharply by the landless and by marginal farmers, and especially by women within these groups.

5   Official claims of a marked reduction in the incidence of rural poverty in recent years, including the beneficial effects of government policies, are examined in greater depth in Chapter 7.

4

AGRARIAN STRUCTURES AND CHANGE: SUB-SAHARAN AFRICA

HENRY BERNSTEIN

About four-fifths of Africa was only colonized some 60 years or more after Latin America had thrown off its colonial status, and when the world economy was already dominated by industrial capitalism. However, well before the 'scramble for colonies' of the late nineteenth century (also the period of the first 'modern' international recession), Africa had been gradually if unevenly drawn into the emerging world economy, not least through the supply of slaves to the plantations of the Americas from the sixteenth to nineteenth centuries.

As the Atlantic slave trade gradually wound down following the successive abolition of slavery in one part of the New World after another (and which in the USA required the cataclysm of the Civil War), it was replaced by a transitional period of 'legitimate commerce' in West Africa in particular, the region that had been most affected. While trade in well-established luxury commodities (notably ivory) continued, the shape of things to come was indicated by the expansion of agricultural exports, above all materials for vegetable oils, demanded in increasing quantities in Europe for processing into soap, lubricants and candles. By the late 1880s Senegal was exporting an average of 29 000 tons of groundnuts a year, and Lagos in Nigeria (still a major slave port in the 1850s) an average of 37 000 tons of palm kernels (Hopkins, 1973, p.128). However, the more comprehensive and systematic orientation of African labour and land to the world market awaited the completion of the colonial system.

Before examining the colonial period, it is important to point out the diversity of African societies before colonialism. They were agricultural societies of different types because of:

1 their adaptations to the various tropical, subtropical and temperate zones of the continent (rainforest and arid savannah, river estuaries and lake shores, the high plateaux and mountain ranges of the east and south);

2 the relative importance within their economies of cultivation and livestock keeping, craft production and trade (as well as hunting, fishing and collecting natural produce);

3 their different processes of agricultural innovation, including the adoption and diffusion of new food staples (maize, cassava and other root crops);

4 their social structures and scale ranging from small classless societies (albeit differentiated by gender and age) to hierarchical empires of peasants, slaves and military aristocracies;

5 their continuing processes of endogenous change charted through migrations, new areas and patterns of settlement, state formation and dynastic rivalries.

In large areas of the interior these processes of change occurred quite independently of external linkages well into the nineteenth century. In other cases they absorbed them into their own dynamics. The Atlantic slave trade stimulated the expansion of warrior states in West Africa, securing slaves through warfare and raiding to exchange for guns, among other imported commodities. With the end of the slave trade the ruling groups of these states often tried to control the production, exchange and profits of agricultural exports, which brought them into increasing conflict with African traders as well as peasant communities. Cooper (1981, p.30) observes that while at the height of the slave trade 'disorder had been the basis of profit', on the eve of colonization 'a new obsession with law and order' was manifest on the part of European merchants who 'feared fighting would cut off supplies from the interior and worried about the security of property within the (trading) ports themselves'. 'Law and order' was to be delivered in the form of the *colonial state*.

4.1 The colonial export economies

The consolidation of colonial rule was no more smooth nor uniform a process in Africa than anywhere else, and campaigns of colonial 'pacifi-

cation' continued at least until the First World War. In southern Tanganyika (now Tanzania), for example, the Maji Maji uprising of 1905–6, sparked off by the imposition of forced cotton cultivation, was only defeated by the Germans through a 'scorched earth' policy with enduring demographic and ecological consequences for the areas concerned.

As 'pacification' was more or less completed ('more or less' because of the various forms of resistance manifested throughout the colonial period), colonial states undertook to harness African labour and land to the requirements of both metropolitan demand for raw materials and their own revenues. This included establishing and operating an administration (at minimal cost) to ensure law and order, collect taxes, stimulate and regulate the development of markets, and build a transport infrastructure to carry export crops and minerals to the ports. Most importantly, colonial administrations had to find means of organizing African labour to build the railways and roads and to produce the commodities they would carry (Figure 4.1).

The making of colonial economies geared to the export production of raw materials usually entailed an initial use of *coercion* whether direct or indirect. Direct coercion included the forced

Figure 4.1 Harnessing labour to build transport infrastructure: clearing ground for the Cape-to-Cairo railway, 1914.

conscription of labour for building roads and railways (as notoriously in the Belgian Congo), for plantation labour, for gathering and delivering natural produce (e.g. wild rubber in the Congo again, and in French Guinea). It also included the compulsory cultivation of specified export crops by African peasants (e.g. cotton in Tanganyika and in neighbouring Mozambique).

Indirect coercion involved money taxation (including the revival of the medieval English poll tax), hence the need for a cash income that could only be acquired by producing crops for sale, and/or selling labour power for wages. Sir Percy Girouard, Governor of the Protectorate of Kenya, expressed this clearly in an article in the *East African Standard* in 1913:

> "We consider that the only natural and automatic method of securing a constant labour supply is to ensure that there shall be competition among labourers for hire and not among employers for labourers; such competition can be brought about only by a rise in the cost of living for the native, and this rise can be produced only by an increase in the tax."
>
> (quoted in Manners,1962, p.497)

Whether colonial regimes were more interested in changing Africans into cash crop farmers or into (wage) labourers, as in Kenya, depended on the patterns of colonial economy that evolved. We can broadly distinguish colonies of *peasant economy* and of *settler economy*, although some territories combined elements of both.

The first were exemplified above all by the French and British colonies of West Africa, where peasants produced the 'classic' cash crops of the forest zones (oil palm, cocoa, coffee) and the savannah (groundnuts, cotton, oilseeds). Here colonial regimes played a key role in connecting the labour of millions of peasant households with the world market through their activities in taxation, transport, and some aspects of marketing and credit (Figure 4.2). Samir Amin (1972) calls the same areas 'Africa of the colonial trade' because key linkages of exchange between peasant producers and the world market were dominated by giant commercial companies like the United Africa Company (later Unilever).

While the need for cash crop cultivation was typically imposed, especially in savannah regions, some West African peasantries pioneered important

Figure 4.2   Linking peasant households to the world market.

This cartoon and those in Figures 4.3 and 4.5 were drawn by Tony Namate for an educational text on agricultural policy and hunger published in Zimbabwe.

cash crops without any 'help' from, and indeed despite, colonial administration. African initiative in clearing and developing new farming areas in fertile parts of the forest zone is well documented in the development of cocoa in Ghana and western Nigeria. African farmers and traders also adapted and extended prior networks of production and exchange of food staples, parallel with and in opposition to those that colonial states tried to impose and regulate.

The activities of colonial regimes were also crucial in much of East, Central and Southern Africa where white settler farming and mining dominated peasant production to varying degrees (Figure 4.3). These capitalist enterprises required a supply of readily available and cheap labour, which in turn often necessitated blocking off ways of earning a money-income alternative to wage employment (e.g. through growing crops for sale). The state was central to securing and regulating labour, even if this proved far less 'natural and automatic' than Sir Percy Girouard's vision of colonial capitalism aspired to. Not least, this was because of resistance and evasion by Africans, reflected in the constant preoccupation of European employers and colonial officials with the 'native labour problem'.

Figure 4.3 The 'settler economy'.

The most evolved settler regime, and the one exhibiting the most widespread and brutal expressions of its fundamental social contradictions, remains South Africa. South Africa's history shares much with those colonial economies where a basic premise of resolving the 'native labour problem' was the movement and restriction of Africans to 'native reserves'. The principal source of money income for people in the crowded reserves, as in other rural areas where land was sufficient but opportunities for cash cropping were lacking, was employment in areas of European capitalism. These were the settler estates and plantations of Kenya and Southern Rhodesia (now Zimbabwe), and parts of Tanzania, Northern Rhodesia (now Zambia), Nyasaland (now Malawi), Angola and Mozambique, and the mining concentrations of Katanga (Belgian Congo) and the Rhodesias, as well as of South Africa which drew on neighbouring countries in a massive regional economy of labour migration. In short, some rural areas functioned as 'labour reserves' for colonial capital.

With the exception of 'native reserve' areas, the most important point of contrast with Latin America and India is that peasants in colonial Africa experienced neither widespread alienation of land nor its conversion into private property as the basis of taxation. In addition, land was mostly plentiful and allocated for use by households and individuals according to various customary principles. Shortages of labour at key moments in the agricultural calendar, rather than shortages of land, help to explain some of the characteristic practices of indigenous farming systems (see Chapter 8). The commoditization of rural life in the colonial economy – principally (if not exclusively) through export crop cultivation and labour migration – was absorbed by, and changed, existing forms of communal and household organization in a variety of complex ways, making it impossible to construct any single 'model' of the African rural household.

The most evident manifestation of the unevenness of commoditization (underlined by the pattern of roads, railways and ports built under colonialism) was *spatial*: the demarcation of areas specialized in producing different export crops, food crops for domestic markets, livestock, and labour for the regions of commercial exploitation. The last reminds us that spatial differentiation is also social. 'Labour reserve' areas supplied not only the mining industry, plantations, and white settler farms (concentrated in the more amenable highland plateaux of the countries noted earlier), but also areas of intensive development of peasant cash cropping, notably in the flow of labour migrants from the Sahelian belt of West Africa south to the forest zones. Otherwise, social or class differentiation of African farmers was relatively limited, mainly because the divisions of labour based on race which were typical of colonialism inhibited accumulation by Africans. One interesting exception was accumulation by those chiefs who combined their administrative offices in the colonial system of 'indirect rule' with a generous interpretation (or invention) of their 'traditional' rights to mobilize the labour of others to carve out special opportunities for themselves in cash crop farming.

It is a reasonable general hypothesis that rural Africans preferred the 'peasant option' (Ranger, 1985): that is, gaining their livelihoods through their own farming when possible, rather than working for colonial capital (Figure 4.4). For example, Sukumaland in Tanzania changed from a labour migration area to a cash cropping area when cotton was introduced, providing cash incomes for paying taxes and acquiring consumption goods as an alternative to wage work in the sisal estates. In Kenya, Mozambique and elsewhere, wage employment was used by some as a means of saving to invest in household farming. Even in the extremely oppressive conditions of South Africa's black 'homelands', labour migrants have often fought to retain rights to land, however unlikely the prospects of securing a livelihood from farming.

Ironically from the viewpoint of today's agrarian 'crisis' (see below), the latter years of colonial rule and the first ten years or so of independence was a period of relatively buoyant economic growth. The 1950s and 1960s (together with the 1920s) were the decades of greatest agricultural growth in twentieth-century Africa, facilitated by favourable world market conditions and a certain amount of productive government investment in rural areas. The latter was provided under the umbrella of the Colonial Welfare and Development Acts in British territories, and similar measures in French colonies, and there were some significant continuities between programmes to 'modernize' peasant farming in the late colonial period and those undertaken by African states following independence.

4.2 Peasant farming and the 'developmental' state

In the post-war world generally there was a much more positive view of state planning and economic management than prevails in today's ideological climate of neo-liberalism (*Hewitt et al., 1992*, ch.5). This was influenced both by the achievements of comprehensive planning in the USSR, of considerable appeal to many newly independent states (as India), and by Western experiences of wartime economic management followed by European reconstruction under the Marshall Plan and Keynesian policies (*Allen & Thomas, 1992*, chs. 6, 11). John Maynard Keynes himself participated in establishing the Bretton Woods system of institutions to regulate the post-war international economy, including the World Bank which was active in promoting national planning in the 1950s and 1960s.

This central role for the state in development (or 'developmental states') had a strong resonance in Africa. First, it was seen as the principal means of controlling the economy, directing it away from the interests of foreign capital and markets towards meeting national needs and aspirations. Second, it was argued that African business and accumulation were so constrained by colonialism that there was no indigenous private sector alternative to state-led development. Third, the state and 'the nation' were virtually coterminous; indeed, 'nation building' was considered as important a task for the state as development (or inseparable from it) – that is, forging united and coherent national societies in the regionally and

Figure 4.4   Contemporary 'peasant options': (above) communal farming in Angola; (right) household maize production in Ghana; (below) pastoralism in Niger.

culturally diverse territories carved out between the colonial powers. Moreover, the potential translation of diversity into division, and division into conflict, was underlined by colonial policies of 'divide and rule' and establishing ethnicity as a principle of local administration ('indirect rule'). Divisions and latent conflict were also reinforced by extremely uneven economic change and the unequal spatial distribution of activities, resources, incomes and opportunities it generated.

The aspiration to economic development included diversification and, in particular, industrialization, which had been limited in the colonial economies to some agricultural processing and a few branches of consumer goods. Industrialization required substantial investment in plant and equipment, infrastructure and energy, hence major increases in imports. Similarly, the commitment to social development on a scale never envisaged by the colonial order – the provision of education, clean water supply, health care, urban housing, and other basic needs – also required substantial investment and expansion of the capacities, hence costs and personnel, of the state. On the other hand, such ambitions had to be pursued, to begin with, on the narrow economic base of the export economies inherited from colonialism. These were typically dependent on a few strategic agricultural commodities, whose output is often subject to major annual fluctuations and whose earnings are often subject to dramatic shifts in world market prices.

What have been the effects of the policies and actions of 'developmental' states for agrarian structures and change, and rural livelihoods? This is an area of intense controversy in the current period of 'crisis' (see below), and one frequently marked by facile generalization and simplistic diagnosis and prescription. The most useful way forward is to identify some key themes and issues, recognizing that these are subject to considerable variation and struggle in different countries and rural areas at different times.

One key theme concerns the experience of many government projects to generate growth by *'modernizing' agricultural production*, whether in state farms (Raikes, 1988, pp.53–5) or, most

widely, in peasant farming through 'packages' of new crops or crop varieties, fertilizers and other inputs, improved cultivation and management practices, credit, and so on. Such projects, of varying scale, almost always involve technical design and funding, and some element of management, by foreign aid donors, sometimes in collaboration with international agribusiness. The term 'state peasantries' has been used to describe those peasants incor-porated into the most highly controlled projects, which require them to follow strict farming schedules laid down by project planners and managers (Barker, 1989, p.96).

Despite some success stories claimed by both governments and aid agencies, the record of agricultural modernization projects is largely disappointing. Their technical design ignores the detailed local knowledge of peasant farmers, and standardized packages of 'improved' inputs and practices are often inappropriate to the variable (and fragile) ecologies of much of rural Africa. Cost–benefit calculations (the economic basis of project planning) are often based on mistaken assumptions about the social organization and objectives of peasant farming systems, and of different groups within them. Peasant resistance to projects imposed on them (as in the colonial period) has been a potent factor in many instances (Box 4.1).

A second theme, relevant to the poor performance of projects but commonly advanced as a general explanation of agricultural decline from the late 1960s, is that of state *'taxation'* of the peasantry, above all through control of crop (and input) *prices and marketing*. Opinions differ as to whether (increasing) taxation:

1 expressed the pursuit of a misconceived model of urban-industrial development, and the interests of a dominant urban class coalition, at the expense of the peasantry; or

2 was a channel of accumulation for a new ruling class emerging within the state itself – a 'bureaucratic bourgeoisie' exploiting workers and peasants (see Chapter 9); or

3 was used to finance more 'traditional' forms of wealth and patronage, i.e. 'political' accumulation

Box 4.1   Development projects and peasant households

In a widely used World Bank text on *The Economic Analysis of Agricultural Projects,* projects are termed 'the cutting edge of development' (Gittinger, 1982, ch.1). This is because projects are selected and designed as interventions to achieve the best possible rates of return to capital. Projects in this sense are above all *investment* activities. The methods of cost–benefit analysis are used to select and plan the investment, in terms of a sequence of targets for expenditure, expected production increases, and income streams, within a given time frame. In turn this provides the basis for project management, monitoring implementation, and evaluating the results. The combination of precise investment planning and control of implementation is why donor agencies favour the project format to channel and target their aid.

At least this is the theory. A voluminous literature suggests a variety of reasons why so many development projects, supposedly conceived in this way, go wrong in practice. It is widely recognized that aid-funded agricultural projects in Africa have a particularly high rate of failure. One reason noted above, and relevant to our concerns here, is their 'mistaken assumptions about the social organization and objectives of peasant farming systems and of different groups within them'.

The assumptions are that peasant households function like any other enterprises in a competitive market: that is, they are the same as a business firm which (a) has a common stock of assets, (b) is unified under a single management, (c) decides the optimum use of its assets, (d) intends to maximize its returns (or profits).

Why might these be mistaken or inappropriate assumptions in planning projects to develop peasant farming?

First, peasant households (especially in much of Africa) do not conform to this model of a single 'firm'. They may contain a number of distinct farming and other enterprises, controlled by different members of the household with different kinds of claims to assets and to the income from their use (see Chapter 5). Furthermore, some activities may be based on co-operation between households, or particular categories of people in different households. Who is assumed to be the household's decision-making agent, its 'managing director' as it were? Implicitly, it is the male 'household head' – there is thus a gender bias in this conception (see further below).

Second, as well as assuming the internal homogeneity of interest of the peasant household as a 'firm', project planning calculates the costs and benefits of a new economic activity on the basis of a 'typical' or 'average' household. This is a sociological fiction. Peasant households even in the same village tend to vary in many ways, not least because of class differentiation, as the example of the Green Revolution in India shows (Chapter 3). Some may possess the resources necessary to benefit from an innovation or project intervention; many others may not, and their livelihoods may suffer as the result of agricultural development projects.

Third, in terms of objectives – and as already implied – the interests and objectives of different household members may diverge as well as converge, notably along lines of gender and generation. And objectives may run counter to maximizing returns on a particular asset or activity (e.g. a new cash crop promoted by a project): for example, maximizing security of subsistence by minimizing risks in farming and/or diversifying the use of resources, especially labour. If circumstances allow, rural households are more likely to pursue livelihood strategies as 'foxes' rather than 'hedgehogs' (Chapter 1), while projects view them as firms specializing in a particular commodity.

and the forging of ethnic coalitions rather than productive accumulation and class formation.

Concerning food crops grown for domestic markets, it is widely argued that governments suppressed producer prices (paid to farmers) in order to provide cheap food to (mainly urban) consumers. Concerning export crops, the contentious issues of price policy also include the rate at which domestic currencies exchange against the currencies used in international trade (notably the US dollar). 'Overvalued' exchange rates discriminate against export producers who receive less in local currency for each US dollar that their products earn in world markets (Box 4.2). Conversely, overvalued exchange rates favour the users of imported commodities. In short, arguments about crop prices, and marketing arrangements and costs, concern their incentive, or disincentive, effects for peasant farmers.

A third theme is social differentiation in the countryside. While 'modernization' projects have been one source of pressure on many peasant farmers (partly because of the increased *risks* they usually entail, see Box 3.1 above), other farmers – especially those with more substantial assets and political connections – have gained

from privileged access to the new inputs and credit that projects make available. The position of the 'progressive' (more commercial) African farmer, a figure idealized by colonial agricultural officers, has often been strengthened by development projects, but not necessarily in ways intended by project planners. Richer peasants often get the lion's share of project resources, not because of their greater economic 'efficiency' but through their ability to influence the local administration of projects.

The nature and extent of class formation in the countryside also qualify the general thesis of state taxation of agriculture and its disincentive effects. Richer peasants and capitalist farmers (for example, in the Ivory Coast and Kenya) are more likely to benefit from 'institutional rents' (see Box 2.4 in Chapter 2), and to exert effective pressure for better prices for the products they specialize in.

The complexity of processes of commoditization and agrarian change in Africa, as elsewhere, therefore includes their differential effects for different rural classes. Among other factors relevant to the 'crisis' that emerged in the 1970s is the one cited from the World Bank's *World Development Report* in Chapter 1, namely growing pressure on land use in many areas as a result of population growth and/or particular types of commercialization and 'modernization', and associated processes of environmental degradation in the form of deforestation, soil erosion and soil exhaustion (see Chapter 8).

4.3 Economic crisis

Table 4.1 gives some economic indicators for sub-Saharan Africa as a whole, for its six most populous countries combined and individually, and for South Asia by way of comparison. We do not have to believe in the precision of the GNP figures to accept the broad truth of the trend they depict: the *negative growth* of African economies in the 1980s. The table also shows, subject to the same qualification, variations between different countries concerning both overall economic performance and food production (which we will come back to).

Box 4.2 Another example of rents

An influential exponent of the 'urban bias' explanation of agrarian crises in Africa refers to state extraction of surplus from peasants as 'bureaucratic rents' (Bates, 1981). This is another application of the concept of economic 'rents' that we encountered in Chapter 2 (Box 2.4).

In this case the state is a rent receiver through its *monopoly* in the marketing of key export crops. The rent effect is enhanced by 'overvalued' exchange rates: when state companies sell these commodities in world markets, peasants receive less in local currency for the dollars (or other 'hard' currency) than the state receives from the sale.

Table 4.1   Sub–Saharan Africa: economic indicators, 1987

	Population (to nearest million)	GNP per capita US$	GNP per capita Average annual growth rate (%) 1965–73	1973–80	1980–87	Average index of food production per capita (1979–81 = 100)
Total sub–Saharan Africa[a]	451	330	2.9	0.1	–2.8	100
Six most populous countries	235	270	3.9	0.4	–0.4	100
Ethiopia	45	130	1.1	0.0	–1.6	89
Zaïre	33	150	0.3	–4.7	–2.5	99
Tanzania	24	180	2.0	–0.9	–1.7	90
Kenya	22	330	4.7	1.3	–0.9	93
Sudan	23	330	–1.7	3.5	–4.3	100
Nigeria	107	370	5.3	1.2	–4.8	105
South Asia	1081	290	1.5	2.0	2.6	109

Source: World Bank (1989a) *Sub-Saharan Africa: from crisis to sustainable growth*, IRBD/World Bank, Washington, tables 1, 8, pp.221, 234–5.

[a] Excludes South Africa.

Table 4.2 suggests there has been little movement away from the heavy export dependence of African economies on primary or raw materials (again contrasting with South Asia). Moreover, the aggregate distribution of export earnings between fuels, minerals and metals and 'other primary' (i.e. agricultural, including timber, livestock and fisheries) reflects the absolute value of petroleum exports from a few countries (Angola, Cameroon, Congo, Gabon, Nigeria) plus minerals and metals from a handful of others (including Zaïre), and obscures the more typical pattern of *agricultural export dependence*. This comes out more clearly in the contrast between Nigeria and Zaïre on one hand, and the other four countries shown whose principal export crops (and their share of total export value in 1987) are coffee for Ethiopia (50%), coffee and cotton for Tanzania (51%), coffee and tea for Kenya (46%), and cotton and sesame seed

for Sudan (48%) (World Bank, 1989a, table 15, pp.246–8).

Deteriorating terms of trade in the world market for many key African exports (not least against petroleum, for the non-oil producing countries), together with declining export volumes in some cases, further constrained the capacity to import and led to increased international borrowing. As a consequence, *external debt* increased massively in the 1980s, one of the strategic macro-economic effects of Africa's crisis (Table 4.3).

It has been observed (Barker, 1984) that the vocabulary of African 'crisis' was adopted and generalized by Western aid agencies and development experts as a perception, in the first place, of a crisis of the 'developmental' *state* which they have advised and assisted since independence. On the other hand, the popular image of 'crisis' is one

Table 4.2 Structure of primary exports, 1987

	Percentage of total exports		
	(1) Fuels, minerals, metals	(2) Other primary	(1) + (2)
Total sub–Saharan Africa[a]	47	39	86
Six most populous countries	72	24	96
Ethiopia	3	96	99
Zaïre	63	31	94
Tanzania	7	75	82
Kenya	21	62	83
Sudan	14	79	93
Nigeria	91	8	99
South Asia	8	28	36

Source: World Bank (1989a) Sub-Saharan Africa: from crisis to sustainable growth, IRBD/World Bank, Washington, table 14, pp.244–5.

[a] Excludes South Africa.

Table 4.3 External debt, 1980 and 1987

	1980 (US$billion)	1987 (US$ billion)	External debt per capita, 1987 (US$)	External debt per capita as % of GNP per capita, 1987
Sub–Saharan Africa[a]	55	133	294	93
Ethiopia	0.8	2.6	58	45
Zaïre	4.8	8.6	265	177
Tanzania	0.3	0.9	36	20
Kenya	3.5	5.9	269	82
Sudan	5.1	11.4	493	149
Nigeria	8.9	29.8	279	75

Source: World Bank (1989a) Sub-Saharan Africa: from crisis to sustainable growth, IRBD/World Bank, Washington, tables 19, 1, pp.252–3, 221.

[a] Excludes Angola, Mozambique, South Africa.

of chronic famine and starvation in Africa, transmitted periodically by the Western media since the early 1970s (see further below). What is not captured by either vision is 'regular, persistent deprivation' (Drèze & Sen, 1989, p.258), the mundane reality of everyday struggles for livelihoods by the poor, rural and urban, throughout the Third World.

The crisis of the 'developmental' state in Africa has various dimensions. Problems of food supply to the cities and rising food prices contribute to fears of political instability. Falling export volumes and values squeeze import capacity, leading to rapid escalation of external borrowing followed by declining creditworthiness (the particular concern of the International Monetary Fund, IMF) (Figure 4.5). Not least, poor or negative rates of return to investment in development projects increasingly frustrated and embarrassed the aid

agencies (especially the World Bank) that were centrally involved in their design, funding and management, which contributed to their conversion to neo-liberalism in the 1980s (Bernstein, 1990).

At the same time, the crisis of the 'developmental' state has far-reaching implications for the 'regular, persistent deprivation' experienced by many Africans. Generally, as the Director General of UNICEF observed, 'it is widely known that the poor have usually gained least in good times and suffered most in bad times' (quoted in Saith, 1990, p.183). More specifically, the ambitions of African 'developmental' states and the extent of their activities (indicated in Section 4.2) affect livelihoods and relative well-being in many ways, directly and indirectly. Most directly, the state is by far the largest employer of waged and salaried labour, especially in cities and towns. Economic

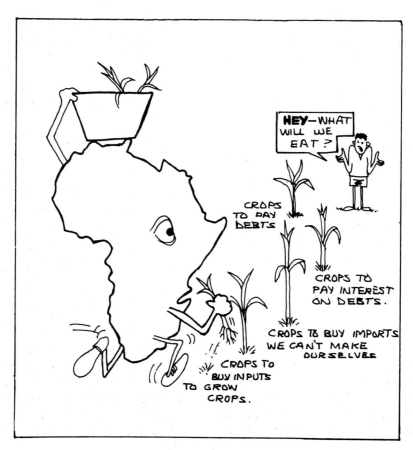

Figure 4.5   The dimensions of crisis.

crisis has tended to reduce the numbers of state employees, and above all has drastically undermined their real incomes through inflation. In the discourse of neo-liberalism, this was 'a brutal but necessary adjustment', even if in some cases 'public sector wages are barely enough for subsistence... the correction has gone too far, and productivity has plummeted as a result' (World Bank, 1989a, p.29).

Declining import capacity has a range of serious effects. It exacerbates the underutilization of industrial capacity due to lack of spare parts and materials, 'knocking on' to shortages of basic consumer goods. It also exacerbates the deterioration of physical and social infrastructure. In many rural areas, lack of farming inputs and basic consumer goods, deteriorating roads and vehicles and increasing transport costs, have contributed more to losses in marketed crops than the effects of producer prices. Shortages of drugs and fuel for generators and water pumps in rural medical centres, of books and writing materials for schools, have undermined the provision of health care and education, inadequate as it usually was.

In short, production and incomes, livelihoods and consumption (of social as well as individual goods and services) are affected by their linkages with state structures, activities and expenditure, and the pressures on them of economic crisis. The extent to which this applies to the countryside as well as to the towns is often underestimated, partly because of persisting myths of rural 'self-sufficiency' and exaggerated views of a rural–urban 'divide', as we shall see when we examine rural livelihoods. Before that, it is worth looking more closely at food 'crisis' in Africa.

4.4 A general food 'crisis'?

We may wonder about the political and international reasons why the crisis of livelihoods and consumption for many Africans is often seen primarily as a food crisis (Guyer, 1987, p.44). The terrible images of famine transmitted by Western media are drawn repeatedly from a few countries

ravaged by internal wars, above all Ethiopia, Sudan and Mozambique. Instances when food shortages are handled effectively by African governments do not receive any publicity (Drèze & Sen, 1989, pp.133–61). This also reminds us that food production is not the same as food availability (production minus exports, and plus imports), and aggregate availability and the ability to acquire food (or food 'entitlements') are not the same, as we noted in the case of the Green Revolution (Chapter 3). If the figure in Table 4.1 above is credible, Sudan has maintained per capita food production in the 1980s while many people have starved.

However, food production data for Africa are notoriously unreliable and also conceal some systematic downward biases – for example, concerning the important root crops and tubers (yams, cassava, sweet potato) mostly grown and traded by women in the densely populated forest zones of West Africa (Guyer, 1983). Moreover, there may be vested interests in exaggerating shortfalls in production, whether by governments seeking food aid or by food exporters (the USA, the European Community) and international agencies that want to supply it (Raikes, 1988), or by those whose proclamation of generalized crisis serves to promote generalized solutions like market liberalization (Berry, 1984).

Nor are increasing food imports a valid indicator of declining food production. They can reflect rapidly rising demand due to population growth *and* a growing proportion of the population which does not produce its own food, rural as well as urban. They can also reflect changes in both international and internal trade patterns and strategies, of which rice imports provide good examples. As the countries of South and East Asia became self-sufficient in rice production, Africa (along with the Middle East) was targeted as an alternative market by the world's two leading rice exporters, the USA (where production and exports are highly subsidized) and Thailand. Rice imports to Guinea have expanded recently with trade liberalization and the return to the country of big traders with strong international connections.

Another important consideration – related to the 'taxation' argument (Section 4.2) – is that it is much more difficult for states to control trade in food (including cross-border trade, or smuggling) than in export crops. For example, in Tanzania, where the state sector had a formal monopoly in marketing food staples until recently, it is estimated that at least 50% of food traded was exchanged illegally in 'parallel' markets. Rising demand, and food price inflation, has also stimulated the growth of food production for the market by peasants and larger farmers in some countries, often at the expense of export crops; nor should we overlook the significance of urban and peri-urban (city outskirts) market gardening, providing many of the vegetables that are a vital part of popular diets (Figure 4.6).

None of this is to deny that *per capita* food production has probably declined in some African countries (especially in the 1970s). Nor does it vitiate the more general point about food availability and entitlements: that adequate aggregate supply does not mean that everyone has access to an adequate diet. In fact, in some instances the commoditization of food production in response to rising urban demand (and prices) has also occurred at the expense of local food markets in the countryside. This further suggests that particular patterns of commoditization (and social differentiation) in the context of economic crisis have complex effects for rural livelihoods, and for the livelihoods of certain people.

4.5 Rural livelihoods

The various factors outlined so far – ill-conceived 'modernization' projects, 'taxation' of agriculture, the environmental effects of population growth and commoditization, shortages of inputs and basic consumption goods, deteriorating physical and social infrastructure – have exerted strong pressures on the livelihoods of perhaps a majority of rural Africans.

Posing the question *'whose* livelihoods?' is to ask about *gender* differentiation as well as class differentiation. Drawing on the analysis by Ann Whitehead (1990) we can sketch some of the elements of an answer, subject to the usual qualification about the complex specificities of different

Figure 4.6   The growth of urban food markets: a marketplace in Burkina Faso.

situations, places and times. First, and by contrast with the 'classic patriarchy' of South Asia, sexual divisions of labour in Africa were historically associated with different spheres of control over land, labour and income. Women had claims to land to farm on their own account, and to dispose of its product as well as of income earned from other independent activities like trading. Secondly, part of women's time was committed as 'family labourers' to farms assigned for common ('household') provisioning, and sometimes to work the personal fields of senior men – fathers or husbands – as well.

Various tensions and processes of negotiation implicit in such arrangements were likely to shift and intensify with the new opportunities and risks accompanying commoditization. This was a more complex process than suggested by the common stereotype of a sexual division of labour in which men specialize in labour migration and cash cropping while women are restricted to 'traditional' food farming (the 'feminization' of subsistence), as Whitehead emphasizes. Pressures on farming and rural livelihoods from commoditization and class differentiation, exacerbated by economic crisis, today frequently assume the form of a 'sex war' in the countryside as men seek to resolve their problems by extending their claims on the labour of women and changing the terms of the 'conjugal contract':

"For African women in peasant households, recruitment primarily as family labourers represents the construction of a hitherto rare form of dependence within and on marriage...As a 'family labourer', a woman may produce more crops and more income; as an unpaid labourer for her husband she may become better off if she helps him become successful. More important, she may also feel that her children's welfare is more secure. However, in addition to a lack of control over spending and welfare decisions, as unpaid workers women do not build up their long-term resources...(which) may hook a woman into a dependence which leaves her very insecure at times of crisis. The growing number of poor female-headed households and the crisis over marriage in Africa...are evidence for this."

(Whitehead, 1990, pp.64–5)

Moreover, the balance of forces in this 'sex war' has been tilted towards men by practices of land allocation and registration that acknowledge only male claims to land, and by development agencies that 'target' inputs and credit on male farmers who try to extend their enterprises by manipulating or redefining customary claims to women's labour. Resistance by women partly explains the failure of some development projects, but may also increase the possibility of abandonment without access to independent livelihood resources, the dilemma for women highlighted by Whitehead and illustrated in Box 4.3.

Box 4.3 Women farmers' dilemmas: a Tanzanian example

Women can no longer tolerate these oppressive conditions. It is so bad that they decide to leave. They leave their children, the husband, the farm, to go follow the life of a prostitute. Look at the women! They have lost weight this season because of work. The woman says, 'I have the right to be self-reliant, but what will my children do?' She agrees to return home.

Women work on the village farm, but very few men do. Women weed the coffee, they pick coffee, pound it and spread it to dry. They pack and weigh it. But when the crop gets a good price, the husband takes all the money. He gives each of his wives 200 shillings and climbs on a bus the next morning... most go to town and stay in a boarding house until they are broke. Then they return and attack their wives, saying, 'Why haven't you weeded the coffee?' This is the big slavery. Work has no boundaries. It is endless.

(statement by a woman activist quoted in Mbilinyi, 1990, pp.120–1)

Development projects which target women to try to increase their productive activities and incomes often make the error of assuming that women's labour time is infinitely elastic. Women's working days, of course, combine food production and income generation with a substantial burden of reproductive work (Figure 4.7).

Given the compound pressures on rural livelihoods in Africa today, albeit unequally distributed by patterns of spatial, class and gender differentiation, we may ask how viable the 'peasant option' is in many areas and for whom. In the 'extreme but not exceptional' case of Uganda (Mamdani, 1987, and Chapter 9 below), the rural

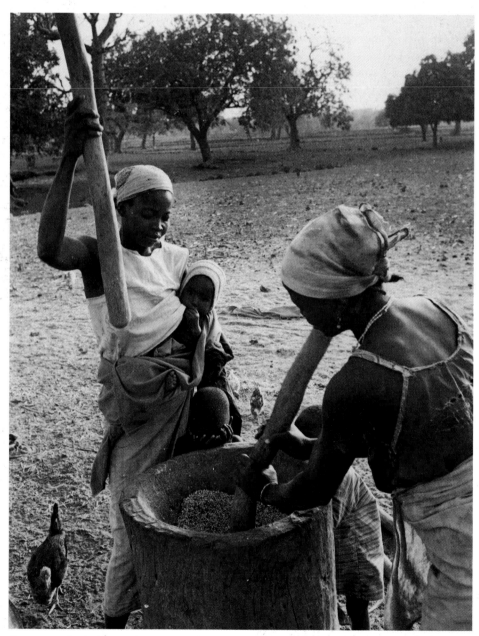

Figure 4.7   (above) Combining childcare and pounding millet in Burkina Faso...

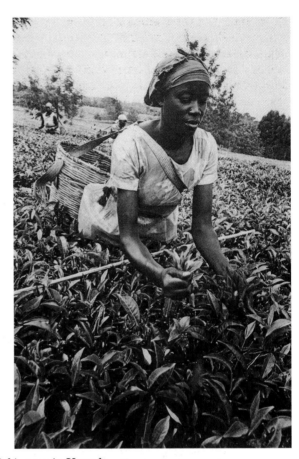

(above left) hoeing fields in Burkina Faso; (above right) picking tea in Uganda.

poor, deprived of access to other resources, intensify their exploitation of such land as they possess and of their own reproductive capacity. Both are rational for *individual* survival in conditions of extreme constraint but with harmful aggregate or *social* consequences. Intensified exploitation of land, without the means to restore its fertility, produces ecological degradation. Having more children to provide additional family labour, and as a source of security for old age, increases future pressures on land that is already unequally distributed and subject to declining fertility. It also has particular consequences for women whose health is undermined by frequent pregnancies, especially when their nutritional levels are low and susceptibility to related illnesses high.

Of equally general significance for the viability or otherwise of the 'peasant option' is that a decreasing proportion of rural households rely on farming as their principal source of livelihood. Sara Berry (1980) suggests that *diversification* of income-earning activities is a key factor because farming in Africa is usually so risky: crop yields are subject to the uncertainties of rainfall and input supply, and farming incomes are subject to the uncertainties of both yields and prices. Diversification is also patterned by social differentiation, however. For richer peasants it expresses a strategy of accumulation by investing in crop processing, trade, transport, urban housing, secondary and higher education for their sons, patronage, and political influence. For poor and many middle peasants diversification is indispensable to their livelihood strategies, pursued through craft production, petty trade and other micro-enterprises (e.g. beer brewing by women), and wage labour – for shorter or longer periods, and whether locally, in other rural areas, in town, or in other countries.

In effect, the income profiles (if not levels) of many African rural households would approximate those of poor peasants in Latin America (Table 2.5 in Chapter 2).

In the conditions of today's crisis, the customary mobility of Africans is even more marked than in the past: 'rural Africa is on the move' (Barker, 1989, p.8). The combination of spatial mobility and social flexibility in pursuit of livelihoods further undermines images of a largely traditional, static and self-sufficient rural existence, and of a sharp rural–urban divide in contemporary Africa. Many rural households depend on intermittent or regular contacts with the urban economy as part of their livelihood strategies, to supplement farming incomes and to obtain basic consumption goods (salt, cooking oil, sugar, soap, textiles, medicines). Reciprocally, some urban residents confronting declining real incomes and employment opportunities rely, at least in part, on rural relatives for their supplies of food staples.

Summary

1 Sub-Saharan Africa was systematically incorporated into the world economy from the late nineteenth century; colonial economies specialized in the export production of raw materials, agricultural and mineral.

2 The labour of rural Africans was harnessed to export production as both peasant farmers and (typically migrant) workers on white farms and mining enterprises.

3 After independence, 'developmental states' attempted to plan national economic development (with the assistance of aid donors), to diversify narrowly based export economies, to undertake industrialization, and to meet basic needs through social provision of education, health, (urban) housing, etc.

4 Relatively buoyant economic growth in the 1950s and 1960s gave way to increasing problems in the 1970s and crisis in the 1980s; the contributory factors included misconceived agricultural 'modernization' projects, state taxation of (especially peasant) agriculture, and the increasing pressures of population growth and commoditization in many rural areas, manifested in problems of environmental and social reproduction.

5 The effects of the crisis are mediated by spatial, class and gender differentiation; they include increased struggles between men and women over the control of assets, work and income within rural households, and the urgent need to find sources of income additional to farming, in order to survive.

Conclusion to Chapters 2 to 4

Chapters 2–4 have surveyed agrarian structures and change, and how they bear on rural livelihoods, in three major areas of the Third World.

The survey has been both broad and historical in its approach, with an emphasis on forms of landed property, processes of commoditization of farming and economic and social life in the countryside more generally, and class and gender differentiation. Analysing social differentiation helps us connect the analysis of 'structures' and structural change with the access to resources and employment that determines the livelihood opportunities of different classes and groups, and of women and men within them. The way to remember this connection is always to ask 'whose livelihoods?', and to link this question to those posed in Chapter 1: who owns what? does what? gets what? what do they do with it?

The survey suggests both contrasts and similarities in the historical paths and contemporary conditions of the three regions. One evident contrast is the *timing* of their incorporation in a world market, effected through colonialism:

* the colonization of Latin America began five hundred years ago;

* India was colonized in the course of the eighteenth century;

* most of Africa, despite earlier links with an emerging capitalist world economy, was formally colonized only in the late nineteenth century.

These differences in timing were also associated with different *types of colonial agrarian structures* and their *legacies*.

While Latin America (with the Caribbean) was the first major site of plantation production for the world market, the *hacienda* (an adapted type of feudal estate) dominated much of its agrarian history, and limited the possibilities of viable commercial peasant farming (petty commodity production). The expansion of *hacienda* production with the great commodities export boom from about 1870 involved a new wave of expropriation of Indian land and coercive forms of labour, along with immigrant contract labour and the beginning of an uneven transition from labour tenancy to wage labour. The conversion of the *hacienda* to capitalist farming (employing wage labour and extensive mechanization) was only completed in the course of the twentieth century. Despite a number of land reforms, an enduring legacy of the *hacienda* is the extremely unequal distribution of land in most of Latin America today.

In India, the key element of change was the establishment by the colonial state of private property rights in land together with taxation in money. This generated a combination of 'parasitic landlordism' and the 'forcible commercialization' of peasant farming. Colonial India's agricultural raw materials and its revenues made a vital contribution to Britain's industrialization in its various phases from the late eighteenth century. The limited land reform following independence facilitated the emergence of rich peasants and capitalist farmers, given a further impetus by the Green Revolution from the later 1960s. Widespread landlessness and marginal farming define the intense rural poverty of India today.

With the exception of parts of eastern and especially southern Africa, the land used by most African farmers was neither expropriated (as in Latin America) nor incorporated in a system of private property rights (as in India). By a mixture of direct colonial impositions and spontaneous processes of commoditization, Africans were integrated in the world economy of the late nineteenth and twentieth centuries as producers of raw materials to meet the massive demand of an already industrialized and urbanized West. They did this both as growers of export crops on their own farms, and as (usually migrant) workers in large-scale agriculture and mining. The relative buoyancy of peasant agriculture, or petty commodity production, in the 1950s and 1960s was highly vulnerable to adverse changes in its conditions (whether through environmental pressures, world market shifts, government 'taxation', aid-funded 'modernization' projects, or increasing social differentiation) as the setbacks of the 1970s and continuing crisis from the 1980s have demonstrated.

What similarities can be perceived in these different histories? One kind of similarity is that the fortunes of the 'agricultural sector' and of the rural people largely (though by no means exclusively) dependent on it, are closely linked to processes and forces outside the countryside:

- the cycles of the world market;

- the extent and types of industrialization within and between countries;

- patterns of urbanization and urban demand;

- the availability of employment outside household farming;

- not least, the effects of state policies and practices (as we have seen in all three chapters).

Another more substantive similarity reflects how capitalism works, with all its specific differences of time and place that have been indicated. This similarity concerns the conditions that define the livelihood opportunities of the rural poor, and the 'survival strategies' they devise. On the former, we have noted that landlessness (especially in Latin America and India) and the 'marginalization' of small farmers (in all three regions) are common characteristics of the rural poor. The rural poor are those for whom the 'peasant option' is not viable, and for whom there is little likelihood of social reproduction as agricultural petty commodity producers at adequate levels of income, consumption and security. In short, within given relations of property and power, the rural poor have to pursue their livelihoods as proletarians or semi-proletarians.

For most of them this means *diversifying* their sources of livelihood, and minimizing their risks as best they can by combining one or more of: farming their own parcels of land if they have them, renting land as tenants or sharecroppers, wage employment and labour migration, petty crafts, processing and trade. This is perhaps the most striking point of convergence of the rural poor throughout Latin America, India, and sub-Saharan Africa today.

We have also noted that the broad lines of class differentiation in the countryside, important as they are, cannot account for all the variations in the causes, manifestations and experiences of rural poverty and insecurity. Gender relations also play a central role. Again, a significant point of convergence is that poor rural women (of all ages) bear the brunt of poverty more heavily, with their typically more constrained access to resources and income, combined with their responsibilities for reproductive labour and household provisioning.

These various kinds of similarities highlight another key theme of this book: that rural society in the 1990s, and the lives of those who create and recreate it through their activity, cannot be understood apart from relations of property and power, and processes of change, in wider regional, national or international settings. This means that we have to abandon inherited images and mythologies of rural life as 'traditional', 'primitive', 'isolated', 'unchanging', and so on. In late twentieth century capitalism, such stereotypes are as inappropriate to the pursuit of livelihoods in the countryside as they are to urban existence.

RURAL HOUSEHOLDS

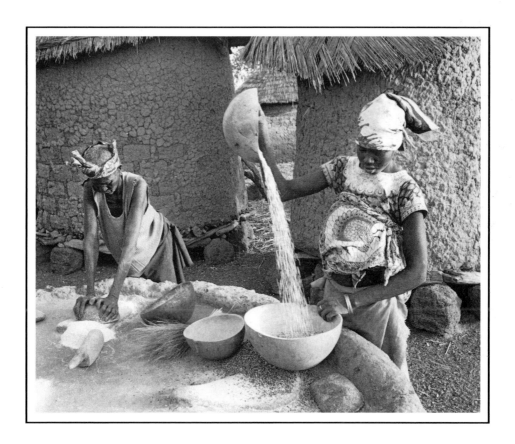

5

RURAL HOUSEHOLDS: MAKING A LIVING

KATE CREHAN

How households make their living and survive on a day-to-day basis and in the longer term are the subject of this and the next chapter. As we shall see, household survival strategies concern many different kinds of activities – producing food, making a cash income, conserving access to resources, bringing up children, negotiating different kinds of social relations within and between households, and dealing with the effects of agrarian change or social and political upheaval. Different themes concerning aspects of household and individual survival strategies are raised in later chapters in the book (8, 9, 10 and 12). Chapters 5 and 6 examine the organization of rural households, their diversity and how they are linked to other aspects of social and economic organization.

> **Q** How are rural families and households organized to enable them to survive, produce and reproduce?

The key issues noted at the end of Chapter 1 (who owns what? who does what? who gets what? what do they do with it?) are central to distinct but related ways of understanding households and livelihoods: how are households located in the wider economy and society? and what are the social relations and dynamics internal to different kinds of households? Chapters 2 to 4 dealt with some important aspects of the wider 'structural' relations of rural households. Without losing sight of the wider context, this chapter and the next use several case studies to look at what happens *inside* rural households. Chapter 5 examines how they function as units of production and consumption. Chapter 6 looks at the reproduction of households and how they are affected by changes over time.

5.1 Studying households

What is 'the household'?

At first sight this question may seem a typical example of the way academics insist on creating problems where no ordinary sensible person would find them. Common sense tells us that in every society there are easily identifiable groupings, 'households', within which people live, and which are recognized as society's basic building blocks. Similarly, even if households in different parts of the world may take very different forms, are they not always based on some kind of family or kinship relations? The problem is that the meanings of family and kinship, and how they are translated into the institutions of everyday life, vary widely between different cultures, and across different social groups. Rural households in the Third World (as elsewhere) come in all shapes and sizes (Figure 5.1), from an aged widow living on her own (also a common occurrence in Britain) to

Figure 5.1   The variety of rural households: (above) a rich Ghanaian farmer with wives and some of their children; (right) a family from Kerala, India, which includes three generations as well as two married sons with families and unmarried daughters.

the large family groupings which the Ghanaian Kusasi describe as their ideal. The examples from Africa in Box 5.1 – to take just one continent – illustrate something of this diversity.

Apart from highlighting diversity, the extracts draw our attention to differences between cultural ideals and actual households. For example, Whitehead refers to 'the *ideal* Kusasi household' – actual households may be different from this ideal. In Britain the monogamous, life-long married couple with two children from the same biological parentage may have been an 'ideal', but divorce rates, remarriage rates, and the incidence of single parenting tell another story.

Why we have these cultural ideals and why reality deviates from them are part of understanding society and social change. The fact that ideals about households and families differ from reality also has important practical implications. For example, in Britain conditions of employment for women and men, wage levels, taxes and benefits are based on the assumed ideal of monogamous

male-headed households rather than the actual realities of women's and men's lives (Barrett & McIntosh, 1982, pp.7–8).

The extracts also show that rural households in Africa have different ways of making a living. They do not necessarily derive their income and livelihoods exclusively, or even mainly, from farming (look at the extracts on Lesotho and Senegal). Compare the descriptions in Box 5.1 with those in Table 5.1 which are taken from a study of rural households in north-west India. Again, we can see that there are substantial differences in both household composition and how household members make a living.

These points suggest that when we see or hear terms like 'households', 'family' and 'kinship' being used, we need to ask exactly what they mean in that particular context. By the same token, we should be wary of references to '*the* household', or '*the* family', as if they describe a single and uniform social reality.

Box 5.1   Some African households

North-eastern Ghana '…the ideal Kusasi household is composed of a male head, his junior brother, both of whom are married with two wives each, an unmarried adult male (brother or son) and able-bodied daughter or daughters, a woman given in pawn and one or more "mothers" ' (Whitehead, 1981, p.95).

Lesotho 'Koali, aged about 30 in 1972, lived with his wife and three young children independently of his father's household nearby, which consisted of his father, his mother and their unmarried youngest son…In July 1973 Kaoli died…his wife had to go to South Africa to seek work in order to support the three young children, two of whom went to stay with her own parents in her natal village some miles away, and the third went to reside in her father-in law's household nearby. Meanwhile Koali's younger brother left his parents' home on his first mine contract' (Murray, 1981, p.106).

Tanzania 'One coffee and maize farmer's household has 17 members. The father and head of the

household has three wives, all of them mothers with children. None of the children is married. The head of the household is 45 years old and the wives are 38, 30 and 25 years old respectively. There are four sons aged 22, 15, 7, and 3; and nine daughters aged 12, 11, 11, 9, 6, 3, 2, 2, and 2 weeks' (Barker, 1989, p.47).

Senegal 'The household is a small one with only four members. The head of the household came to the Birkelane area from Guinea (Conakry) a number of years ago as a *surga*, or dependant attached to a household who works for the head of the household in return for the use of some land on his own account. Recently he married a woman from Guinea and began a part-time business as a small trader. They have an infant child. Although he still lives in the concession with his former *nyatigu*, or master, he is now regarded as a head of household with a separate production unit… now he has as his own *surga* the young son of a prominent man in the village' (Barker, 1989, pp.50–1).

Table 5.1   Some north-west Indian households

Household member	Age	Occupation
Household 1		
Nand Ram	63	Farms family land
His wife	58	Farms family land; domestic work
Son (1)	28	Army
Daughter-in-law (1)	22	Domestic work; farms
Grandchildren	1, 3	
Son (2)	25	Navy
Daughter-in-law (2)	17	Domestic work; farms family land
Son (3)	20	Road mender
Three other sons are married and live separately		
Household 2		
Chandu	32	Farms family land; agricultural labourer
His wife	30	Farms family land; agricultural labourer, domestic work
Children	1, 2, 4, 5	
Wife's brother	25	Agricultural labourer
Wife's mother	55	Domestic work

Source: Adapted from Sharma, U. (1980) *Women, Work and Property in North-west India*, Tavistock Publications, London and New York, pp.68, 72–3.

Why study 'households?

While there may be no neat *formal* definition of 'the household', nonetheless the basic units within which people live in the Third World, as in our own society, are households. The important point to remember is that the nature of these households, and the kind of things they do, varies enormously. For instance, in modern industrial societies the household is normally a unit of consumption and only rarely a unit of production producing goods for sale. In the rural areas of the Third World, however, as well as being units of consumption households often *are* units of production (Box 5.2). Another area of life for which households are responsible, although again in different ways and to different degrees (as we will see in Chapter 6), is the biological reproduction of the next generation and its socialization – in other words, having children and bringing them up. In fact, one of the

few general characteristics 'the household' does appear to have is that it is a unit within which a range of *different* activities are organized – moreover, activities which are absolutely central to individuals' lives.

A good starting point, then, is to ask:

- What do households do?
- How do they do it?
- Why do they organize their activities in particular ways?
- How do they survive over time?
- How and why do they change?

These questions involve examining the physical and human resources that households command to secure livelihoods and reproduce themselves.

Box 5.2   Units of production, consumption and reproduction

The use of the term 'unit' here and in other chapters needs explanation. It usually refers to an identifiable social institution within which particular activities take place. For example, we often talk of 'production units', by which we may mean factories, farms or workshops. In many studies of the organization of rural life in the Third World, 'the household' is often referred to as a 'unit' in which both production and consumption (as well as reproduction and residence) take place. But the extent to which these activities actually occur within households varies, as does how 'the household' is defined, as Chapters 5 and 6 show (see also Box 4.1).

They also involve exploring social relations – or relations between people – both within and between households.

Two key dimensions of social relations in households are age and gender. Both affect access to, control over, and use of resources (land, tools and labour), output and income. Furthermore, people's responsibilities, activities and daily life experiences are also different depending on their age and gender, particularly as divisions of labour are often based on both characteristics (see Table 5.1 for example). Because of this, the concept of 'producer' is not a straightforward one: how (younger and older) women, men and children gain access to resources (who owns what?) is likely to be significantly different, as are their roles in production (who does what?) and their rights over what is produced (who gets what?). With respect to gender in particular, a simple concept of 'producer' is likely to run into all kinds of problems, most commonly that 'producers' become synonymous with 'male producers', while female producers are relegated to some shadowy background existence, or subsumed within the category of family labour. Unless we understand the gendered nature of social relations, and the powerful role played by relations between men and women in shaping society, we cannot understand the shape of a society as a whole.

5.2 Households as units of production

The first case study of these chapters draws on an account of life in the Senegalese village of Kirene by Maureen Mackintosh (1989) in the book *Gender, Class and Rural Transition*. Mackintosh wanted to understand the village farming systems and their relationship to the local economy. This involved researching *social relations*: in the wider context of the villages, and between and within households. Her investigation of this 'web of domestic and social relations' led her to discover that assumptions often made by economists (e.g. that the organization of small-scale agriculture is based on single family units with male household heads, using family labour) were quite erroneous in the West African context.

So how was production organized? What were households like?

In Kirene, households were based essentially on residence and consumption:

> "the smallest residential unit was the household, varying in size from 2 to 37 people. The household was composed of a group of people who ate meals together, jointly consuming the cooking of some of the women in the household. Most households had a common granary, and each had housing set closely together within some form of marked boundary."
>
> (Mackintosh, 1989, p.47)

So households were where people lived and ate together. But were they also units of production? To find out, we need to see how people gained access to land, how they organized their labour, and what relations they had with the wider economy.

Land

Land is not the only resource required for agricultural production and the following questions could equally be asked for other means of production, such as tools or draught animals, and would help to give information on how production is organized:

- Does the household hold land as a unit, or do individual members hold land separately?

- How do different household members gain access to land?

- How is land used? Who makes decisions about it?

Access to land

In Kirene, as in much of rural Africa, there was no market in land. Land was acquired in two ways: first, inherited land which passed between the male kin of the mother's household (in earlier times, from the mother's brother to her son, or uncle to nephew), and secondly, through well-established use rights which based on residence. These customs meant that men generally obtained land from the households in which they grew up. For women, the situation was rather different. Although land inheritance was maintained through the male relatives of the female line (known as **matrilineal** inheritance; see also *Wuyts et al., 1992*, ch.10), on marriage women moved to their husbands' households (known as **patrilocal** marriage or residence). Married women could therefore expect to obtain land through their husbands. However, if the husband had insufficient land, the woman could turn to her mother's brother, or to the household in which she grew up. In general, men and women tried to marry close to home to 'keep land in the family'.

While people could expect to obtain the use of land in different ways, the household head was the person who allocated different pieces of land between household food crops, the head's own cash crops and other household members' cash crops. A key point about this kind of land tenure system is the relative looseness of the ties between particular individuals and particular plots of land. This is a different kind of system from notions of private property in Western capitalist societies where land owners have exclusive rights over use and

Matrilineal: Descent and inheritance through the male relatives of the female line.

Matrilocal: Residence in the house or village/community of the female line.

Patrilineal: Descent and inheritance through the male line.

Patrilocal: Residence in the house or village/community of the male line.

Household arrangements do not have to be matrilineal *and* matrilocal, or patrilineal *and* patrilocal. For example, in Kirene inheritance was matrilineal (customarily) but marriage was patrilocal.

disposal, except where there are legal constraints (see Box 9.1 in Chapter 9).

The land tenure system in Kirene, as in many parts of Africa, was one that had developed over a long historical period during which land was in relatively abundant supply. Consequently, people still tended to view land in the vicinity of their community as something to which all its able-bodied inhabitants had a certain right. This was not necessarily a clearly defined right, and there may well have been fierce and bitter disputes over a particular tract or boundary. But customarily, use rights and their transmission had been a relatively flexible and fluid business, with household heads borrowing land from each other without payment.

Thus, the Kirene system of rights over land is based on a diffuse and generalized cluster of norms and expectations. An analogy in British society would be the norms and expectations involved in a marriage contract. Most Britons would probably agree that husbands and wives have certain rights to each other's worldly goods, but exactly what these rights encompass and how they should be translated into actual transactions in specific cases is far more problematic, as often becomes evident in the case of divorce.

The problem with a system of land tenure based on these kinds of rather diffuse rights (which depend to a large extent on the enforcement of *moral* norms rather than legally enforceable contracts) is that it tends to work only as long as there is no land shortage. Once there is a shortage of usable uncultivated land, those in the community who have *de facto* control over existing cultivable land become less willing to make it available to others, with the consequence that access to land is likely to become increasingly skewed, with certain groups losing out.

Land use

The customary flexibility of land *allocation* was mirrored in land use *practices*. At the time of Mackintosh's fieldwork the villagers' basic crops were millet, grown for consumption not for sale, and groundnuts sold to the government buying agency. Other cash crops were also grown, such as various vegetables and tree fruits, henna and cassava. This mixture of food crops for direct consumption and crops that generate a cash income is characteristic of many rural areas in the Third World (Figure 5.2); many of the problems village economies face stem from the need to maintain a proper balance between these two needs. Mackintosh gives us the following information about the way control over individual plots of land was exercised in Kirene:

> "Each farm plot had someone who 'farmed' it, in the sense that they were responsible for the work on it, oversaw the disposal of the crop and appropriated the revenue if it was a cash crop. Each household…had one and generally only one millet farmer: normally the household head. In addition, most households belonged to what I have called a household grouping. These groupings were organized as a single household during the wet season. In addition to the millet fields of the separated dry season households, the senior household head would also plant a common millet field. The crop from this common field was stored separately and kept for common consumption during the following wet season.
>
> …Most adults in Kirene farmed cash crops on their own account. The household head in

Kirene had considerable influence over the cropping decisions of all household members… This influence was exercised not by giving orders, but through land distribution and access to seeds. Most land used by the household members was held and distributed by the household head. If land had to be borrowed to allow dependent members of the household to farm, this borrowing was usually done by the household head…

> The cash crops were grown by the household heads, by other adult men in the household, and by older married women. Newly married young women rarely had plots, unmarried girls never, but young men would be given plots for cash crops from their late teens."

(Mackintosh, 1989, p.55)

Households and land

This discussion of villagers' access to land and land use reveals that there is no simple answer to the question: is the Kirene household a unit of production? Although the household in Kirene does sometimes operate as a unit of production, particularly as regards the cultivation of the bulk of the staple, millet, a household can also be seen as containing different units of production within it (the separate cash crop enterprises of individual family members, for instance), and equally from time to time as part of a larger unit of production (when it combines with other households to form a single joint unit of production). What this description shows is that to understand how production is organized, we have to look at the different ways in which individual members of households, groupings within households, and households as a whole, are involved.

Labour

When economists and policy makers talk about households as units of production, they usually assume implicitly that there is a (male) household head who controls the organization of labour as well as land and other resources (see Box 4.1 in Chapter 4). If there is no simple answer to whether Kirene households were units of production based on access to land, did they operate on the basis of family labour directed towards producing household food and income? The kind of questions we need to ask of the Kirene households are:

- Who controls the labour of household members: is it the household head, or do individual household members determine how they use their labour?

- What rights do household members have over each other's labour or the labour of others?

- What sexual divisions of labour are there and are they household based?

Figure 5.2   A mixture of food crops and cash crops in Senegal: (left) winnowing millet for household consumption; (above) the tomato harvest – a source of cash income.

• What rights do household members have over output and income?

Controls and rights over labour

With respect to the role of the household head, Mackintosh tells us:

"The household head...had considerable capacity to manage and influence farming activity. But that capacity was based, first, on the head's ability to fulfil demands such as that for land and, second, on the options open to other members of the household, including the women. It is a serious mistake, in analysing West African farming systems, to assume that male household heads always have an unchallengeable right to the unpaid labour of women or other members of the household."

(Mackintosh, 1989, p.29)

So what happened in practice? While household heads could regulate which crops were produced by allocating land (and seeds), they did not necessarily control the actual organization of labour. Each plot would have someone in charge of it and that person might have access to other labour from inside and outside the household. Co-operative work on individual plots was frequent. So, for example, men within the same household or across a household grouping (or compound) might work together (Figure 5.3), not just on common millet fields but on individual groundnut fields too (and might help on women's as well as men's fields).

Another type of collective labour was the *dahira* which was a male work group based on age and residence (rather than kinship), with money paid for the labour going to the mosque (see also Chapter 9 for other examples of work groups). *Dahira* labour was usually used to harvest groundnut fields (including those belonging to women) as this was particularly hard work. There was also a certain amount of paid labour performed by migrants, although villagers did not tend to pay each other for work done. In addition, the co-operative labour provided across households acted as a kind of insurance if individual farmers were ill or unable to carry out their tasks. The only crop

Figure 5.3   Men working together on a groundnut crop in Senegal.

which involved a high level of individual labour and control was vegetables, although women farmers might have had help from other women in the same household.

With the exception of vegetable growing, there was a clear sexual division of labour *by task* in agricultural work. Thus, for example, on married men's millet plots, women from the same household would transport the harvest but men (individually or collectively) did all other tasks. Likewise in men's groundnut fields (Figure 5.4), women did most of the winnowing while men did the rest of the work. Married women did not grow millet on their own account but had groundnut fields. While men were largely responsible for clearing the women's land, seeding and early stages of hoeing, women did most of the other tasks, although *dahira* labour was often used to lift the groundnuts (ibid., tables 3.1, 3.3).

Thus, who worked for and with whom, how labour was organized, and what sexual divisions existed, involved wider forms of co-operation than just household units. Being able to call on family or household labour at certain times and for certain crops and tasks was obviously important, and so were the forms of collective labour that stretched across and outside households,

Figure 5.4   Weeding groundnuts in Senegal.

particularly in groundnuts, the main marketed crop.

The degree of control exercised by different household members over other members' labour for any given activity is therefore always something that has to be investigated. Even where (according to local norms) male household heads are expected to have total authority over others' labour, especially that of their wives, there are limits. The following account taken from the biography of a black South African sharecropper, Maini Kas (1895–1985), illustrates how such limits can operate.

Van Onselen (1990) describes Maini Kas as a patriarch who, *formally* at least, exercised complete control over his large polygamous household, deciding what crops were to be grown, how much was to be sold and appropriating the cash earned in this way. At one point Kas decided to begin growing sunflower, a profitable cash crop; as usual, Kas's wives provided the main labour force. Sunflower is a demanding crop and harvesting the heads and removing the ripe seeds is a particularly unpleasant task. After the experience of the

first year, Kas's wives realized that not only was it hard work, but since it was a cash crop, it gave them nothing in the way of food to meet their obligations of feeding the household. Kas, however, was delighted with the profit the sunflower had brought him and determined to increase the acreage the following year. Unfortunately for him, just when the sunflower was almost ready to harvest he had to make a journey to the local town and was away for a few days. Taking advantage of his absence, the women went to the sunflower fields and proceeded to walk up and down the rows hitting the stalks of the plants with large sticks, and scattering the seed. On his return Kas was met with stories of a terrible attack by birds that had destroyed most of his sunflower harvest. There was little Kas could do: even if he had his doubts about the great bird attack, he had no real sanctions he could apply given his dependence on his wives' labour, and to challenge the story would merely risk a loss of face. The next year, however, the amount of sunflower planted was reduced.

This example illustrates how the stated norms of a society – what people *say* they do – do not necessarily reflect what people *actually* do. This is particularly a problem in the case of households. The relationships between members of a household, unlike those between employer and employee, are not based on a clearly specified contract, but tend to involve a diffuse and rather generalized set of rights, which are translated and negotiated in the course of day-to-day living into more particular obligations and expectations. Getting at these substantive realities is always difficult because if asked who decides how household labour is used (especially if asked by an outsider), people tend simply to reiterate the accepted norms of their society: what they perceive as the way things *ought* to be done. Thus it is necessary to distinguish between the *ideal definitions* of what constitutes a household and how households function, and households as *substantive realities*, i.e. what actually happens in them.

Types of work and sexual divisions of labour

There are two useful distinctions concerning the *types* of work carried out in rural households. The first is that between productive and reproductive work (*Allen & Thomas, 1992*, pp.297–300); the second is that between monetized and non-monetized activities.

The distinction between productive and reproductive (or domestic) work is often difficult to make in practice. For example, while house cleaning may be seen as reproductive or domestic work, how might we categorize food storage or food processing (Figure 5.5)? Both food storage and food processing are simultaneously an extension of crop production and part of preparing food for consumption (hence the reproduction of household members). Also, what is defined as reproductive work or domestic labour often varies widely between societies and in the same society over time. Although in almost all societies women seem to be responsible for 'domestic work', we should not regard this as 'natural' or unalterable. Not only does the content of 'domestic work' vary, it also varies in the way it is shared within households between women and men, and, importantly, among different women. Equally, things are not fixed and timeless; the shape and nature of this work

undergoes all kinds of changes and even radical transformations, if these are sometimes masked by claims that nothing has altered.

In Kirene, for example, the burden of domestic labour fell on women, but not equally on all; their responsibility depended on their position within the household:

> "Domestic work – especially cooking and fetching water and firewood – was hard physical work which was organized entirely on the basis of a division of labour among women within the household, with virtually no household co-operation. A young unmarried girl in a household would assist her mother with her domestic work. After marriage, when she went to live with her husband, she would take over the husband's mother's work completely…By contrast, adult female kin of the household head had no domestic work responsibilities; for example, a divorced sister of a household head was doing no cooking except as an irregular favour to another woman. The domestic responsibilities resulted from the marriage tie and kin relations between men."

> (Mackintosh, 1989, pp.48–9)

The degree of monetization of different activities is important because it affects the basis on which goods and services (including labour) are exchanged, and the extent to which individuals and households require cash incomes to carry out productive and reproductive work. In rural households, domestic or reproductive work is generally not monetized: unless people have paid servants, these activities are part of the unpaid family labour involved in household survival (even where distant kin are involved – see Chapter 6). Other activities may not be directed to market exchange, like growing subsistence crops (although inputs used in their production or storage may well be purchased, requiring cash income from other work). The extent of monetization of any activity needs to take into account whether money is used in production (to buy inputs, pay labour etc.) and whether the product is sold.

Figure 5.5 *Productive or reproductive work?*
(below) Senegalese women selecting
groundnuts for the next season's seeds;
(left) pounding grain for domestic
consumption.

To obtain an overall picture of the activities of households, both productive and reproductive, we therefore need to look at different types of work, who does them, what is produced, and under what conditions. We also need to know how different types of activity are valued: what cash income they bring (and to whom), what they contribute to household consumption (and who gets what), and how activities (and their products) are viewed, i.e. what importance they are given by different household members.

It will have become apparent that a crucial part of understanding these complexities centres on sexual divisions of labour. Whatever other sexual divisions of labour exist within particular societies, the bulk of what is commonly termed domestic labour (i.e. cooking, cleaning, childcare and all those other tasks that service the needs of household members, but do not produce saleable commodities) are generally the responsibility of women rather than men. The ability of peasant households to produce commodities for sale (including selling labour) is dependent on their ability to mobilize and organize labour (mostly female) to carry out the domestic work necessary for their reproduction both on a day-to-day basis and in the longer term. That women are willing and able to carry out domestic work tends to be taken for granted but, like the idea that peasant farming represents a static 'traditional' way of life, it is wrong to assume that domestic labour somehow *naturally* falls within the female sphere.

Rights over output and income

The distribution within households of the goods and services produced (both individually and collectively), and any income (whether in cash or kind) often involves co-operation and negotiation. In Kirene, different individuals within the household had their own separate fields. In addition, the household as a whole would periodically join together with other households to form what Mackintosh terms a household grouping. This household grouping would constitute a single 'unit of production' in relation to a particular set of fields. But none of these arrangements involved payments for labour in cash or kind. 'Money did not act as a medium of exchange within the household boundaries.'

Within a household, the main staple, millet, was normally produced under the supervision of the household head, who also allocated millet from the household granary. Crops produced on an individual basis by household members could be disposed of by the producer. Again, any cash income obtained was usually used to buy essential household goods that were not produced locally or to pay for *outside* labour, or given to needy relatives.

Outside village farming, both young men and young women frequently migrated to do wage work, although this tended to cease on marriage. The incomes earned by young migrants were regarded as their own: for the boys and young men, income went into a fund for marriage payments; for the girls and young women, cash incomes helped to buy clothes and other essentials, also for when they married.

Households and labour organization

This discussion of the use and organization of labour in productive and reproductive work in Kirene has shown that the household is not the only basis for work organization. In summary, household heads could determine to a large extent what household members did with respect to crop production through land and seed allocation. But the actual organization of work in crop production, and the distribution of benefits, did not just depend on (or accrue to) the household head. There were different types of labour organization, involving individual and joint labour, that operated within and across households. Among the villagers of Kirene none of this labour was paid in monetary form but reciprocal arrangements ensured that each person's crops were farmed and harvested. Cutting across these types of labour organization were sexual divisions of labour which functioned at two main levels: in crop production (Figure 5.6) and in the divisions between productive and reproductive work. There were also different arrangements for distributing what was produced and for the use of cash income.

Figure 5.6   Senegalese women weeding around tomato plants.

The wider economy

Final questions we need to ask of productive activities in Kirene are:

* What are the relations between household members and the wider economy?

* Do household members sell or otherwise contribute their labour to production units outside the household?

* How do these wider relations affect the productive activities of household members?

The organization of productive and reproductive work in Kirene, as well as who benefited from it (and how), was the result of long-standing processes of historical change and adaptation to ecological conditions in Sahelian Senegal. While relations within and between households were based largely on forms of reciprocity, they were not untouched by the wider economy and commercialization, evident in the commoditization of groundnuts (since the colonial era) and vegetables, and in the cash income from migration. Futhermore, while farming systems were organized in response to delicate and variable ecological conditions, and were directed to forms of distribution within and between households to cope with the risks and vulnerability inherent in such an environment, households were facing increasing pressures. Arrangements about land and labour use originally based on an abundance of land were being gradually squeezed as land became more

scarce, and patterns of inheritance and access to land were also becoming rigid. Reinforced by changes in land tenancy laws enacted by the state, the fluid nature of a system in which land could be distributed flexibly according to needs began to break down.

Kirene was thus not an isolated, static or self-contained community. But the arrival of an agribusiness transnational corporation in 1972 had a critical impact on different households in the village, reinforcing existing processes of commercialization, tendencies to individual farming and gain, and increasing the pressure on the land. Two large commercial farms were established by an American company, Bud Antle. The Senegalese operation was called Bud Senegal – referred to here as Bud. Bud's farms were to produce fruit and vegetables for the European market and the workforce was to come from nearby villages, including Kirene.

Understanding the internal complexities of Kirene households and how they relate to household and community survival strategies is critical for understanding the impact of Bud, which was qualitatively and quantitatively different from previous commercial links. As noted, the villagers of Kirene had long migrated to towns in search of wage work. 'What was new was the presence of wage employment within walking distance of their own homes.' Bud's arrival offered the people of Kirene a new and different kind of opportunity to earn a cash income, but it also created new pressures on existing village production.

Prior to the scheme, the Bud management argued to the Senegalese government that establishing commercial farms would benefit local communities. The Bud farms would provide a new source of income in two impoverished areas, and since the farms would be based on irrigated cultivation in the dry season, when local farmers did not need to work on their own farms, employment by Bud would not compete with the local agricultural system. In the event, the project did not succeed as a commercial enterprise, nor did it create the envisaged happy symbiosis between the local farming system and the Bud farms. While some people gained, there were also many problems – for individuals, for households, and for local farming. One problem, for instance, was that although the farmers in the localities close to the Bud farms did indeed work *mainly* on their farms in the wet season, 'the dry season was not seen as time "free" for migration in search of work.' The farmers 'saw the dry season work including hedging, house maintenance and vegetable growing, as important' (Figure 5.7).

Three main ways in which the arrival of a large commercial enterprise can affect local livelihoods are: (1) through changes in local land availability and use; (2) through changes in markets for local produce; and (3) through changes in the use and organization of labour. Given the rather delicate balance of local farming systems and their dependence on very particular land and labour arrangements, the extent to which such changes

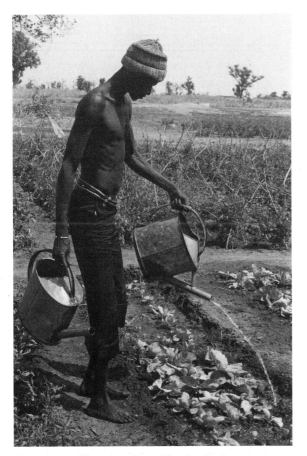

Figure 5.7   Watering vegetables in the dry season in Senegal.

either enhance or jeopardize particular people's livelihoods is obviously an important issue.

Land

With respect to land, Bud appropriated some of the best low-lying land which had previously been used by villagers for their own vegetable crops. The households most affected by this loss were those in closest proximity to the Bud farm, and particularly those that had already been relatively land-short. Some households managed to borrow land from others in the village subsequent to Bud's takeover, but this redistribution did not fully compensate for their losses, mainly because land distribution mechanisms tended to work within household groupings (kin networks) rather than within the village as a whole.

Heads of households that lost land continued to control decisions about land use and decided to prioritize subsistence food crops as part of a 'risk aversion strategy'. However, for some households it became increasingly difficult to assemble land for collective millet fields. Collective fields were also affected by conflicts over labour use (which we will return to below). Some fields were abandoned, and the customary rationale for household grouping based on exchange of labour, equipment, and other forms of reciprocity, was undermined.

In addition to the effects on collective fields, vegetable production was also undermined. The area appropriated by Bud had been prime land for households' own vegetable production because it remained humid even in the dry season. Diminishing availability of land as well as competition for use of labour by Bud, and the effects of weather changes and drought, reduced the viability of vegetable crops.

Vegetable markets

What about market competition between Bud and local vegetable growers? The Bud farms were set up to grow crops for export. At first none of its produce was sold locally, mainly because, given the tiny local market and Bud's relatively high production costs, such sales were unprofitable, and the Senegalese government was worried about the impact of these sales on local farmers. Produce of inadequate quality for export was simply dumped or ploughed back into the land. In subsequent years, however, an increasing amount of this sub-standard produce was sold locally, which pushed down prices and made vegetable growing less viable for local producers. Interestingly, though, Bud's increasing local sales were 'in part a response to pressure from the local population, who saw food rotting, and in particular from the local traders who saw profit opportunities in buying and reselling the produce.' Some local women, for instance, 'bought the red bell peppers, unknown in the local diet until then, cooked them with a little tomato, and sold them as "tomato paste", a product widely used in cooking in the area.' This illustrates an important point about how markets can work to undermine existing productive enterprises: once cheaper goods become available, local people are likely not only to buy them, but actually to demand them.

Labour

What about the effect on labour? To what extent did Bud compete for labour with individual and household production? We will see that Bud's impact was different for different kinds of households.

For households that lost land, Bud offered some 'compensation' through wage work while other households hoped to increase their overall incomes by working for Bud. Most millet farmers (men) worked at Bud but all combined it with their own crop production. Although young men worked for Bud, they also continued to go to Dakar to work, as before. For women, working at Bud was particularly attractive to those who did not have large domestic workloads – older married women with married sons (and therefore daughters-in-law to help) and unmarried adult women living in the household, although other women also found ways to combine farming, domestic work and wage labour. For men, employment at Bud competed with crop production to a certain extent because of overlapping agricultural cycles, particularly with the groundnut harvest and with land clearance and millet seeding. There was also a clash between the vegetable harvest (now largely in the hands of women) and Bud's demand for labour.

This competition for labour was an additional factor in the disintegration of collective millet fields, in which there was increasingly uneven participation by members of household groupings. With respect to the *dahira* collective labour, there were two effects: men employed by Bud could afford to pay the fine due on dropping out of the *dahira*; men with improved incomes from Bud could also afford to pay for a *dahira* instead of carrying out their own (or other) field work. Thus these men did not contribute to the *dahira* work either. Mackintosh concludes: 'Bud was thus an agent of disintegration of mutual assistance in farming within Kirene. Through this effect it was reducing the commitment to food production within the village, despite farmers' attempts to give priority to millet growing, and increasing the vulnerability of any household to economic disaster in any year by reducing the "safety nets" embodied in the collective labour system.'

Labour organization in Bud itself established new sexual divisions of labour in estate work. To some extent these mirrored – and were influenced by – existing sexual divisions of labour in village households:

> "All of the permanent workers were men, as were all but three of the regularly employed non-farming seasonal workers (clerks, transport workers, local sales people, bricklayers, watchmen). Among the farming jobs, all those recognized by the management as skilled or semi-skilled and requiring training were done by men: tractor driving, plant protection, mechanics…Of the jobs classified as unskilled, certain were done exclusively by men: irrigation work, hand hoeing, melon and pepper harvesting. The others, packing produce and bean harvesting, were done mainly, but *not* exclusively, by women. The men's teams specifically excluded women; the jobs 'where the women are' were sometimes done reluctantly by young men who could find no other work."
>
> (Mackintosh, 1989, p.172)

Mackintosh states that 'within three years, that familiar category of "women's work" had been created' – sexual divisions of labour associated with differences in pay and working conditions had been established.

Wider implications

In summary, the income changes by category of producer were as follows: the incomes of farming men with sufficient land increased; those with insufficient land declined; the incomes of independent, unmarried men increased as a result of wage work; women's cash incomes increased in general even though they did not earn as much as men, although older women who lost vegetable land also lost income. Even though some women earned more money by working for Bud, Mackintosh explains how their responsibilities for food provisioning increased with declines in household food production, because they had to buy food with their wages.

Lessons from Kirene

So what can be learned from the relations between Kirene households and the wider economy? How was the social organization of production affected, and what was the impact on rural livelihoods? Changes brought by increasing commercialization and a squeeze on land meant that forms of production that had previously been based on reciprocity were becoming 'individualized', that is, becoming based on individual access to resources, jobs and income. Different households were affected differently by these processes, as were women and men within them. Mackintosh concludes:

"Several reasons suggested that the decline in food farming and the individualization of the economy were not likely to stop in 1975. The strains which were pulling apart the collective labour institutions would continue, and were likely further to damage labour co-operation among households. The physical toll of doing two, or in the case of women three, competing or overlapping types of work would continue, and women and young men, in particular, seemed likely to progressively drop out of farming. The

withdrawal of dry season work from farming would in the long run reduce the productivity of the system, reducing land preparation, manuring and fencing; faced with a choice, the married men would sacrifice their cash crops to Bud where Bud provided a higher cash income, which was in a majority of cases. The division between men with the better paid seasonal jobs at Bud, and those with only casual work, seemed likely to harden…and to put further strain on the collective labour system."

(Mackintosh, 1989, p.85)

This case study shows that simplistic ideas about the household – as a single unit of resources, income, responsibilities, and interests – are untenable. The social organization of Kirene illustrates different patterns of land holding and use, of labour, and of control over produce and income, all of which combine individual, household and collective dimensions. The case study also shows that understanding these social relations and divisions is critical to understanding the effects of increasing commercialization on rural livelihoods.

5.3 Households as units of co-residence and consumption

Having examined the complexities of households as units of production, we now explore the meaning of households as units of co-residence and consumption. That households have these functions may seem obvious: from our own experiences, households are generally where people live and eat together, and share everyday domestic life in different ways. This idea of households presupposes that resources, food and incomes are somehow shared between household members, but as we have seen from the case study of Kirene, there may be different forms of provisioning, and access to food and income, within the household itself. Who lives with whom, who provides consumption needs, and who consumes what, are all subject to different norms, values and customs.

Co-residence

Thinking of a household as a unit of co-residence says little in itself about who resides with whom. The terms of co-residence are a fundamental part of social organization in rural communities. They also affect the impact of agrarian change on the lives, needs and interests of individuals, and how individuals as well as households try to bring about, adapt to, or resist change (see Chapters 9, 10 and 12). Knowing what is meant by co-residence – what happens in practice as well as in its 'ideal' form – is also important for any social and economic policy directed to rural areas.

Look back at Box 5.1 (and Figure 5.1) for a reminder of how varied household composition can be. However, knowing who constitutes a household or a household grouping does not tell us directly who lives with whom and under what conditions. For example, in **polygynous** households where there may be more than one wife, each wife may have her own house and individual responsibilities for particular people or dependants. Or, in the case of Kirene, Mackintosh says that households could comprise from 2 to 37 people residing together, but most common were houses of 6 to 9 people constituted through marriage (women joining their husbands' households). Most children lived with their parents, although boys were occasionally sent to live with their maternal uncle. Girls married early and moved out while boys might stay in their parental households until their thirties. Critical to this pattern of co-residence was the structure of domestic work which was done by wives, girls and unmarried young women. Adult female relatives of the male household head who also lived in the household would not, however, be obliged to do domestic work. Mackintosh concludes: 'It was... the preparation and common consumption of food, organized on the basis of marriage relations, which constituted the household.' We return to the relationship between co-residence and consumption shortly.

Comparing the different descriptions of African households in Box 5.1 and Kirene with the household compositions of Table 5.1 again shows that 'co-residence' is relative within and between cultures. The households in Table 5.1 are also formed

> **Polygynous:** Having more than one wife. *Also:*
>
> **Polyandrous:** Having more than one husband.
>
> **Polygamous:** Having more than one wife or husband.

by marriage, with women going to their husbands' households (hence the absence of adult daughters), but in terms of residence they may comprise a range of kin (again, except adult daughters). Both households in the table contain three generations.

Now compare the composition of those households with the household in Table 5.2. In many ways, the household composition is similar – there are three generations living in the same household based on patrilocal marriage. The main superficial difference between this household and those described in both Box 5.1 and Table 5.1 is that the household head is a widowed woman.

While this may seem a minor point, the question of household headship in co-residence (and in consumption, as we shall see) is an important factor in household composition and dynamics. As was noted in Section 5.1, it is often assumed that households are headed by men who are also responsible for the distribution of goods and resources. (This assumption has often resulted in rural policy going badly awry – see *Allen & Thomas, 1992*, ch.15.) While male household headship is in practice common (remember the role of senior male compound heads in Kirene) there can be no automatic assumptions about their roles – either about the extent of their control, or whether it results in an internal distribution of household resources which benefits all members equally. Rural policy based on such assumptions often misses the unequal relations that exist in some households as well as the importance of women's areas of control in others. As we have seen, women and men may have different access to land and other resources, as well as independent income streams, and collective labour and resources may cut across individual households based on marriage.

Table 5.2   Another north-west Indian household

Household member	Age	Occupation
Household 3		
Sarla (widow)	60	Farms family land
Son (1)	37	Shopkeeper
Daughter-in-law (1)	30	Domestic work; farms family land
Grandchildren	2, 7, 10	
Son (2)	30	Clerk in Punjab
Son (3)	20	Labourer in mill in Punjab

Source: Adapted from Sharma, U. (1980) *Women, Work and Property in North-west India*, Tavistock Publications, London and New York, p.71.

However, the issue of household headship has another dimension. Assuming that household heads are men ignores the large number of households that in practice are headed by women. This fact brings us back to the question: who lives with whom and on what terms? Women may head households for a number of reasons: they may have been deserted or divorced; partnerships may be rather fluid; or women may be effective (rather than actual) heads of households because their husbands or partners are migrant workers (Figure 5.8). How male migration can affect the functioning of households as units of co-residence is described in the following example from Lesotho, taken from a case study by Colin Murray (1981) called *Families Divided*.

Although Lesotho is nominally an independent nation state, its economy is totally dependent on that of its giant neighbour, South Africa, for which Lesotho is primarily a labour reserve. Although in the past Lesotho had a thriving agriculture, nowadays it is virtually impossible for its people to make a living from farming, and large numbers of them, especially men, have to migrate for work to South Africa where they work mainly in the mines. A survey of one area in 1976 found that 62% of all males aged between 18 and 59 were away working as migrant labourers, while for the age group 20 to 39 the percentage rose to 77 (quoted in Murray, 1981, p.40). The money sent back by these migrants is crucial for the survival of those left at home. Official rhetoric in Lesotho may claim that agriculture is the 'backbone of the economy', but according to Murray, this is only true in the residual sense that there is little within Lesotho in the way of other income-earning possibilities (Figure 5.9), so that its population today 'is aptly described as a rural proletariat which scratches about on the land.'

For the predominantly male migrant labourers of Lesotho, the hope is that eventually the years spent working in South Africa will enable a man 'to establish his own household…to retire from migrant labour and to maintain an independent livelihood at home'. In other words, the idea is that through migrant labour a man can earn enough money to set up a viable production enterprise, normally a farm. This involves the purchase of land, livestock and equipment, and in addition the acquisition of cattle, or their equivalent in money, with which to marry, since in order to establish his own independent household a man needs a wife. The Basotho (the people who live in Lesotho) are a patrilineal people and marriage depends on the payment of bridewealth by the groom to the bride's family. In fact, the amounts of money that can be

Figure 5.8 *Women may actually, or effectively, head households for many reasons: for instance, in Namibia (right) many men migrate to work in mines.*

Figure 5.9 *The difficulties of earning income in Lesotho: building a road as part of a UN/FAO food-for-work scheme.*

earned through migrant labour are almost never enough to achieve all this. Most men 'must commit themselves in their declining years to dependence on the remittances of sons or of other junior kin who in their turn engage in the oscillating pattern.'

Just to persist in 'scratching about on the land' usually requires continual inputs of cash income to buy fertilizer, hire oxen for ploughing and so on, but at the same time farms seldom produce enough even to satisfy a household's basic consumption needs. And even if people were able to produce enough food for themselves, they would still need clothes, blankets, cooking utensils, and a whole range of other goods they cannot produce themselves and must buy. Given the uncertainty and lack of profitability of farming in Lesotho, why do so many people continue to invest so much of their time and energy into it? A factor stressed by one analyst, Spiegel, is the importance of farming in terms of providing people with a sense of belonging to their local community, and therefore having some kind of claims within it as regards security in the future. '[M]igrants still aspire to be allocated land and continue to invest in agriculture and livestock, even when they cannot ensure a worthwhile return in the short term, in order to demonstrate their long-term commitment to the rural social system' (Spiegel, 1979, quoted in Murray, 1981). In most Third World societies the state provides little in the way of welfare systems so that the household and wider kinship networks are crucial in this respect. Such longer term needs can also be seen as part of what makes up household provisioning.

Husbands and wives living in Lesotho's rural areas are subject to two contradictory pressures; on the one hand, if husband and wife are to maintain an independent household, then it is normally necessary for the man to engage in migrant labour while the woman stays in Lesotho and runs the household. On the other hand, the prolonged separations of husband and wife, sometimes for years on end, put enormous strains on the marriage and separation and divorce are common.

This example demonstrates some further issues about households as units of co-residence. The pervasiveness of migration as a form of semi-proletarianized labour (Chapter 2) means that many households in Third World economies are relatively fluid units in terms of who actually lives together at any one time. Although, in Lesotho, migration is a male phenomenon, in other contexts other adult members of the family may come and go at different times in their lives. (Migration is not the only reason for fluid co-residence. Children may also move, or be moved, between different sets of kin, especially where there is financial hardship, or emotional or physical disruption.) While temporary (as well as permanent) migration may be part of economic survival strategies, it has an effect on how households are organized. In many instances, women effectively become responsible for farms and for most aspects of household maintenance – whether there is cash income coming from their migrant husbands, sons or daughters, or not. As Murray suggests, the stress can take its toll. However, in the Lesotho case, there was still an 'ideal' of household composition and character to which male migrants aspired: to establish independent households (which they 'head') based on their own land and an economically viable farm. The reality was in fact rather different, as we saw.

Consumption

The Lesotho example also draws our attention to the relationship between co-residence and how households meet their consumption needs. In this case farming alone did not meet such needs, and, indeed, to maintain investment in farms required cash income from migration. The result was two rather independent but connected units of residence and consumption – that of the wife who ran the farm and that of the husband who may have lived collectively with other miners in South Africa.

The living arrangements of migrant miners in South Africa and other parts of the world are of course not households constituted by marriage or male–female partnerships, and they emphasize again varied forms of co-residence, histories and rationale. Seasonal migration to do wage work in rural areas (often by whole families) can also involve temporary collective housing. Co-residence does not necessarily presuppose a particular kind of consumption unit – who provides what and who consumes what still need to be established.

The Lesotho case also shows the links between consumption and production: income from migration was critical to maintain consumption and investment in continued production on the farm. However, farm production by women also provided consumption needs, and women's management of provisioning was obviously a key part of household reproduction.

Household provisioning includes food, but it also requires securing and maintaining a house and other means of daily life. For most households, domestic work is a fundamental component of provisioning and consumption. Note how Mackintosh defines the unit of co-residence and consumption for the villages she studied in Senegal:

> "The domestic unit…is defined, not simply as a food consumption unit, but as the unit of consumption of domestic work [Figure 5.10]. By this is meant the unit within which a division of labour is established, generally between women, for the performance of the classically female tasks of cooking, cleaning, childcare and care of the sick; the unit also includes the people who benefit from their output. By association, this 'household' is generally, in the societies studied here, the unit of consumption of meals in common – but not necessarily a unit of common food stocks."

> (Mackintosh, 1989, pp.29–30)

We have seen from Kirene itself and from Lesotho that household provisioning may be a complex process in terms of who produces (or provides) what and what divisions of labour are involved, between women and men, and between women. Men are the main providers of the food staple, millet, in Kirene, but in many parts of Africa, farming and growing food crops for household consumption is carried out by women. This provisioning is also intimately linked to internal distribution of food in households, as shown by the following example from a study in Malawi by Megan Vaughan.

In southern Malawi, where Vaughan carried out a survey in the 1970s, the people are matrilineal. Most of the women Vaughan interviewed lived in the villages of their matrilineal relatives (men normally move to their wives' villages on marriage, i.e. marriage was matrilocal). With respect to food distribution Vaughan was told that:

> "Ideally, each adult woman has access to land on which she grows the food to feed her immediate family. She will own a grain bin in which she stores her food, and which is hers alone. A group of sisters [one of the closest and most powerful ties for matrilineal people] living in neighbouring huts will each possess their own grain-bins, and it is apparently unheard of for one sister to take grain from another's bin."

> (Vaughan, 1983, p.277)

In this case, while in any formal or legal context the head of household will be perceived as being a man, the distribution of basic food staples is controlled by women. If there is more than one adult woman in a household then each woman will control the distribution of grain from her grain bin and will be responsible for feeding her husband and children. This is not the whole story, however. What is partially masked by this ideal picture is a continual flow of food transfers both within and between households. Thus the unit of consumption is not necessarily, or always, the household. When the *household* is not the unit of consumption, what is? And how *are* consumption units related to the household?

In the villages Vaughan studied it was common for a group of matrilineal relatives living within the same village to eat communally, i.e. the women and children would eat together in one group while the men also ate together but in a separate group. The norm that each woman should be self-sufficient is very powerful. 'Women continue to own and repair grain-bins which they are only able to fill to a tenth of capacity. Not owning a grain-bin amounts to admitting destitution.' Consequently:

> "Food transfers are, in fact, disguised as much as possible. Where a number of sisters,

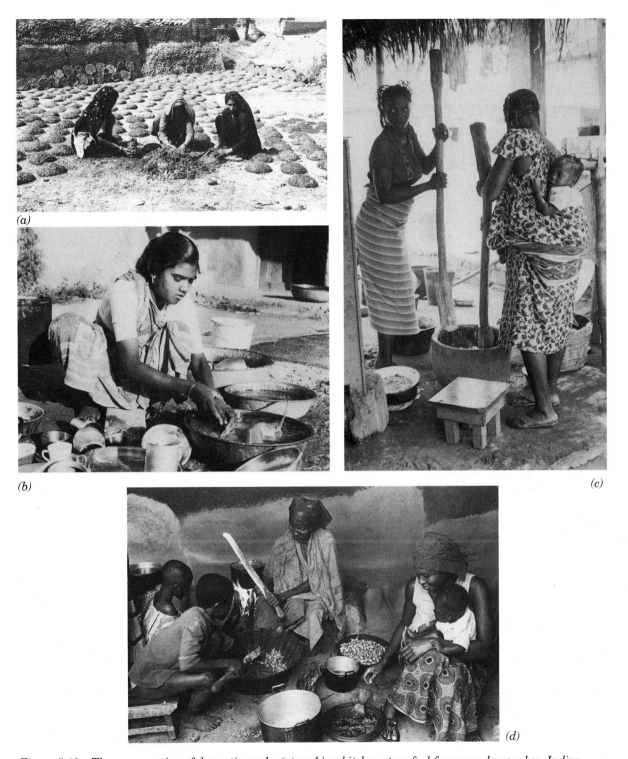

Figure 5.10 *The consumption of domestic work: (a) making kitchen stove fuel from cow dung cakes, India;
(b) washing up with mud, ash and water, India; (c) pounding yam, Nigeria; (d) combining cooking and
childcare – a courtyard kitchen in Ghana.*

or sisters and other maternal relatives, eat communally, each woman cooks her family's food and brings it to one of the houses where the meal is to take place. Food is transferred at the stage of eating, and people do not perceive this as 'food sharing'. As the male members of the group are served separately from the women and children, the transfer of food between households is even less apparent to them."

(Vaughan, 1983, pp.277–8)

As Vaughan points out, it is very difficult to know what is going on unless one is able actually to witness the business of cooking and eating! But such food transfers may be very important. They act as a household survival mechanism and they can also be a clue to whether or not there is increasing differentiation in the community. Vaughan found that 'the poorest households (often headed by women alone) were heavily dependent on food transfers from their more prosperous relatives.'

Various informal systems of sharing, whereby food and other goods circulate very freely among a large group of kin and/or neighbours, are common. Sometimes these transfers are the result of mutual obligations between kin sanctioned by powerful moral norms. But again, as in the case of rights to land described by Mackintosh, these norms tend also to be rather general, or diffuse. The idea that matrilineal kin must help one another does not specify in any precise way just what transfers should be made. Transfers may take the form of gifts or loans, but discovering exactly what happens is always problematic, particularly as this sharing (often involving very small amounts but taking place almost continuously) is frequently so woven into the texture of day-to-day life that those within the community are scarcely aware of it. In fact, such forms of redistribution *depend* on people not being too conscious of each individual transaction, but rather having a sense that in the long term there is a kind of reciprocity. Once people do become aware of individual transactions it tends to mean the system is under pressure as constant need replaces temporary fluctuations in the patterns of distribution.

A different picture of household provisioning and arrangements over intra-household distributions is presented from studies in parts of South Asia. For example, a study of villages in Bangladesh shows considerable male control over resources and their distribution (Kabeer, 1990). Kabeer states:

"The primary domestic grouping in Bangladesh is the family-based household...Men tend to control most of the household's material resources, including the labour of female and junior members of their households, and also to mediate women's relations with the non-familial world. Women...are generally reluctant to seek incomes outside the socially sanctioned relationships of family and kin, first because there are few options to do so and second, because they could forfeit the support of their kin."

(Kabeer, 1990)

The sexual division of labour takes the form of an 'inside/outside' division: 'field-based stages of production have to be carried out by men; those activities located in or near the homestead are the preserve of women, along with various domestic chores' (Kabeer, 1990).

As Kabeer points out, social relations within these households are asymmetrical or unequal. But as in most domestic units, they imply a process of negotiation between the various members of the household concerning who has access to what and who consumes what. In the context studied by Kabeer, women submit to male authority for their own security but they also find ways of negotiating access to resources for themselves and their dependants through informal networks and exchanges with kin outside the immediate domestic unit and neighbours (see also Chapter 12).

Another example of how intra-household distribution and control of resources for domestic consumption needs are organized is illustrated in Table 5.3 for household budgets in north-west India. While there may be many small variations within each category, depending on the cash income coming into the household and the extent to

Table 5.3 Control of household budgets in two villages in north-west India

Types of control	Numbers of households
A Man gives fixed allowance to wife or mother from monthly earnings for household expenditure only	15
B Man gives all or almost all monthly cash earnings to wife or mother for most kinds of expenditure	4
C Man and wife each keep respective earnings and share responsibility for budgeting	8
D Man keeps all the cash and does all kinds of budgeting himself	0[a]
E There is no male member of household; the woman receives all cash and budgets for herself	1
Total	28

[a] There were no cases of this type of household in the sample, but they were not totally unknown in the areas studied.

Source: Adapted from Sharma, U. (1980) *Women, Work and Property in North-west India,* Tavistock Publications, London and New York, p.106, table 3.

which women earn cash income, men tend to control domestic expenditure and women have little access to resources for their personal consumption.

Other manifestations of asymmetry can exist in intra-household distribution and consumption of food. Data from India and Bangladesh have shown that the satisfaction of food requirements tends to be higher for men and boys than women and girls, especially among poor households. Population statistics for India also show higher sex ratios for males relative to females, especially in the north and north-west (Agarwal, 1986). These data are thought to represent the effects of unequal access of males and females to health care and other resources, as well as food (Drèze & Sen, 1989, p.52; see also *Allen & Thomas, 1992*, pp.29, 31).

This section has shown that the organization of living arrangements – who lives with whom, who consumes what, who controls access to income and other resources for consumption – are, again, highly varied. Standard patterns cannot be assumed, but discovering what people do in practice is a fundamental part of understanding and explaining the dynamics of rural life.

Summary

1   Rural households organize their domestic lives in many different ways. 'The consumption of domestic work' is closely linked to production and distribution, even if the units in which these activities take place are not identical and may change in composition over time. Relations within households are not necessarily equal either in access to resources, control over income, or consumption of basic needs. Production, provisioning and household management are subject to many types of division of labour and forms of negotiation.

2   Households might not be neatly bounded entities; people may move easily in and out of them. Our analytical definitions can also vary depending on the particular questions being asked.

3   Households may have many different activities which may or may not be organized jointly by household members. To understand the organization of rural livelihoods, we should not lump households simply into groups of autonomous individuals, all with their own particular interests and all competing with each other. Nevertheless, it is equally important not to subsume the interests of different household members into a single overarching entity with a single decision-making head.

6

RURAL HOUSEHOLDS: SURVIVAL AND CHANGE

KATE CREHAN

As we have seen in Chapter 5, households are affected by social and economic changes in the wider society. In addition, the reproduction of households from generation to generation is not an automatic process. This chapter focuses on how households come into being, persist, change, and are reproduced over time (or not). The chapter is divided into four main sections: the first is concerned with having children, and the second with children's role within households and how they are socialized. The third section looks at economic and other kinds of differentiation between, and within, households, focusing particularly on those households, or household members, who are not simply surviving but are actually prospering. The final section turns to households in crisis – those for which day-to-day existence and survival itself are a struggle.

6.1 Having children

When, as in many peasant societies, households are the key units of production, their survival depends on their ability to reproduce themselves and maintain their viability. On one hand they must contain, or have access to, the minimum number of adults needed to constitute a unit of production, and on the other, they should not have a too high ratio of dependants (small children, the old and the sick) to active producers. The primary mechanisms through which peasant households acquire new members are marriage and the biological **reproduction** of existing household members, and recruitment on the basis of other kinship links. The enormous value peasant societies characteristically attach to children is based on the material importance of children as a future labour force for the household and for its reproduction over time. Children are not, as in capitalist societies, simply 'unproductive' consumers who must be supported until they are able to become wage earners in their own right.

Marriage necessarily involves the coming together of two sets of kin: the husband's and the wife's kin. The question of to whom the children belong depends on the nature of the kinship system. In

> **Reproduction**: All the processes by which the inputs of production are themselves produced. The 'production of the producer' involves *biological reproduction* (childbearing), *generational reproduction* (childrearing) and *daily reproduction or maintenance* (provision of human needs like food, shelter, etc.). This chapter looks at aspects of biological, generational and daily reproduction or maintenance.

patrilineal systems children belong to their father's kin group, in *matrilineal* ones to their mother's, and in *bilateral* systems for certain purposes they belong to their father's and for others their mother's.

The rules governing descent, however, do not tell us with whom a married couple will live. Sometimes a new couple may form an independent household, but often, maybe until one or two children are born, either the wife is incorporated into the household to which the husband belongs, or the husband may join his wife in her family's household. Usually a society has rules about where couples should live, but it may well be that in practice these rules are not always followed, residence depending on the particular circumstances of individual households. In addition, the household as the site for raising children is not necessarily synonymous with the household as a unit of residence. In some peasant societies, children may not live in the same residential unit as their biological parents. They may live with grandparents or other kin and in the course of their childhood they may move freely between different kin.

Such movement between residential units can help modify the arbitrariness of biology and demography. Some households may have too many small children, or only boys, or only girls, while others may have no children, or a household may be suffering from economic hardship and may need to reduce the number of unproductive dependants.

Although kinship may be the dominant mechanism by which peasant households recruit new members, it is not the only one. As units of production, households may employ some wage labour. There can be various other practices which are not so easy to classify. For instance, it is common for richer households to incorporate young women or sometimes men who are related in some way to other household members, and receive their board and lodging in return for performing various household tasks (often the relatively invisible tasks of cleaning, childcare, etc.). The powerful ideology of kinship means that such people are often described by other household members as

kin, and it needs careful observation of what they actually do, and whether or not they have reciprocal claims on other household members, to determine whether or not such people are better termed 'employees'.

Although kinship is not the only mechanism by which households recruit new members, it remains the most important one, in practice and with respect to what people believe about it. These ideological dimensions tend to make it appear as if households are *determined* by kinship, coming into being and reproducing themselves quite separately from wider economic and social relations.

An area that illustrates how the reproduction of households is shaped both by wider economic and social relations, and by individuals' decisions made in the course of their day-to-day lives, is that of fertility, or the number of children that women have. Much popular writing on the Third World suggests that one of the major problems is 'overpopulation' – Third World peoples are poor because they have too many children, and what is needed is education about family planning and the provision of birth control (this view is criticized in *Allen & Thomas, 1992*, ch.4). An alternative view is posed by Mahmood Mamdani in his study *The Myth of Population Control*, in which he says that poor people, rather than being poor because they have large families, 'have large families because they are poor' (Mamdani, 1972, p.14).

Mamdani studied Manupur, a village in India which had been one of the 'test' villages for a major birth control programme funded by the USA and the Indian Government in the 1950s. The region selected was one with a very high population density in which the fragmentation of land, because of its subdivision through inheritance, was recognized as a growing problem. Despite the massive funds poured into this project, the results in terms of reducing the birth rate were negligible, although it took some time before the villagers' reluctance to use the contraceptives provided became apparent. The following two quotations explain why. The first is an explanation one man gave to Mamdani as to why villagers accepted the contraceptive tablets, although most had no intention of using them:

> "But they were so nice, you know, and they came from distant lands to be with us. Couldn't we even do this much for them. Just take a few tablets? Ah! even the gods would have been angry with us. They wanted no money for the tablets. All they wanted was that we accept the tablets. I lost nothing and probably received their prayers. And they, they must have gotten some promotion."
>
> (Mamdani, 1972, p.23)

Later Mamdani visited another house together with a villager, Asa Singh, who had also taken but not used the tablets:

> "Gradually, my eyes got used to the faint light, and I saw small rectangular boxes and bottles, one piled on top of the other, all arranged as a tiny sculpture in a corner of the room. Along with the calendar prints of gods and goddesses, movie stars, and national leaders, it decorated the room. This man had made a sculpture of birth control devices. Asa Singh said: 'Most of us threw the tablets away. But my brother here, he makes use of everything."
>
> (ibid., pp.32–3)

So why had the villagers of Manupur been so reluctant to use the contraception they were given? Was it that the villagers were simply obeying the dictates of their 'culture' or their 'tradition'? According to local norms, families were expected to have many children, particularly boys, and underlying this expressed preference for large families were sound economic reasons for having children: they provided the labour crucial for a family's survival.

Although landowning and landless households may have different labour needs, both indicate why children are crucial:

> "Of course, I am worried about the fragmentation of land. But even before I worry about my land being divided up tomorrow, I must worry about making a living on it today. Just look around: no-one without sons or brothers

to help him farms his land. He rents it out to others with large families. Without sons, there is no living off the land. The more sons you have the less labour you need to hire and the more savings you can have. If I have enough, maybe we will buy some more land, and then fragmentation will not matter."
>
> (ibid., p.78)

Another villager without land, Fakir Singh, whose former livelihood as a water carrier had been undermined by the increasing number of electric water pumps in the village, told Mamdani how indispensable his eleven children were to him. He was prevented by caste norms from doing any work other than watercarrying, but the restrictions did not apply to his children. He claimed all his sons are an asset:

> "The youngest one – aged five or six – collects hay for the cattle; the older ones tend those same cattle. Between the ages of six and sixteen, they earn 150 to 200 rupees a year, plus all their meals and necessary clothing. Those sons over sixteen earn 2000 rupees and meals every year. Fakir Singh smiles and adds: 'to raise children may be difficult, but once they are older it is a sea of happiness'."
>
> (ibid., p.109)

The need for security in old age is another important reason for having children. As one man put it:

> "You think I am poor because I have too many children. If I didn't have my sons I wouldn't have half the prosperity I do. And God knows what would happen to me and their mother when we are too old to work and earn."
>
> (ibid., p.111)

Mamdani's study focuses mainly on the male household heads and their attitudes to fertility and birth control, but what of those who actually bear the children, the women: how do they feel? A study by Naila Kabeer in a village in neighbouring Bangladesh tried to answer this question. In the village she studied, as in Manupur, marriage is *exogamous*, i.e. people marry outside their kin group; in this case women would leave their natal village to live with their husband and in-laws (Figure 6.1). Kabeer sketched out such a young bride's future:

"As a young and inexperienced bride, she enters a strange and hostile household where her behaviour is viewed with suspicion until she has been successfully integrated into it and has learned to identify with its interests. The practice of village exogamy reinforces her isolation, depriving her of the possible support of her own kin. Her position in the household is subordinate to her mother-in-law and senior sisters-in-law, and she assumes much of the domestic drudgery. Her status in this hierarchy will only improve with the birth of children, specifically of sons, because of their importance in perpetuating the lineage. Barrenness, on the other hand, brings social disgrace, and can be grounds for divorce or polygamy. There is further improvement in her position when her sons come of age and start to contribute to the household income. Finally, when they marry, she is able to obtain the only position of authority open to most women: that of the mother-in-law. Women's own interests therefore predispose them to high fertility rates, independent of the need of the household economy to build its labour force and perpetuate itself. In the early years of her marriage, children are a young wife's only solace; as they get older, they help her around the house; and in her old age, control over her daughters-in-law reinforces her status in the household, and assures her of some freedom from drudgery."

(Kabeer, 1985, pp.98–9)

Figure 6.1   A woman leaves her natal home: a marriage procession in Bangladesh.

Both in the village studied by Kabeer and in Manupur, having sons was valued more highly than having daughters – a preference reflected in statistics showing more acute malnutrition among young girls than young boys (Bangladesh Nutrition Survey 1976–77, quoted in Kabeer, 1985, p.85), and 'the much higher infant death rate among females' (Mamdani, 1972, p.141). However, in both villages, provided some sons had been born, girls were also valued, particularly by women who depended on daughters for help with domestic and other 'female' work.

Mamdani's and Kabeer's studies illustrate how the ideological stress on having many children is founded on sound economic reasons. It is when, and if, these change that peoples' desire to have large families may change rather than through the prompting of birth control advocates (Figure 6.2).

For instance, in Manupur something that did have an effect on the birth rate was a rise in the average age at which girls married, from 17.5 years in 1956 to over 20 in 1969. But this rise was the result of various technological changes in farming which made the labour of young girls more valuable to their families. To put it another way, it is not having many children that explains the conditions of villagers' lives, it is the realities they face that shape the decisions they make about family size.

The economic rationale for family size is also influenced by other pressures, particularly on women, to protect their own positions and interests. Kabeer noted that women in the village she studied tended to want to have many children 'independent of the need of the household economy to build its labour force and perpetuate itself.' For

Figure 6.2 *A family planning worker demonstrating an intra-uterine device to a group of women in India.*

these women, children gave additional forms of security in terms of status, company, moral and physical support, as well as assistance with daily work and financial security in old age.

6.2 Reproduction: socialization, education, and providing household labour

This section focuses on the place of children within households, where much of their socialization takes place, as well as their general education when there is little formal schooling, as for many rural children in the Third World. In peasant households children can be a vital source of labour, whether for work in the fields or various domestic tasks, such as caring for younger siblings. Children may also be expected, or may choose, to sell their labour outside the household.

The upbringing of children and their socialization into different social roles is also the best way in which individuals 'obtain' the expected characteristics of their genders. The structures of power and authority in households that assign junior members to particular tasks in productive and domestic work are also 'gendered', and affect how children are brought up, as well as the role models they seek or have to follow. How children became adults, and what sort of adults, how they learn their roles at home and in society, are all critical to family and household survival as well as the social cohesion of the community.

These points are illustrated in the following short account of the role of children in Mukunashi, a village in north-western Zambia where I carried out research between 1979 and 1981.

"Participation in the labour process begins at an early age. Almost as soon as a child can walk, he or she begins to be expected to help adults in their daily tasks, work growing organically out of play. Tasks are mastered gradually and informally, first as play, but, as soon as their physical capacities allow, play turns into work. The acquisition of skill takes place almost by a process of osmosis

from the largely female world in which small children live. In the course of their general socialization, therefore, children of both sexes acquire the basic skills of cultivation, food processing and all the essentially female tasks. But whereas small girls help their mothers more and more as they grow up, small boys become increasingly unwilling to undertake 'women's work'... from the age of about nine or ten, boys begin increasingly to move into the male orbit of their mothers' brothers [this is a matrilineal society in which the mother's brother is a central figure] and their fathers. It is only then that they begin to learn the specifically male skills."

(Crehan, 1985, p.87)

The north-western region of Zambia was incorporated into what was then the British colony of Northern Rhodesia in the early part of this century, and with colonialism and the missionaries came schools modelled on Western lines. At first these were scarce and provided an extremely rudimentary education. But their number grew, particularly after Zambia gained its independence in 1964, and although at the time of my fieldwork there were many Mukunashi children who did not go to school, the state school system is now an important part of the reality of Zambian villagers' lives (Figure 6.3). The schools seem to offer a path to well-paid formal employment and the modern world of urban prosperity, even if nowadays those from the remote rural areas find this path increasingly blocked.

Whereas in the past, therefore, Mukunashi children were socialized and equipped with all the skills they needed to take their place in adult society through the kind of informal process I have described, now there is something of the same split as in modern industrial societies between two distinct spheres of education: the informal socialization that takes place within the family and under its control, and formal education, conducted in school and under the control of the state. Although the particular elements of socialization which belong to the different spheres are often a matter of dispute, there is a broad consensus within industrial societies that the 'moral' side of

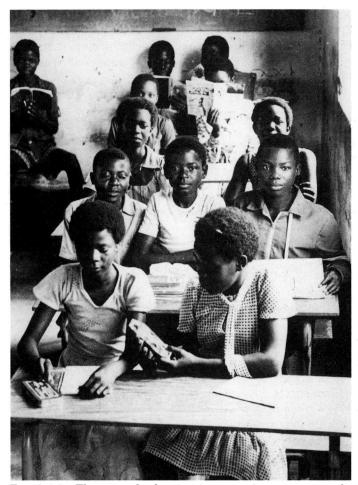

Figure 6.3   The state school system is now an important part of the reality of villagers' lives in Zambia.

socialization is the responsibility of the family, while the school is responsible for equipping its pupils with the necessary skills to become 'productive' – that is, wage earning – members of society.

Under the older system, which was an organic part of the village community, children's socialization and their participation in the labour process were intertwined in a way that is not possible with the modern school. The old system could at times treat children harshly and certain children might well have been exploited (the plight of step-children is a common theme in local folk tales) but there was not the radical separation between education and work of modern industrial society. A frequent consequence of this new separation is conflict over how children should spend their time: at school, or working for their parents and other kin in the fields, looking after younger children, doing various kinds of domestic work and so on. At the time I was in Mukunashi, for instance, one of the tasks expected of children was scaring the birds away from the ripening sorghum crop. Sorghum was the main staple in this area and was vulnerable to attack by birds. Preventing the birds alighting involved long hours of shouting and banging drums at a period of the year when children were supposed to be at school. However high a value rural people put on formal school education, as children in Mukunashi explained to me, priority must be given to producing the family's food.

There is often also a new disjuncture between the kind of socialization children receive within the household, and the education they are given at school. For instance, in no sense is the local school in Mukunashi controlled by, nor does it belong to, the local community; its ties are with the wider state structures which train, appoint and pay its teachers and devise its curriculum. The school has not grown from the community where it is sited. While children continue to receive much of their basic socialization within the household and local community, if they are to have any hope of participating in the wider world beyond the village (and in particular the urban world of wage labour, to which so many rural peoples aspire or are driven) they need a 'modern' education, something that in industrial societies is seen as the business of schools not households.

In industrial societies characteristic of the developed world, 'childhood' is a distinct category, clearly marked off from adulthood. The definition of childhood varies but tends to rely heavily on exact age in years during which a person cannot be employed, vote, get married and so on. Children have neither the responsibilities nor rights of adults: childhood is a time specifically devoted to equipping the child with what is needed to enter the adult world. In many peasant societies (and in the 'informal' sectors of urban life in the Third World) things are seen rather differently: everyone, men, women, young, old, are seen as having a role to play in production. Certain tasks are often allotted specifically to children – the complaint of the childless that they have no one to run errands for them is a common theme in many folk tales and in daily life. There is not the same rigid separation between work and a child's education and socialization, and children are seen as learning the skills they will need as adults by helping adults in their work (Figure 6.4). It is therefore taken for granted that, in so far as their strength and skill permits, children will take part in the normal production processes (see also *Wuyts et al., 1992*, ch.2).

Whether or not this means that children are exploited depends on the context. If children belong to households which face a struggle to survive (for example, having to farm increasingly exhausted land where output can only be maintained by using more labour, or where households face falling prices for their crops), then children may well be pushed to the limit of their endurance. Another important factor may be a child's relationship to the household in which he or she lives. For instance, a distant relative may be treated more harshly than the head of household's own child. (For comment on use of children's labour and that of other kin in a Ugandan context, see Chapter 9.)

Whether children do domestic labour or weed cash crops, the rights and obligations of children are still primarily determined by the norms of kinship. It is a child's relatives – most commonly one or both parents – who have rights over its labour, but are also responsible for its well-being: a child has a duty to help produce the food it eats, but also has a right to a 'proper' share of the household's resources as well as a right to whatever kind of socialization and education is seen as 'normal' in that particular society. Once again, these are not clearly defined contractual rights so much as diffuse and implicit, part of the unstated, taken-for-granted assumptions that shape everyday life, and which are only stated explicitly if it is felt that they have been transgressed.

Children's labour is not only used within the household. It may be lent or hired out to other households, or children may also do paid work employed outside the household. On South African farms earlier in the century it was common for the white farm owners to employ whole families. In a study of child labour on the Natal sugar estates, William Beinart wrote:

"Labour tenancy was becoming the predominant social relationship on white owned farms in the early decades of the twentieth century and the transfer of the labour of the children to the landlord was part of many such agreements... herding, domestic service, hoeing and weeding could all be done by African boys and girls or youths. Masters and servants legislation, and court cases testing this, allowed parents to bind their children under the age of sixteen as part of a contract for the family as a whole."

(Beinart, 1990, p.18)

Figure 6.4 Learning skills by helping adults: (above) pounding grain in Senegal; (left) cleaning maize cobs in Zimbabwe.

These kind of family labour contracts, still common throughout much of the Third World, can be seen as a shift from an informal, non-contractual system of children's labour within households to a contractual system in which children's labour power has become a commodity. Within Western industrial societies, the categories 'child' and 'employee' are seen essentially as mutually exclusive, and part of the process of becoming an adult is taking on the responsibility of paid employment: children should not work, adults must work. However, the struggles in early industrial society over child labour bear witness to the fact that modern Western definitions of childhood are not 'natural' or universal; they came into being at a particular historical moment in response to particular social forces. In societies in which other ideas about childhood prevail, children tend not to be excluded from the adult world in the way they are in industrial society, nor does there tend to be the same rigid distinction between education and employment. The idea of young people, maybe as young as 12, selling their labour does not therefore transgress accepted norms in the way that it does in the West. However, when young people are employed in this way within capitalist-based enterprises, which lack the informal – but crucial – restrictions on the use of such labour characteristic of family-based enterprises, young people are often vulnerable to extreme exploitation.

A key difference between children working within a household setting on the basis of kinship ties and children working for wages is that the former usually involves some sense of a longer term reciprocity. In the future children will benefit from the work they are contributing now: the boys herding will be given the cattle they need to marry, the girls the marriage goods they need, while the older generation expect support in the future from younger generations. In societies without state welfare provisions, children are the basic insurance for old age and for sickness. Also, if children run away, they cannot be replaced in the way that deserting wage labourers can. All in all, there are objective forces, quite separate from the normal feelings of family affection, which tend to act as some restraint on abuse of children's labour. While it is important not to romanticize peasant life, it is also important to recognize certain fundamental differences in the role of children in the labour process and how they relate to their socialization and education.

6.3 Households and differentiation

Differentiation between households

A common stereotype of rural areas in the Third World, whether in Africa, Asia or Latin America, is of poverty-stricken peasants. Television documentaries, newspaper reports and appeals by charitable organizations continually reproduce and reinforce the image of an undifferentiated mass of suffering humanity, victims of famines, disease and ignorance (see Chapter 1). This picture obscures the complex economic and political reasons for such rural poverty, and also ignores the often significant degree of differentiation in rural communities. While differences may apparently be small in absolute terms, they are significant in a relative sense: most people may be poor, but some are much poorer than others, and some, rather than struggling to survive, are prospering and accumulating. This section examines household differentiation and accumulation. Accumulation means that certain individuals, households or other groups are able to acquire extra resources that enable them to raise themselves above their fellows on a basis that can be reproduced and transferred to the next generation (see Chapter 2).

How do we know whether a process of social differentiation is taking place or not, and if it is, how can we explain it? As I have already stressed, *all* peasant societies are characterized by differences in income and levels of wealth, which are significant within communities even if they may appear very small to outside observers. Whether or not these differences are evidence of a process of *social differentiation* is a different question, that has given rise to considerable debate. On one hand, it is suggested that in capitalism rural communities experience the creation, or strengthening, of *class* divisions: some people manage to establish themselves as rich peasants or capitalist farmers (see Chapter 9), while others lose their rights to land and become landless labourers. On the other hand, considerable differences in wealth at any given moment do not necessarily indicate an *irreversible* process of differentiation: if we were to look at the community over a longer time span, we might see that the fortunes of individual households rise and fall as they move through the domestic life cycle.

The latter possibility starts with the recognition that the amount of family or kin-based labour available to households varies over time: it may initially consist of a single married couple with no children, or only very young children, and then grows as more children are born and if the household attracts other members. At a latter stage the household may shrink again as children grow up and move away to form their own independent households, or join the households of others, often as spouses. When the ratio of active adults to dependants (children, the old, the sick) is high, the household, or at least the more powerful members within it, may be able to achieve a relative degree of prosperity, but this tends to be vulnerable and short lived. When adult children move away to form their own households, leaving ageing parents, the ratio of active adults to dependants is likely to fall, and the household may sink into poverty. Even if a household has managed to acquire certain assets, such as land, on the death of the parents these might be divided among a number of different heirs. The key question, then, in determining whether or not accumulation and social differentiation is taking place is: can households use the surplus they produce to secure a basis for future surplus production that can be *reproduced* and passed on to the next generation so as to create a potentially *permanently* more wealthy stratum?

Discovering whether or not social differentiation is taking place in any given case depends on investigating the following kind of questions:

* What differences are there between households' material circumstances?

* What are the factors underlying different levels of prosperity and poverty?

* Is accumulation possible?

To explore this set of questions, I use a historical example from nineteenth-century South Africa in a study by Colin Bundy (1979) called *The Rise and Fall of the South African Peasantry*. This study charts the brief flowering of a relatively prosperous peasantry in the course of colonial incorporation which ultimately transformed pre-colonial pastoralist–cultivators into 'sub-subsistence inhabitants on eroded and overcrowded lands, dependent for survival upon wages earned in "white" industrial areas and on "white" farms' (Bundy, 1979, p.xv).

The discovery of diamonds in the Cape Colony in the 1860s led to something of an economic boom. Among other things this resulted in an increased demand for agricultural produce of various kinds: 'prices of meat, draught animals, dairy products, grains, fruit and vegetables all rose sharply, and encouraged increased agricultural output.' According to Bundy, this led to:

"a virtual 'explosion' of peasant activity in the 1870s which affected the lives of the great majority of the Cape's Africans... for smaller numbers of Africans access to capital and to larger landholdings, and the successful adoption of new productive techniques, among other factors, created a class of small commercial farmers and large peasants...All too often the period of prosperity was brief; black Africans were driven onto ever more crowded 'tribal reserves' as more and more land was appropriated for use by white colonists. Within 50 years there were few areas in which prosperous black peasants were to be found, and in many of the 'reserves' survival depended on the annual exodus of contract and seasonal migrant workers to mines and factories and farms. "

(Bundy, 1979, pp.67, 146)

But the trajectory of this rise and fall was not exactly the same for all the Cape's African inhabitants. 'Within the overall pattern of decline and proletarianization, there persists the phenomenon of a small stratum of successful and moderately well-to-do farmers.' A key dimension of Bundy's argument is that 'the success of a stratum of large peasants is not only compatible with but is a predictable feature of the underdevelopment of a peasant community as a whole.'

If we direct the questions posed above to Bundy's study, what answers do we get? First, what differences were there in the material circumstances of households? Bundy gives this description of African 'progressive farmers' whose number was increasing during the 1870s and 1880s, although they never constituted more than a small minority:

> "They were men who had consolidated early peasant success or who invested income from other sources in agriculture...Their farms – almost always on land they had purchased outright and held as individual proprietors – might be quite large, and were distinctive for the amount of re-investment of capital in the shape of fencing, walling, irrigation, improved stock breeds, and for the adoption of mixed farming."
>
> (ibid., p.92)

As well as being farmers, members of this more prosperous stratum often diversified into other economic activities, particularly transport, using their carts and wagons for carrying crops and all the other goods which were being bought and sold. In general, these progressive farmers, Bundy tells us, adopted a way of life which in its material and ideological aspects closely resembled that of advancing farmers of other races in South Africa.

Bundy's preferred designation for these 'progressive' farmers is small-scale commercial farmers, and his discussion of what differentiates them from peasant farmers brings us to the second question I posed: what basic factors underlie the marked differences in wealth and prosperity?

The mode of production of these small-scale commercial farmers was, Bundy tells us, 'more capitalist than peasant'. The differences here included the following characteristics:

Scale They were wealthier and held larger units of land.

Social relations They extracted surplus from tenants and wage labourers outside the extended family.

Market relations They were predominantly cash farmers, not subsistence farmers who sold a small surplus.

A key technological change at this time was the adoption of the plough in place of the hoe, which made cultivation on a larger scale possible and allowed the deliberate production of an agricultural crop for sale. Private ownership of land was introduced where previously there had only been communal land ownership with usufruct rights, making it possible for a small minority to lease parts of their land to tenants, labour tenants and sharecroppers. At the same time the presence of those who had neither *de jure* nor *de facto* rights to land and other means of production meant that there was a pool of people who were obliged to become wage labourers. Sufficient numbers of potential wage labourers for *white* employers was of course a continuing concern of the Cape's colonial government. In terms of attitudes and prejudices, the small élite of African commercial farmers were often closer to local white farmers than to their fellow, but poorer, Africans. Their prejudices were in fact very much those common to the owners of landed property generally. One African farmer, for instance, who owned 500 cattle, explained why he did not want tenants who had been to school: 'The educated man wants higher wages: I endeavour to get the cheapest labour.'

Becoming one of this African élite was open to very few. The ploughs, wagons and draught animals required represented a considerable investment. The account of Stephen F. Sonjica gives an insight into the emergence of a 'progressive' farmer. As a young man Sonjica realized the potential of acquiring a freehold title to land, an aspiration not shared by Sonjica's father, but Sonjica was determined and in the 1870s left his home to find work which would enable him to save the necessary capital:

> "While ostensibly sending home my month's earnings to my father in the usual Native custom, I cunningly opened a private bank account into which I diverted a portion of my savings without the knowledge of my father. This went only until I had saved eighty

pounds… [I bought] a span of oxen with yokes, gear, plough and the rest of agricultural paraphernalia."

(quoted in Bundy, 1979, p.94)

He then left his job and devoted himself to farming. Within a few years he had made good, bought himself a wagon and had accumulated £1000 in savings. He then began buying land: first a small farm and then various other plots.

So the answer to my third question (is accumulation possible?) is clearly affirmative. Indeed, at the end of the nineteenth century it might have seemed as if the 'class formation and differentiation among African agriculturalists would lead to the emergence of a class of black farmers, a diminishing "traditional" peasantry, and a growing permanently proletarianized urban working force.' But this process was aborted; the growth of a powerful lobby representing white settler farm interests ensured that aspiring black capitalist farmers such as Sonjica who represented unwelcome competition were squeezed out. As well as white farmers' fears of competition, there was also the demand of white employers for African labour.

"Mine owners and white farmers alike sought legislative measures designed to dislodge labour from African areas; they vied for fierceness in their railing against an image of idle young men lolling at home amidst fields of corn and herds of cattle."

(Bundy, 1979, p.240)

Legislation was passed to exclude African farmers from good agricultural land, which was reserved for white settlers. African cultivators were increasingly confined to areas which, if not poor land to begin with, became so as population densities rose and forced an overexploitation of the land. And then there was the question of access to markets. As Sonjica stressed, this was a key factor: 'No native farmer is worth calling a farmer who has no agent in a big town through whom he may dispose of his produce at market prices.' The fortunes of all commercial farmers, however small or large their scale, are linked to their access to markets in which they can sell the agricultural commodities they have produced, and to the level

of the prices they can obtain in those markets relative to their production costs. But the terms on which different producers compete within the market are also related to securing support for their interests within the political arena, where crucial decisions concerning the provision of basic infrastructure such as roads, the nature and levels of subsidies and taxation, and so on, are decided. Only the white farming areas were supplied with the roads and other infrastructure necessary to engage in production for the market. Similarly, black producers were subjected to relatively higher rates of taxation. These and other measures had the effect desired by the mine owners and white settlers of forcing African labour onto the market and, in particular, since it had no other option, to work for pitifully low wage rates.

Differentiation within households, and household and kinship networks

So far, our discussion of social differentiation and accumulation has been directed to households and their relationship with wider social and economic processes. To look at how accumulation relates to differences *within* households we need to ask a further question:

• To what extent is accumulation based on some household members appropriating what is produced by others or having access to the labour of other household members or kin?

Let us remain for the moment with Bundy's case study which was written in the 1970s before the impact of feminist scholarship; consequently almost all the actors who figure explicitly in his story are men. The accumulating élite he describes are essentially male heads of households who, in the archive sources on which Bundy relies, were seen as also representing wives and other dependants. Here the patriarchal Cape peasants and the colonial government saw eye to eye. But however invisible these dependants may have been in the arenas of official and public life, their crucial value in other spheres is captured in a statement by a colonial magistrate who explains that the problems of labour recruitment lie in the attitudes of local peasants. In the magistrate's words, these (male) peasants ask: 'Why should we work? Is not the country ours, and have we not lots of land and many women and children to cultivate it?' Behind all the Sonjicas and other male small-scale commercial farmers, there were always other household members on whose labour the male household head depended (see also Chapter 9).

It is useful to remember the point made at the beginning of Chapter 5. *Ideologies* of kinship – 'ideal' values of kinship including reciprocity – are likely to diverge from (and to obscure, if we are not careful) how family and household relations work *in practice*. An example from Latin America shows how kinship ideology can mask exploitation and accumulation within households behind notions of reciprocity. In an article entitled *The Elementary Strictures of Kinship*, Tom Brass (1986) examines how a system of 'fictive kinship' (see below), common throughout much of Latin America, operated in a rural area of Peru during the 1970s.

La Convención is a relatively remote province in the north of Peru, well suited to the cultivation of cash crops such as sugar, cocoa, tea, coca and coffee (Figure 6.5). In the earlier part of this century the region's large landowners began leasing plots to peasant migrants from the south, who in return for land, had to work for their landlords for a certain number of days. Until the 1940s this labour rent was generally low, but as the world demand for commodities such as coffee, tea and cocoa rose, landlords began to increase their demands for labour, part of which they used on their cash-crop farms. Their tenants in turn subleased some of their holdings to subtenants who would work the tenants' land and also perform the tenants' labour service. In most cases tenants and subtenants were kin, and most of the labour tasks (work on the landlord's fields and domestic work in his house, as well as work for the main tenant) were in fact carried out by the female members of subtenant and tenant households: mothers, wives and daughters.

Interestingly, the development of capitalist agriculture in the province led to labour by kin becoming more rather than less important. On the basis of credit obtained from local merchants, there emerged 'a small but economically powerful

stratum of rich peasants', who began to accumulate capital through the production of cash crops, and particularly coffee, for the export market. Coffee, however, is a relatively labour-intensive crop and La Convención is a region where labour is in short supply. Labour became even more scarce since the 1969 land reform which gave land to many of those who would previously have had to work for others. Prior to 1969 a standard method of labour recruitment was *debt bondage*, whereby both local poor peasants and 'landless migrants contracted from outside the province' performed labour service to pay off previously contracted debts. Since the land reform, however, rich and middle peasants have replaced such bonded workers with local workers, who are in the category of 'fictive kin'. According to Brass, the system of 'fictive kinship', at least in La Convención, provides richer peasants with an important source of cheap labour.

What is *'fictive kinship'*? Fictive kinship involves richer peasants becoming godparents to a child at the time of his or her baptism, confirmation or marriage, a relationship which also makes them 'co-parents' of the child's parents. Child, parents and godparents are then fixed in a continuing relationship which is seen as entailing the same highly patriarchal and hierarchical pattern of authority as biological kinship. Godparents are expected to make gifts to their godchildren, and the co-parents and/or the child are expected to reciprocate. According to the ideology, there is no accurate accounting of just how 'reciprocal' the exchanges are; the value of the various prestations is not reckoned in economic terms. In reality, however, this kind of sponsorship of labourers and poor peasants often involves them receiving cash sums which they repay (or reciprocate) in the form of labour service. Most importantly, these are not reciprocal 'gifts' between *equals*, but between clearly defined superiors and inferiors. 'The authoritarian structure of fictive kinship is an explicit aspect of this relationship.' Also, fictive kinship:

> "is not limited to a restricted triad of god-child, parents and godparents but extends laterally and vertically to include all the

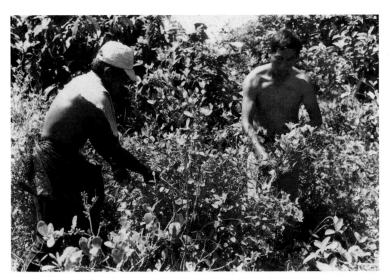

Figure 6.5   (above) Picking coca leaves; (right) combining coffee and bean cultivation (Peru).

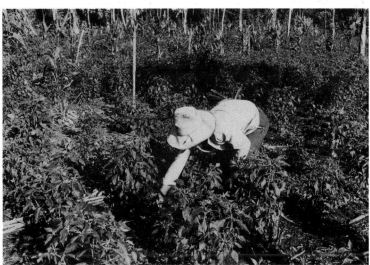

brothers and sisters of the godchild's parents as well as its grandparents... Accordingly, the authority of fictive kinship may be invoked by a rich peasant in order to mobilize local opinion against an indebted poor peasant co-parent (or member of the latter's family) who fails to fulfil labour service obligations...Idioms of patriarchal authority, duty, obligation and reciprocity can be utilized to control a workforce of debt bonded labour composed of godchildren, parents and classificatory co-parents while at the same time disguising the class base of this control."

(Brass, 1986, p.63)

The 'kinspeople' who worked for rich and middle peasants were therefore mainly 'indebted poor peasant co-parents together with their domestic kin and classificatory co-parents', whose own smallholdings were next to the more extensive landholdings of their creditor-employers, who were also their godparents or co-parents. Their ensnarement in relationships of debt bondage, meant that these fictive (and sometimes actual) kin were unable to take advantage of employment opportunities elsewhere offering higher wages and better conditions, and provided cheap labour force for rich and middle peasants:

> "This enabled [rich and middle peasants] to cultivate cash crops for the national and international market, to reinvest in trading activities, to purchase agricultural inputs and buy land elsewhere in the province. In short, it enabled them to transform themselves from simple commodity producers into small capitalists."
>
> (ibid., p.65)

Although Brass's study is concerned primarily with how *fictive* kinship works in La Convención, his argument raises interesting questions about accumulation and kinship relations in general. In looking at any specific instance of accumulation by peasant producers, we need to examine the role of kinship in mobilizing labour, appropriating surplus and so on. Ideologies of kinship, however different in other respects, all tend to stress reciprocity rather than exploitation, mutual supportiveness rather than possible oppression. And while it would be wrong to go to the other extreme and see all kinship relations as disguised class relations based on exploitation and oppression, it is always necessary to look at the realities behind the ideology, as Brass does.

6.4 Households in crisis

All forms of economic relations involve satisfying various human needs. At the same time, there is typically some tension between short-term and long-term needs: the need for grain to eat versus the need for seed corn later; the need to feed children versus the need to pay for their education. Most rural people in the Third World have no access to state welfare or security as provided in industrialized societies. When people are sick, when they grow old, when their farms are hit by natural disasters, they look to their kin for support. An important reason why peasants often redistribute their surplus to various kin members is to ensure that when they themselves are in need, those kinship ties can be relied on. One of the pressures that ensures kinship obligations are met is the unspoken knowledge that, at some time or another, virtually everybody is likely to suffer some kind of hardship. If certain individuals or households can escape the need to redistribute their surplus to kin, and are able instead to use it to buy labour or land, or to invest it in expanding their productive base, their actions could have serious repercussions for those to whom they would otherwise have distributed their surplus.

This final section looks at those whose struggle is not to accumulate but simply to survive. We focus on the poor as active subjects trying as best they can to shape their own lives, albeit with very limited access to resources and often confronted with impossible choices over satisfying immediate and longer term needs. The two basic questions we look at are:

- What survival strategies do people use?

- Are the causes of their poverty temporary, short-term ones, or structural and more long term?

In trying to understand what happens to households in crisis, we again need to decide what our unit of analysis should be. Official statistics give little choice because they tend to use 'the household' as their basic unit and tend to have little information on intra-household relations. The unstated assumption is that the household is a homogeneous unit within which there is an equitable sharing of resources. Yet the way resources are distributed within households – which may be far from equitable – is a crucial factor in understanding the specific effects of poverty in any given instance.

There are always various conflicts within households, whether these are explicit or implicit, acknowledged or unacknowledged. They may be between the interests of individual household members or between groupings within the household. On one hand, such strains are likely to be exacerbated when times are hard and the 'cake' which is to be divided among household members is too small, regardless of how it is cut. Such strains may tear households apart, when those members – often men – able to leave to try their chances elsewhere do so, abandoning women and children for whom such mobility is more difficult. On the other hand, privation may draw household members together, the pooling of certain resources becoming a vital strategy for survival.

People can be poor for many different reasons: they may not have enough land, they may lack the resources to make use of their land, or, if they wish to sell their labour, they may not be able to find work, and so on (see Chapter 1). Whatever the particular reasons, poverty tends to be *gendered*; that is, access to employment, to productive resources, legal rights, etc. all tend to be different for women and men. For poor women it is important whether they obtain goods and services in their own right, or are dependent on a male head of household. How poverty affects men partly depends on the kind of unpaid resources they can obtain from wives and other household members.

Bina Agarwal (1990) illustrates some of the survival strategies used by the rural poor in India (see also Chapter 1, Boxes 1.2 and 1.4 for Bangladesh). Agarwal's concern is primarily with basic food security. Two of the questions she sets out to answer are: 'How do poor agricultural families seek to cope with the problems of food insecurity associated with seasonal troughs in the agricultural production cycle?' and: 'How do they cope with calamities such as drought and famine?' Agarwal uses the terms household and family interchangeably and to mean units of consumption. In general, it is useful to distinguish, as Agarwal does, between the problem of (regular) seasonal shortages and inherently unpredictable calamities.

Seasonal shortages can be the result of the household not being able to produce enough food from one harvest to carry them through to the next. If, as is common in much of India, the household is dependent on some or all of its members working as casual agricultural labourers, then there may be shortages at those times of the year when wage employment is not available. Rural households have various ways of coping with these

times of shortage which Agarwal divides into five categories:

1 diversifying sources of income, including seasonal migration;

2 drawing upon communal resources – village common lands and forests;

3 drawing upon social relationships – patronage, kinship, friendship – and informal credit networks;

4 drawing upon household stores (of food, fuel, etc.) and adjusting current consumption patterns;

5 drawing upon assets.

This list indicates the wide range of strategies employed. Differences within households, especially those of gender, affect their adoption and outcomes. Women and men are often differently placed in adopting survival strategies and, in general, women tend to be in a weaker position. The following discussion of some of the strategies listed by Agarwal highlights this point.

Wage employment is crucial for many rural households in India (Figure 6.6). This may mean local employment by richer farmers or, particularly in times of shortage, may involve migration on a seasonal or longer term basis. Such employment tends to be casual and dependent on the fluctuating needs of the employer rather than anything permanent and secure (see Box 1.2). Sometimes, a whole family will be employed: 'in Gujarat, the labour hired seasonally for sugar-cane harvesting...has to work in teams – the male cutter followed by a helper (usually a wife, sister or daughter) to clean the stalks of leaves, followed, in turn, by a child to bind the cleaned stalks together.' But individual employment is more common, in which women agricultural labourers seem to be disadvantaged in relation to men. Drawing on a wide range of data from different regions in India, Agarwal identifies a number of the disadvantages women face:

> "[because of] much greater seasonal fluctuations in employment and earnings than men due to the greater task specificity of their work, [women] are noted to have

Figure 6.6 Wage employment in India: (above) men cutting cane; (right) women stripping and gathering cane.

sharper peaks and longer slacks than men in the irrigated rice regions of South India, and have a lesser chance of finding employment in the slack seasons; ...are much more dependent on wage labour than men, have lower average days of annual employment (and more days of involuntary unemployment), and lower daily wages (often even for the same tasks), which makes for considerable gender differences in annual real earnings; ...within agricultural wage work, are usually only employed as casual labour, typically men alone being hired on long-term contracts, ...have lower job mobility due to their primary and often sole responsibility for childcare, the ideology of female seclusion, and their vulnerability to class/caste-related sexual abuse."

(Agarwal, 1990, p.350)

While women agricultural labourers normally earn less than men, their contribution to household earnings may well exceed that of male household members. One study of 'landless and near landless agricultural labour households in 20 sample villages' found that in more than half of the villages, 'although the wife's earnings from agricultural wage work were typically half or one-third of the husband's, in absolute terms her contribution from her earnings towards household maintenance was [equal or] greater than his.' And 'the wives typically contributed 90–100 per cent of their earnings and the men rarely gave over 60–70 per cent of theirs, keeping the rest for personal use.'

The second coping strategy of poor households Agarwal lists is making use of communal resources, i.e. forest and other uncultivated land owned by the state or held in common by the local community. These communal resources 'serve as a source of various types of food, medicinal herbs, fuel, fodder, water, manure, silt, small timber, fibre, house-building and handicraft material, resin, gum, spices, etc. for personal use and sale, especially for the landless and landpoor' (Figure 6.7):

"Typically it is women and children who play a primary role in the collection of CPR [common property resources] and forest produce, ...it has been argued that poor peasant and tribal women as the main foragers and gatherers have a special detailed reserved knowledge of edible forest produce that can help tide over prolonged shortages."

(ibid., pp.357, 361)

An important point mentioned by Agarwal here is the frequent 'divergence between male and female priorities in CPR and forest development – with men typically favouring commercial varieties and women opting for trees and plants that fulfil the subsistence needs of food, fuel and fodder.' The general and rapid decline in these communal resources would seem likely to hit women particularly hard, given their greater reliance on such resources.

Another example of the gendered nature of coping strategies is in connection with 'adjusting current consumption patterns' (the fourth strategy). In times of shortage it is often women rather than men who have to do the belt tightening:

"among the tribals of Orissa, during months of low food availability, children get first preference, then men, and finally women. During the month when ploughing has to be done, food is allocated in favour of men, and women either borrow from neighbours or go hungry. There is no mention, however, of any reverse adjustments by the men when women have a heavy load of work, as during the rice transplanting season."

(ibid., p.369)

In general, the kind of survival strategies that people use to cope with regular seasonal shortages are also those they use in times of exceptional calamities. As Agarwal puts it, 'seasonality, drought and famine may be seen as three points in a continuum, representing increasingly severe threats to the food security of the family.' In fact, one way of defining famine is the point at which the various survival strategies people use are no longer adequate, and people die due to lack of food. It should be stressed, however, that famine is never simply a case of an absolute shortage of food (as explained in *Allen & Thomas, 1992*, ch.1). To quote Amartya Sen:

"Starvation is the characteristic of some people not *having* enough food to eat. It is not the characteristic of there *being* not enough food to eat. While the latter can be a cause of the former, it is but one of many *possible* causes."

(Sen, 1982, p.1)

People acquire food in many different ways: they may produce it, buy it, be given it. In conditions of famine, the mechanisms whereby some people

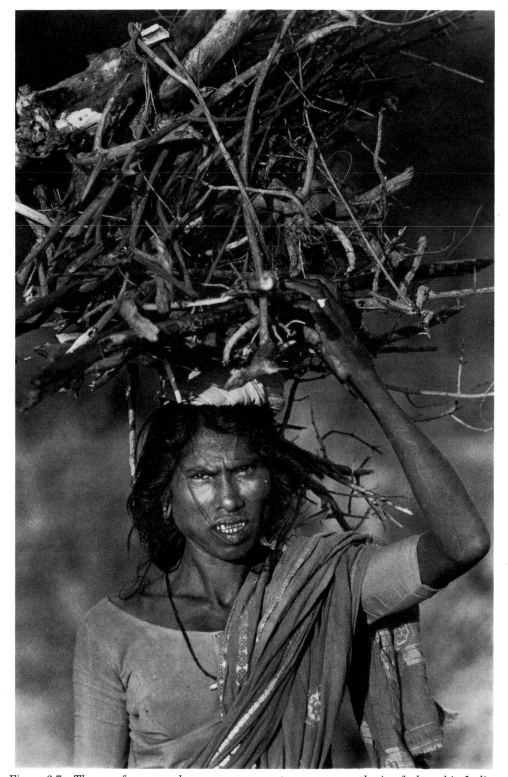

Figure 6.7   The use of communal or common property resources: gathering fuelwood in India.

obtain food have broken down or in some way become inadequate. Characteristically different groups suffer in famines to very different degrees. In the devastating 1943 Bengal famine, for instance:

> "the worst affected groups seem to have been fishermen, transport workers, paddy huskers, agricultural labourers, …craftsmen, and non-agricultural labourers, in that order. The least affected were peasant cultivators and sharecroppers."
>
> (Sen, 1982, pp.71–2)

Agarwal quotes another statistic of the same famine which highlights a different contouring of suffering:

> "57 per cent of males compared with only 16 per cent of the female recipients of relief [according to a survey based on the Bengal Famine Committee's records] were able to pay cash for the food received, …the least excess mortality (in absolute terms) occurred among men of 20–40 years of age."
>
> (Agarwal, 1990, p.391)

Are the causes of poverty temporary, short-term ones, or structural and more long-term? In his book *The African Poor*, John Iliffe (1987, p.4) quotes a useful distinction between '*structural poverty*, which is the long-term poverty of individuals due to their personal or social circumstances, and *conjunctural poverty*, which is the temporary poverty into which ordinary self-sufficient people may be thrown by crisis.' What are the links between these two kinds of poverty?

First, in times of crisis, such as drought or famine, it is those who are already in a vulnerable position, or those who are just managing to get by but have little in the way of reserves to fall back on, who are likely to suffer the most. Agarwal's fifth coping strategy, drawing upon assets, is one that tends to become crucial at such times (see also Box 1.4). At first households draw on

> "stocks of fodder, fuelwood, grain, animal products, dungcakes [for fuel], etc., stored for the lean period. Subsequently household goods such as ropes, cots, utensils and jewel-

lery – the non-productive assets – are disposed of. Some small livestock may be sold…Finally draught animals and land are disposed of – the former being sold outright, the latter first mortgaged where possible."

> (Jodha cited in Agarwal, 1990, p.382)

Those who have few assets in the first place are likely to be the hardest hit. Agarwal also points out that while it is entirely rational for households to try everything before relinquishing control over key productive resources, 'the assets which are the first casualty – namely household utensils and jewellery – also happen to be those typically owned and controlled by women.' Often these are the *only* assets women own.

The second set of linkages between structural and conjunctural poverty has to do with the way crisis situations may drive 'ordinarily self-sufficient people' into conditions of permanent structural poverty. Figures from the 1943 Bengal famine illustrate this process:

> "between April 1943 and April 1944, one-fourth of all families (numbering 1.6 million) owning paddy land before the famine had either sold it in full or in part, or mortgaged it. Sales in full were highest among those with less than two acres: these, constituting 6.1 per cent of all families, thus lost their main source of livelihood. Those buying land were largely based outside the village, many of them urban dwellers. Thirteen per cent of plough cattle was also lost (due to sales or deaths) and only 25 per cent of it could be replaced later."
>
> (Agarwal, 1990, p.382)

As regards assets disposed of in times of crisis, 'once mortgaged or sold the chances of these assets being redeemed in full in the post-calamity period are often slim, especially for the small farmers and landless agricultural labourers.'

In rural areas, lack of access to land is a crucial factor in structural poverty, particularly when there is a shortage of opportunities for wage labour. Although there are variations between different regions, generally rural women in the Third World have systematically weaker rights to land than men, and their access to land tends to depend on their links with male household heads, most commonly husbands. As customary usufruct rights are supplanted by individual private ownership of land, women tend to lose the old *de facto* rights which they may have had. Other productive resources, such as livestock and ploughs, and so on, also tend to be owned by men rather than women, and by household heads rather than more junior men. Two groups that are likely to be disproportionately represented among the structural poor are, therefore, women who are not attached to a male household head, often referred to as female-headed households – a term which implies that 'normally' household heads can be assumed to be male – and younger unmarried men. It is important to remember, however, that these are broad contours of structural poverty and can never be taken as axiomatic; not *all* female-headed households are poor. To get beyond such general contours we need to analyse the specifics of individual cases.

Summary

1   Decisions about having children and family size are affected by economic needs as well as ideologies and values.

2   Families and households are important sites of socialization and informal education for children. Childhood tends not to be a distinct category – children have obligations to perform household activities from an early age but also have rights to household resources, and will benefit as adults from the work they contribute as children.

3   Social differentiation involves the creation of class divisions. Households' fortunes may also rise and fall over time with changes in domestic composition and access to resources. Social differentiation may result from changes in the context of production (such as new market opportunities) and depends on the ability to acquire resources and command the labour of others. Command over labour may involve family members or wage workers. It may also involve an 'unequal exchange' in hierarchical patterns of authority where labour is provided in return for favours made.

4   Poor households adopt a wide variety of survival strategies; diversifying income, drawing on communal resources, family and other networks, running down household stores or reducing consumption, and drawing on assets. Women and men tend to be differently placed with respect to these strategies.

Conclusion to Chapters 5 and 6

One of the popular stereotypes of rural areas in the Third World is of 'traditional' villagers living a life which has remained essentially unchanged since time immemorial. Chapters 5 and 6 have shown how false this stereotype is – how much the lives of those who live in the rural areas of the Third World are shaped by, and themselves help to shape, wider social and economic processes. While it is true to say that rural producers (whether in Africa, Asia or Latin America) live in 'households', the nature of these households is enormously variable.

What we have seen is:

1 how 'the household' is best regarded not as a specific entity, but as a *site*, within which a wide range of activities – production, consumption, childrearing – are carried out, often in vastly different ways;

2 how the relationships within the households are also highly variable, even if they are frequently lumped together under the single term 'kinship'.

While 'kinship' may be popularly assumed to be the cement which binds households together, 'kinship' can be and can do many things. It should have become clear how important it is to examine closely the dense ideological veils of 'natural' sentiments and moral imperatives with which household and 'family' relationships are enshrouded.

POVERTY AND CHANGE

EMPLOYMENT, ENVIRONMENT, DIFFERENTIATION AND HEALTH

7

POVERTY AND EMPLOYMENT IN INDIA

JAYATI GHOSH AND KRISHNA BHARADWAJ

India is a late industrializing country and has experienced high rates of economic growth (*Hewitt et al., 1992*, ch.11). Yet India contains the largest number of poor people in the world – a function of its social and economic development as well as the size of its population.

Q Why is national economic growth combined with rural poverty in India?

Q Are the conditions of the rural poor improving or getting worse?

Q How have economic change and state intervention influenced rural livelihoods?

There are diametrically opposed views about trends of poverty in India. Official household surveys suggest that income distribution has improved and there has been a marked decline in poverty. If this is true, there are potentially significant lessons for the rest of the Third World. Other evidence suggests that income inequalities have increased and levels of poverty and hunger have not been significantly reduced, despite marked economic change. This chapter examines this evidence and suggests that processes of development taking place in India do not appear to be improving livelihoods for the poorest. It then looks at why hunger and poverty persist in India despite significant economic growth.

An explanation of the coincidence of growth and poverty involves many different factors. Earlier chapters have examined historical processes affecting agrarian structures and rural livelihoods (Chapters 2 to 4), as well as the importance of understanding social relations within and between households in making a living and coping with change (Chapters 5 and 6). This chapter examines the role of the wider economy and the state in India's continuing rural poverty.

Problems of rural poverty and the generation of employment do not result from processes in agriculture alone: they are a result of uneven patterns of growth and change within the economy as a whole. The persistence of poverty in India is a result of a skewed pattern of development and a basic 'failure' of the state's development strategy. This 'failure' refers to specific government policies directed at poverty and unemployment and also to broad macro-economic trends which only partly result from state policies. These trends are intimately related to changes in the character of production in different geographical areas and sectors and the formation of different kinds of markets for different goods (crops, land, labour and money). The links between these changes and their wider economic impact affect the livelihood possibilities of the rural poor.

In India, poverty stems from a failure of employment generation and adequate wage levels. The survival strategies of poor households have given rise to a proliferation of low-paying and low-productivity employment and self-employment.

This chapter traces the relationship between poverty and patterns of employment generation. It then describes how state intervention in the economy and changes in the structure of the economy have contributed to the failure to generate adequate levels of employment. Finally, it shows how government policies directed specifically at poverty alleviation and employment generation have been inadequate for the task.

Section 7.1 describes the results of official surveys which indicate a decline in poverty and contrasts this evidence with macro-economic indications of growth and changes in income distribution since the early 1970s. Section 7.2 examines aspects of poverty and the relationship between poverty and economic growth. These two sections suggest that there has been little reduction in poverty, despite economic growth.

Section 7.3 explains how the economy has failed to create productive employment: unemployment and underemployment have been growing, and living standards have been stagnant or declining for large sectors of the population.

Sections 7.4 and 7.5 examine the role of the state in seeking to direct economic development and show why employment generation has been inadequate and poverty alleviation ineffective. Section 7.4 describes three main phases of state economic intervention since Indian independence and how employment generation and poverty alleviation were envisaged in each phase. This section ends by suggesting that state intervention, particularly during the 1980s, came to reflect an increasing division of the Indian economy which has left the majority of the urban and rural poor economically 'disenfranchised'. Economic development and state intervention have responded to the spending power of upper income groups, but that development has failed to create the employment and the types of products required by poor people. Without productive livelihoods, the poor lack spending power to influence economic change.

Section 7.5 reviews the sequence of programmes which successive governments have implemented to reduce rural poverty and promote rural livelihoods. It describes how initial policies of agrarian reform and planned employment generation have

gradually been transformed into what are effectively welfare programmes which seek to alleviate poverty rather than remove its causes.

7.1 Poverty and income distribution

Direct evidence on poverty decline – the official view

Household surveys conducted periodically by the Indian government suggest that there has been an absolute and a relative decline in poverty, and that income distribution has also become more equal. Greater diversification of employment and the proliferation of non-agricultural activities even in the rural areas is seen as evidence of a new and positive dynamism in the countryside.

Official estimates suggest that between 1978 and 1990 the rural population below the poverty line fell absolutely from 253 million to 169 million, and fell as a proportion of rural population from 51% to 28% (Figure 7.1). This is a 'headcount' estimate based on a defined poverty line (which in this case was Rs 40 per person per month at 1972–73 prices, corresponding to a daily calorie requirement of 2400 calories per person). Urban poverty levels are estimated to have declined similarly. The World Bank states that the degree of inequality of rural consumption expenditure also appeared to have declined marginally by the mid-1980s (World Bank, 1989b). The average per capita consumption expenditure of the lowest 40% of rural households has also grown slightly. Even the wages of agricultural labourers are estimated to have increased in some regions and remained constant in others, according to another source, Jose (1987).

Official estimates of poverty decline are doubted, however, by several researchers. Minhas (1990) suggested that the extent of rural poverty in 1987–88 could be more than 50%, a dramatic increase over the official figures. In addition, official surveys do not necessarily fully capture changes in the material living standards of poorer groups. Poverty is multidimensional (see Chapter 1), reflecting not only an inadequate intake of

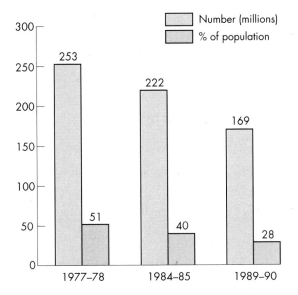

Figure 7.1 Official estimates of the incidence of rural poverty. [Source: Planning Commission (1990) 'Employment: past trends and prospects for 1990s', Planning Commission Working Paper, New Delhi.]

foodgrain, but also the inability to obtain other essential food and clothing, a minimum level of acceptable housing, and access to basic medical and educational facilities. It is thought that the majority of the Indian population has inadequate access to these basic necessities of life.

If we look at the macro-economic indicators of income, employment, consumption and assets, the official view of poverty decline is also thrown into doubt.

What do macro-economic indicators suggest?

The available evidence on aggregate growth, income distribution and poverty in India presents a contradictory picture. While not as exceptional as that of some of the East Asian economies, several indicators suggest that India's overall growth performance has nevertheless been high. In the 1980s overall growth of national income rose to 5.5% per annum. The fastest growing economic sectors have been services and manufacturing industry, but agricultural output too appears to have crossed an important threshold. In the mid-1980s India achieved 'self-sufficiency' in foodgrains in the sense that imports were no longer required to make up shortfalls in domestic production.

However, this overall growth has led to an increase in external debt and the rapid growth of government expenditure, also financed increasingly by loans. Furthermore, income-generating opportunities have been growing very slowly relative to employment needs.

Rural incomes: stagnation

While agricultural output has shown a low but fairly consistent growth of just below 3% per annum over the past three decades, the rural population has grown faster than output. Increasing numbers of people have been forced to seek their livelihoods from agriculture, and overall per capita income in agriculture has stagnated. Meanwhile, the gap between agricultural and non-agricultural incomes has been increasing. In the late 1980s, non-agricultural income per capita was well over three times the level of per capita agricultural income. This stagnation of per capita agricultural incomes results from two major factors: first, inadequate growth of output and productivity within the agricultural sector, and secondly, the inability of other sectors of the economy to absorb labour productively and at a rate sufficient to reduce the pressure of population in agriculture.

Employment: inadequate, insecure and poorly paid work

The inadequacy of employment generation is increasingly apparent in all parts of the economy. There is not enough employment being created, the type of work is low-productivity work in the informal or **unorganized sector**, and the wages are low.

Employment has grown very slowly, and in agriculture the rate at which new jobs are created by growth has been falling. Employment in the **organized sector** has actually fallen over the past decade. The largest growth of employment has been in unorganized services and manufacturing, with poorly paid workers forced to seek low-productivity jobs (Figure 7.2).

Figure 7.2 Non-agricultural rural production may provide employment in rural areas, but it is often low-productivity work for low returns: (left) throwing pots, Jandiali village, Punjab; (below) weaving cotton cloth in Sultanke village near the Pakistan / India border.

Organized and unorganized sectors: The *organized or formal sector* of the economy consists of larger scale production using wage labour, and it 'functions within the framework of civil laws and administrative regulations. This sector comprises joint stock companies, public sector enterprises, multinational companies, large proprietary concerns and commercial plantations to name but a few…' (Jagannathan, 1987, p.4).

The *unorganized or informal sector* uses mostly family labour for smaller scale production, and it is less effectively regulated. It 'covers most rural markets and the peripheral but growing urban informal sector…[it is] at best a loose way to describe a wide range of economic activities by small firms, households and individuals which are to varying degrees integrated with organized sector markets. At one end of the spectrum a family firm, for example, may be closely integrated with a large corporation in the organized sector through industrial subcontracting. At the other end, agricultural labourers and slum or shantytown dwellers could be securing their income-earning opportunities solely within the unorganized sector' (Jagannathan, 1987, p.4).

Employment has also become increasingly insecure. The proportion of casual labour in total employment has grown in the last decade. Employment of casual labour varies with seasonal fluctuations in demand so the growing use of casual labour indicates a worsening of conditions of the labouring classes because of the greater uncertainty of income flows and reduced real incomes over the year.

Consumption: national self-sufficiency with hunger

Increasing casualization of labour and worsening conditions of employment also help to explain trends in total consumption in the economy which do not fit the broad picture of a general reduction in poverty given by official statistics. Two indicators of overall consumption are particularly relevant.

The first is the demand for foodgrains relative to availability. National self-sufficiency in foodgrains has been achieved over a period when per capita consumption has actually *fallen*. One-third to a half of the population lives on a nutritionally inadequate diet.

The second indicator is the structure and composition (reflected particularly in the industrial sector) of trends in aggregate consumption. The fastest growing items of industrial production have been consumer durables (such as cars and televisions) within the reach of the upper and middle income groups, constituting at the very most a quarter of the population. Even among other commodities, the growth items are those which are demanded more by the better-off, such as refined sugar and artificial fabrics (while overall cotton cloth consumption per capita has declined). Similarly, the other growth area of services (such as banking and insurance, hotels and tourism) caters to the upper 20% of the population.

Assets: increasing landlessness

Data on rural assets indicate growing inequality. Landholdings have fallen in size on average, largely because of an increase in the number of 'marginal' holdings (which are considered to be unviable because they are less than 1 hectare). The proportion of the rural population without access to land has also increased considerably. Paradoxically, the *average* value of household durable assets in the rural areas (essentially agricultural assets) has increased significantly (Dantwala, 1987). This can be explained by growing income inequalities in the countryside – it is not a general increase in assets for *all* rural households.

The macro-economic indicators above point to the conclusion that income inequalities have increased in India. While there has been significant growth, particularly in some sectors, the benefits of this growth have not really 'trickled down' to reach the mass of the people. Rather, a significant proportion of the population appears to have been excluded from the beneficial effects of higher income levels in the aggregate, and in some cases their standard of living may even have worsened due to the greater precariousness of means of earning a livelihood.

Although there have been some improvements in material conditions, there is evidence to suggest that the pattern of development has been skewed in favour of the better-off sections of society, and the relatively disadvantaged have not been able to improve their position to any significant extent.

Regional variation in poverty and growth

Studies that disaggregate national data by region also show considerable variation in levels of growth and degrees of poverty. Table 7.1 shows the overall per capita *output* generated in the major states (Figure 7.3), as a percentage of the all-India average, for the period 1986–89. It can be seen that the northern and western regions have greater output. Furthermore, the richest state, Punjab, had a per capita output nearly six times that of Madhya Pradesh, the poorest state. Not only is per capita output widely divergent between states, but differences have been increasing over time. This has also led to increasing inequality between regions.

Regional differences in *agricultural growth* follow a similar pattern to differences in per capita output. Agricultural growth has been highest in the northern and western regions at 3.75% per annum in the period 1970–73 to 1980–83 (Bhalla & Tyagi, 1989). The central region has had crop output growing at 3.0% over the same period. The southern and eastern regions have much lower growth rates, at 1.5% and 0.9% respectively. In both the eastern and southern regions, the agricultural growth rate has also been significantly below the rate of population growth.

Regional differences are equally marked in the prevalence of rural *poverty*. In 1988, the eastern and central regions together accounted for nearly two-thirds of the rural poor in India, and their share in total rural poverty has actually been increasing since the early 1970s. The western and northern regions account for the smallest share of rural poor, around 14%, while the remainder are in the southern region. Consumption expenditure data for the bottom 40% of the rural population across regions reflects this increasing disparity: states such as Bihar show average expenditure per household considerably below the all-India average, while others such as Punjab and Haryana

Table 7.1   Per capita output of states (average for 1986–89 at 1980–81 prices)

State	Percentage of all-India average
Northern region	
Punjab	174
Haryana	143
Himachal Pradesh	97
Jammu and Kashmir	35
Western region	
Maharashtra	143
Gujarat	112
Central region	
Uttar Pradesh	75
Rajasthan	71
Madhya Pradesh	32
Eastern region	
West Bengal	94
Assam	78
Orissa	70
Bihar	53
Southern region	
Karnataka	99
Tamil Nadu	98
Andhra Pradesh	79
Kerala	72

are very much in excess of it. However, the relationship between agricultural growth and rural poverty is complex and is mediated by many other factors, such as the nature of state intervention (see Section 7.2).

Studies of agricultural growth (for example, Rao, Ray & Subbarao, 1988) indicate that fluctuations in both year-to-year as well as regional growth rates of crop output have intensified in the past 20 years. In the low-growth areas, the population is less equipped to cope with such fluctuations than in high-growth areas.

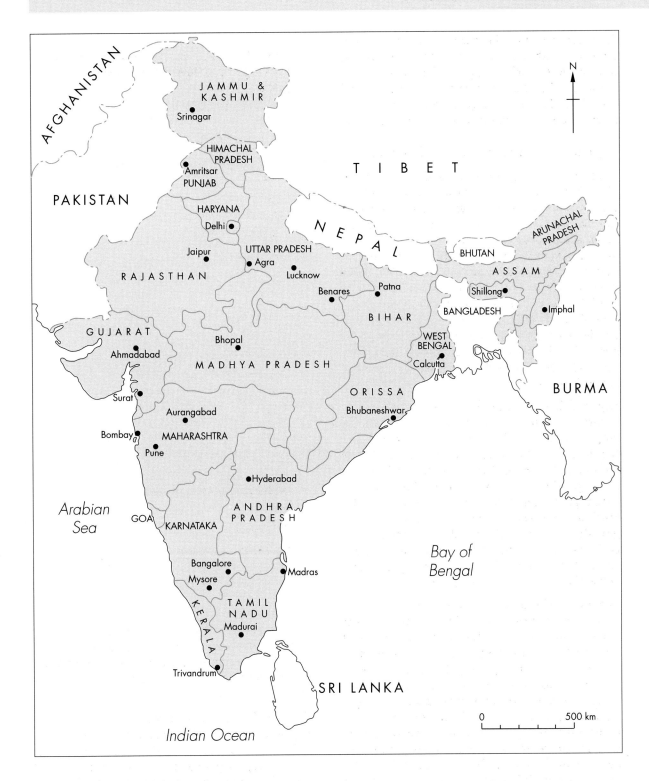

Figure 7.3   The major states and cities of India.

7.2 Aspects of poverty

Poverty and growth in agriculture

Output growth, particularly for foodgrains, shows some relation to the extent of rural poverty. Where per capita foodgrain output has declined, the scope for improvement in the conditions of the poorest is limited. However, the experience of one state, Kerala, indicates that state intervention and popular mobilization can make a difference even in constrained conditions (see*Wuyts et al., 1992*, ch.10). Kerala has experienced a sharp decline in per capita foodgrain output, but the extent of poverty (in terms of the headcount ratio) is contained because of a strategy of providing public goods and a superior system of food distribution in the rural areas. Thus, while growth in itself does not necessarily lead to a 'trickle down' as shown by continuing poverty in several high-growth areas, in low-growth areas the pressure of poverty can be ameliorated to some extent by purposeful and effective state intervention.

While output growth as such need not be related to the extent of poverty, there does appear to be a strong negative association between labour productivity in agriculture and the incidence of poverty. Some studies (Mahendradev, 1986; Parthasarathy, 1987) show that regions with the highest incidence of poverty had either negative or very low rates of **labour productivity** (and were also the regions of high unemployment), while relatively high rates of growth of labour productivity (above 3% per annum) were found in regions with the lowest incidence of poverty.

Labour productivity: The quantity of goods and services that someone can produce with a given expenditure of effort, usually measured or averaged out in terms of time spent working or labour time. It is the ratio of the amount produced to the amount of labour put in, measured as product per person-hour or person-year.

Malnutrition

The prevalence of malnutrition is usually a good indicator of the actual incidence of poverty, and a number of studies conducted in the past decade show that malnutrition is widespread, particularly among the landless and small farmers in rural areas, and unorganized sector unskilled workers in the urban centres. A survey in Bihar (Prasad *et al.*, 1981) found that nearly 90% of the landless labourers interviewed felt that they did not get enough food, and over half suffered physically from malnutrition in varying degrees. Studies in the semi-arid regions found that between 25% and 50% of the children under 6 years in the survey villages were undernourished and thus at risk of various physical debilities (World Bank, 1989b). Other data suggest that in all states except Kerala between about 33% and 50% of all children show nutritional inadequacy on a weight-for-age basis. In some states such as Gujarat (with one of the highest growth rates of agricultural output) the proportion of such children increased between 1979 and 1982. The data also show that females are typically the most nutritionally disadvantaged in all age categories, and malnutrition is particularly marked among them.

Mortality and variation by region, gender and ethnicity

Poverty is also typically reflected in high mortality and morbidity (illness) rates because of poor nutrition as well as inadequate health care and lack of access to amenities such as basic sanitation and clean potable water. Such problems are often more evident for the urban poor, but are widespread in rural areas as well. Mortality rates have declined in recent years, possibly as a result of the spread of community health programmes and other medical facilities, but they remain high by international standards, and vary regionally as do other poverty indicators. Mortality rates remain much higher in the central and eastern regions than in the rest of India.

Again, women are especially vulnerable. Rural women in poor social groups tend to have special problems, not only because of social norms and customs which reduce their effective consumption and ability to increase incomes, but also because the nature of property rights and landholding patterns in India restricts their access to these fundamental rural assets. Nutritional inequalities within households, adversely affecting female members, are by now well documented (Drèze & Sen, 1989, ch.4; see also Chapter 6 in this volume). Additionally, female access to other basic needs, such as health and education, is more restricted than that of males in the same socio-economic groups. Female members of landless labourer and marginal farmer households are thus the most disadvantaged in Indian rural society. Once again, a state like Kerala has been able to reduce the degree of gender inequality in material terms through programmes of mass literacy and person-specific food targeting, but no other states have taken comparable action.

Further, the incidence of poverty is higher in the tribal areas and among the lowest castes in the social hierarchy. To some extent, this reflects a decline in the availability of common property resources, such as grazing and fodder land, access to wood, fuelstuffs and water, all of which have become increasingly privatized, especially in the last decade or so (Figure 7.4). The poor typically are much greater users (proportionate to income) of these common property resources (see Chapters 6 and 8). One estimate (Jodha, 1986) found that in semi-arid tracts of seven major states, such common property resources had declined by 26–62% in survey villages over the period 1952–82, and the bulk of newly privatized resources had been appropriated by the upper-income rural classes. The rapid privatization of such common resources, along with the depletion of forests (mainly for commercial use), generates ecological degradation as well as losses in the real incomes of the poor. It also increases the labour used in obtaining access to fuel and water, which places heavy burdens in particular on rural women.

Figure 7.4   Privatization of common property resources affects the most vulnerable groups: collecting scarce woodfuel in Bihar.

Urban poverty

While the proportion of the poor in urban areas has been declining in the past decade, the absolute numbers of urban poor have increased by well over one-third since the early 1970s. Urban poverty is less regionally concentrated than rural poverty, but is still most marked in the towns and cities of the central region. The poorest groups in the urban areas are rarely better off than their rural counterparts – in fact, they are often recent or temporary rural migrants. They tend to suffer greatly from the inadequate growth of urban infrastructure which has not kept pace with the rapid expansion of urban population even in the smaller cities. Poverty is most evident for the unorganized workers in manufacturing and services (the informal sector labourers), who are typically migrants from rural areas. Studies indicate that undernutrition, inadequate sanitation and health facilities, and lack of access to education, are possibly as prevalent among this group as they are among the rural poor.

It is apparent that poverty characterizes the less skilled labouring groups, both urban and rural, who are forced to seek low-paying, low-productivity jobs simply to achieve a precarious survival (Figure 7.5). The inadequacy of job availability is an important reason for this – there is a clear association between unemployment and poverty (Visaria, 1980). The numbers of such labouring groups have increased as a proportion of the working population as well as in absolute numbers. The expansion of productive employment opportunities is thus the crucial issue in addressing the problem of poverty.

Figure 7.5   Low-paying, low-productivity work in Tapin Colliery, Bihar, in 1991.

7.3 Employment and unemployment

Employment generation is central to any development strategy for two reasons. First, unemployment signifies the waste of the productive resource of labour. Second, it generates poverty, as we have just seen. Unfortunately, growth in employment has been one of the major failures of the Indian development experience. Problems of employment generation have been especially acute in the last decade.

Changes in the structure of employment

In the period since the early 1970s, employment is estimated to have grown at an average of 2% per annum, below the rate of population growth of around 2.4% (Planning Commission, 1990). Disaggregating this national average, we find that urban employment has grown much faster, at a rate of 4%, whereas rural employment has only grown at a rate of 1.75%. But looking at Table 7.2, we can see that, even within an average overall growth rate of employment of 2%, there

has been a declining trend, from 2.82% to 1.55%. Furthermore, the sectoral nature of employment generation has been changing, with the lowest growth of employment in agriculture, which still employs at least two-thirds of the workforce.

We have already mentioned that differential growth rates in the various sectors of the economy are associated with other changes in the structure of employment. To summarize, these include: the stagnation of agricultural employment despite output growth; the growing predominance of casual labour; the tendency, observed in recent years, of diversification of rural employment from agriculture to other sectors (particularly services), together with the increased importance of the informal or unorganized sector, which has become the major employer for all branches of productive activity. We now look more closely at these trends.

Why is agricultural employment stagnant?

Agriculture employs two-thirds of the labour force but recently its ability to generate additional jobs has declined. The rate of job creation in agriculture has been falling continuously (Bhalla & Tyagi,

Table 7.2   Growth rates and structure of employment (% per annum)

Sector	1972–73 to 1977–78	1977–78 to 1983	1983 to 1987–88	% share of sectoral employment in total, 1987–88
Agriculture	2.32	1.20	0.65	65.5
Mining	4.68	5.85	6.16	0.7
Manufacturing	5.10	3.75	2.10	10.8
Construction	1.59	7.45	13.69	3.7
Electricity, gas, water	12.23	5.07	4.64	0.3
Transport, storage, communication	4.85	6.35	2.67	2.5
Other services	3.67	4.69	2.50	16.5
Overall average growth of employment	2.82	2.22	1.55	(total) 100.00

Source: Planning Commission (1990) 'Employment: past trends and prospects for 1990s', Working Paper, Planning Commission, New Delhi.

1989), or, as economists would say, there has been low elasticity of labour demand to output growth (Box 7.1).

Much of the land-augmenting technological change in agriculture (embodied in Green Revolution methods) has been simultaneously labour-saving (Box 7.2). Thus, even though there has been an increase in the cultivation of more labour-intensive crops, rapid mechanization and increases in the use of chemical fertilizers and pesticides have meant that total labour absorption in agriculture has either been stagnant or may even have fallen in several of the most technologically dynamic states (Bhalla, 1987) (Figure 7.6). Aggregate employment need not fall if cropped area and output expand, but in some areas (for example, Kerala, Tamil Nadu and possibly Bihar), even the absolute quantity of agricultural employment seems to have fallen. The decline in labour absorption has been related to the pattern of labour demand: the desire to reduce the seasonal peaks of labour demand created by biotechnological innovations has led to the displacement of unskilled agricultural labourers by either fewer skilled workers or casual workers hired on a short-term basis.

The growing importance of casual labour

The proportion of rural households who receive the major part of their income from wages has increased from 25% in 1965 to 40% in 1988. Of men and women employed as wage labourers, 75% of males and more than 90% of females in the rural areas were employed on a casual basis by 1988 (Vaidyanathan, 1988; *Sarvekshana*, 1990).

Figure 7.7 shows the change in employment patterns in rural areas since the early 1970s. These figures are on an all-India basis, but the trends are similar for the regions taken separately. In some states, such as Kerala and Tamil Nadu, nearly half of all the male workers (both self-employed and hired labour) in the rural areas are

---

Box 7.1   Elasticity of labour demand

The notion of 'elasticity' is frequently used in economics to describe the relationship between two elements of economic activity. It is a measure of the extent to which changes in one factor are associated with changes in another. The elasticity of labour demand is a measure of the rate at which jobs have been created as economic growth (growth of output) occurs. Labour absorption, total labour use per unit area, is a related concept which shows the total labour (household plus hired) requirement in agriculture.

Example: We want to know if economic growth increases the number of jobs. The answer to that question depends on how the growth occurs. If economic growth arises from the application of more human labour to an activity, new jobs may rise at the same rate as economic growth. This would show a high elasticity of labour demand. If, on the other hand, increased output results from the introduction of machines, few if any jobs may be created as economic growth takes place. Then there is a low elasticity of labour demand and little effect of economic growth on the growth of employment.

Box 7.2   Technological change in agriculture

Changes in the technology of agricultural production are called 'land-augmenting' if they add to the productivity of land or to the gross sown area. A typical example is irrigation, which can allow for more intensive cropping or changes in the cropping pattern. Such changes usually imply an increase in total labour requirement per hectare, and therefore absorb more labour. By contrast, 'labour saving' changes refer to those forms of mechanization which substitute machines for human labour without increasing the cultivable area, such as tractors and harvester–thresher combines. In many of the recent Green Revolution areas, land-augmenting changes have been accompanied by labour-saving techniques which reduce problems of supervision and control over hired labour.

*Figure 7.6
Mechanization of
agriculture: tractor
pulling manure in
Jandiali village,
Punjab.*

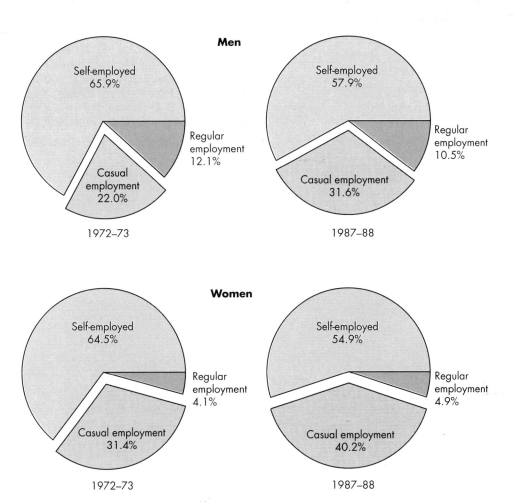

*Figure 7.7 Distribution of the rural workforce. [Data from National Sample Survey
Rounds, in* Sarvekshana, *various issues, Ministry of Planning, New Delhi.]*

employed on a casual basis. In other states the proportion is lower but everywhere the tendency towards casualization of labour has been marked, especially for men. For women the trend is less strong, but a larger proportion of female labour has always been casual.

This process is significant because it implies a definite worsening in labour conditions. Casual workers in the rural areas are typically better paid per day than equivalent permanent workers doing the same jobs. But casual wages are still very low in most states relative to subsistence requirements. Added to this is the insecurity of finding employment, with the near certainty that it will not be easily available every day of the year (Figure 7.8).

Although daily rates may be higher for casual workers, the annual real incomes of casual labour households are generally no higher, and often lower, than those of permanent labour households in rural India. Casual workers increasingly receive payment largely in cash, thus exposing them to the effects of price fluctuations, unlike those who receive part of their wages in foodgrains or other payments in kind. Casual workers are thus much more affected by rises in foodgrain prices in lean seasons when work opportunities are in any case limited. Money wages do not usually increase at the same rate as commodity prices or there may be a long time-lag between wages catching up with prices. Thus a rising trend in the prices of foodgrains or other necessities affects casual labour more directly than the self-employed and others who receive part of their income in kind. Along with this, the short-term nature of work contracts means that casual labourers receive no other benefits or help from their employers, for example in periods of sickness or even for work-related accidents. (These last problems are even more apparent for casual workers in urban areas, who receive none of the benefits that permanent workers in the organized sector consider their due.)

There are several reasons why casualization of labour has proceeded at such a pace. One major cause is the rate and pattern of commercialization

Figure 7.8   Casual farm labour in Jandiali village, Punjab.

in the rural areas and in many regions and sectors where subsistence production was common until fairly recently. It is not only the growth of markets in themselves but *how* these markets have evolved that changes the nature of work contracts and renders them more transient as well as offering less security for workers. Factors that have contributed to this process include the monetization of most transactions even in the rural economy, the fact that both goods and services are now available through cash exchange transactions, and the specialization of production leading to the displacement of many traditional crafts by modern industrial output.

Commercialization is in turn linked to the growth of **reserve armies of labour**. Employers prefer to take advantage of freely available labour at market wage rates depressed by the presence of large numbers of people looking for employment. Thus those regions marked by the highest rural unemployment rates (such as Kerala and Tamil Nadu) are also those with the highest proportion of casual labour in the total workforce. By contrast, in labour-tight situations employers find it useful to tie down labour with permanent contracts, which gives workers greater security (even if typically combined with greater personal dependence).

Does employment diversification into non-agricultural activities indicate dynamism or stagnation?

The contribution of agriculture to rural employment is declining, shown by the growing importance of non-agricultural employment in the rural areas. This diversification of rural employment is

> **Reserve army of labour:** This term was first used by Karl Marx to refer to the body of unemployed labourers in capitalist economies. He argued that such unemployed workers played an important role in keeping down wages of those employed, ensuring worker 'discipline', and providing a cushion of available labourers for periods of increased labour demand.

also shown in official survey data which indicate that by the late 1980s 25% of men and 15% of women employed in rural areas were involved in non-agricultural activities. Most of these are in the services, construction and transport, although manufacturing is also high on the list.

The bulk of the increase in rural non-agricultural employment is related to the rise in casual workers, indicating that the additional employment generated is of a short-term, fluctuating nature. Indeed, the spread of non-agricultural employment may well reflect agriculture's inability to generate enough employment over the year, since casual agricultural workers reported a very high incidence of non-agricultural work in a secondary or subsidiary capacity. There are also seasonal fluctuations in rural non-agricultural employment which can also be explained by changes in the employment structure of rural casual labourers, who shift back and forth between agricultural and non-agricultural occupations depending on job availability.

Migration patterns (particularly seasonal or circular migrations within regions) reflect this tendency. We have already suggested that much of the labour force in the urban informal sector represents the result of urban migration from villages in search of jobs. In some sectors, such as construction, this migration is very short term in nature and organized by contractors who hire for specific projects. Elsewhere there have been movements of labour between agricultural regions, such as the flow of workers from Bihar and Orissa to Punjab, Haryana and west Uttar Pradesh, and circular movement within states such as Gujarat and Maharashtra between the more dynamic and stagnant areas. However, such migration has not provided much relief because of limited employment opportunities relative to the size of the workforce, and because even in areas of high output growth, labour demand has not grown commensurately. Rural labour markets have evolved in a largely localized fashion, notwithstanding examples of interregional mobility.

Some economists might argue that the diversification of rural employment is part of a positive dynamic whereby economic growth entails a shift in employment from agriculture to industry and then to services (Kuznets, 1965). However, in the Indian case, the spread of non-agricultural employment reflects the growing desperation of the rural poor for income-generating opportunities. Non-agricultural employment arises from the survival strategies of rural households unable to obtain employment or self-employment in agriculture. It is a last resort rather than an attractive alternative livelihood.

The expansion in non-agricultural employment is mostly in low-productivity activities which require little capital. Nearly half of such enterprises are in the textile and food processing industries (Bharadwaj, 1989) while some are in low-grade services (such as domestic service). Little has changed in this pattern since the early 1960s.

Thus, almost all the increase in casual non-agricultural work, whether in the rural or urban areas, has been in the unorganized sector, which employs 90% of all Indian workers. Most unorganized sector workers are in agriculture, but, as we have seen, recent increases in employment are essentially in some services and in manufacturing. The unorganized sector in 1987–88 accounted for 86% of the employment in construction, 76% in manufacturing and 51% in transport, storage and communication. Not only are most Indian workers already outside the scope of effective legislation to protect their minimum rights, but also in several parts of the economy informal sector labour has been increasing as a proportion of the total workforce.

Neo-liberal economists argue that this makes the Indian labour market more 'flexible' and 'competitive' and thus ensures efficient allocation of labour as well as more labour-intensive methods of production. (Similar arguments are also made by neo-liberal economists about the British and other industrialized economies that reduction of workers' legal rights will facilitate job mobility and labour productivity – see also *Hewitt et al., 1992*, ch.5.) However, we have seen that in agriculture precisely the opposite is the case: increasing numbers of available workers have still had to face widespread labour-saving mechanization and declines in rates of labour absorption. In the

economy as a whole, this 'flexibility' which is embodied in minimal wages paid to non-permanent labourers has still not been able to ensure any reasonable growth of employment opportunities. Rather, it appears that underemployment has spread across sectors and that some disguised unemployment has even been replaced by open unemployment.

Underemployment: the growth of low-productivity work

Two important trends are discernible in employment and unemployment in the period 1973 to 1988. One is that the proportion of the labour force chronically unemployed (that is, available for work but unable to get work for a major part of the year) has doubled for rural men and increased by more than six times for rural women. By any standards these are dramatic increases in a 15-year period. A growing proportion of the workforce is unable to find productive work for a major part of the year. Much of this is related to the collapse of agricultural employment growth mentioned above.

In practice, the poor cannot remain without income: they have few assets to maintain themselves, their access to credit is limited and their kin are severely constrained in their ability to maintain non-workers. They are thus forced to seek work for however short a period and in whatever activities are available. This explains the second trend: that is, the proliferation of low-paying and low-productivity work, both within agriculture and outside it (see also Box 7.3). As capitalist agriculture expands, the scope for widespread underemployment in agriculture diminishes. Agricultural workers are thus forced into various types of non-agricultural work in efforts to get some income, however meagre.

7.4 State intervention and economic change

We now turn to the process and effects of state intervention in the Indian economy, specifically examining the relation between government policy and employment generation. Sections 7.4

Box 7.3   Gender and employment

'Evidence is overwhelming that in Indian life, access – who gets what – is closely tied to gender. Access is connected to the very meaning of male and female: part of the culture's definition of female is her association with the "inside" – the home and the family. By contrast, men belong to the "outside" where livelihoods are earned and political and economic power is exercised. Because of this dichotomy in the socially accorded domains of men and women, governmental policy aimed at improving the lot of women has viewed them largely as passive beneficiaries of social services and anti-poverty programs...[Yet] their economic productivity is particularly critical for...households below the poverty line: studies show that the poorer the family, the more it depends for its survival on the earnings of a woman' (World Bank, 1991b, p.iii).

This relationship between gender and class reflects two tendencies. More prosperous rural families withdraw women from agricultural production to show their wealth and prestige, and their ability to hire labour from poorer regions. By contrast, women in poor families are increasingly forced to take paid labour in order to meet their survival needs: 'Thus class and regional differences have been widening as a consequence of agricultural development...The growth in the female agricultural labour force has taken place largely among low caste and tribal women who are the most impoverished and disadvantaged groups in the population. They have been increasingly employed in formerly "male" tasks as a direct consequence of their cheapness. The resulting increased competition between men and women for jobs has led to a decrease in male–female wage differentials – not by raising female wages, but by depressing male wages. The sexual division of labour in agriculture is thus weakening with the growth of wage labour' (Standing, 1992).

and 7.5 argue that, by the 1980s, the combination of uneven economic growth and ineffective state intervention had created a situation where the majority of the population were economically powerless.

Growth in India has been very uneven. Although the range of goods and services produced by industry has been widening, technological change has been slow, and there have been uneven growth rates between different parts of the economy and a low and falling profitability of traditional industries catering to mass consumption needs. This uneven growth increasingly separates the leading organized sector (and its beneficiaries, the rich and middle classes) and the poor, even though these two groups are linked to each other through production and trade.

The period of systematic state planning: 1950s to mid-1960s

In the early years after independence, the goal of government planning was rapid industrialization. India was a populous, predominantly agricultural economy with a limited industrial sector, backward in technology and infrastructure and having few linkages to the rest of the economy.

The key emphasis of government intervention was to increase production. Employment creation was to be met by the growth of small-scale and household industries, particularly in consumer and intermediate goods. A 'trickle down' was therefore anticipated as a result of accelerated public investment and the pattern of industrial growth (in consumer and intermediate goods), thus generating income and employment as well as demand for these goods.

However, the pattern of growth of small- and medium-scale production was given little attention. Instead, while government investment strategies focused on large-scale and heavy industry, infrastructure and agriculture, the aim of productive employment generation in the small- and medium-scale production was never fulfilled, and was gradually replaced by welfare-oriented policies to support household and small enterprises. These policies were more in the nature of relief operations to assist the survival of the poor rather than policies generating viable productive units.

The period of support for agriculture: late 1960s to mid-1970s

In the mid-1960s, the economy faced severe drought, foreign exchange crisis, border wars and rising inflation. These crises were reflected in rising food prices, which threatened famine, and government attention was directed to the need to raise agricultural output. This was the beginning of the era of the Green Revolution (see Chapter 3, Section 3.3), which selectively encouraged 'progressive farmers', or rich peasants, to adopt modern techniques. This was a shift away from the previous emphasis on community-based investments in public irrigation, land reforms, co-operatives, and community development projects. The goal was to achieve national food self-sufficiency and to increase employment within agriculture (Figure 7.9).

By the mid-1960s, a growing budgetary deficit led the state to reduce direct public involvement in industry. For about a decade, public investment in industry grew at a much slower rate, and the emphasis shifted from direct government investment to providing credit and subsidizing private investment. This was true in both agriculture and industry.

This was a period when differentiation and commercialization in the countryside accelerated. Some of the differences in agricultural growth rates (noted in Section 7.1) can be attributed, at

Figure 7.9 Labourers hand weeding fields of High Yield Variety of wheat in the Punjab.

least in part, to regional differences in commercialization. Two main types of commercialization can be identified: a 'normal' process in which the market distributes a growing proportion of surplus output; and 'forced commercialization' (see Box 3.2) in which peasants are impelled to sell a large proportion of their output to get cash for repayment of loans or to pay taxes. Peasants may be forced to sell their output at harvest, when prices are low, and purchase grain (using loans if necessary) for subsistence later in the year when prices are high. In the high-growth region of north-west India, normal commercialization is much more common. In the eastern and southern regions of the country, where growth is much slower and poverty levels higher, forced commercialization is more prevalent.

Commercialization of output and the evolution of other markets in land, labour and money have occurred in ways that have accentuated differentiation in production and consumption, and reinforced power structures. These processes have put severe limits on the bargaining power of the poor and imply that the dominant groups are able to benefit from a many-faceted control over the labouring groups as well as over the production and distribution of agricultural output.

The period of liberalization: the late 1970s to present

From the mid to late 1970s there was a third pattern of government involvement in the economy. Calls for 'liberalization' and reduction of bureaucratic control led to a decline in state intervention, but at the same time government expenditure grew to the extent that it accounted for half the total expenditure in the economy. Some of this expenditure was in productive investment in infrastructure, particularly in the field of energy, but the greatest expansion was in expenditure on subsidies, public administration and defence. 'Trickle down' had not generated adequate employment, and government increased its own employment generation and gave income subsidies to particular groups in the population in an attempt to compensate for this failure. These subsidies were primarily in food and fertilizers, and transferred income to urban organized workers

through lower food prices, and to middle and rich peasants through cheap agricultural inputs. The growth of 'welfarist' schemes was also characteristic of this phase. They attempted to contain the material miseries and political resentments of the poorer groups through various 'targeted' schemes. However, the scale of poverty makes welfarist solutions impossible in a purely budgetary sense: the resources required to bring all the working population to a minimum standard of living would require expenditure of a scale beyond the capacity of the state. A solution can only be found if anti-poverty schemes increase productivity to the extent that they can be self-financing.

Economic disenfranchisement of the majority

The pattern of growth described above has led to the systematic deprivation and economic disenfranchisement of the majority. That this now co-exists with the growing awareness of political enfranchisement has contributed to increasing political instability in the country as successive governments are found wanting (Figure 7.10).

Since the 1980s, the development of the organized and unorganized sectors of the economy, and the

Figure 7.10 *'I am the most experienced chap here. I have been removing poverty and unemployment for nearly 30 years now!'*

role of state intervention in relation to each of them, has left the majority of the population economically powerless.

The organized sector of the economy has faster rates of capital accumulation and greater technological dynamism than the unorganized sector, and it also generates a relatively rapid growth of incomes. But the process of growth has created neither sufficient employment nor income in the unorganized sector. The growth of incomes in the organized sector, arising from technological change and expansion, is aided by the fact that wages are higher for those employed by government, in services, and by large corporations. As a result, those associated with the organized sector are able to save money, and they have purchasing power (called 'effective demand') with which to influence the sorts of goods produced in the economy. However, such effective demand is small relative to the total population, and thus has little influence on the modern manufacturing sector.

Since job creation in the organized sector is limited, people are forced to seek low-paying low-productivity employment in the unorganized sectors of agriculture, industry and services. But such employment is threatened as even the limited effective demand in the economy shifts in favour of goods produced in the modern sector. Even in the rural areas people increasingly buy urban mass-produced goods rather than local handicraft production. The needs of the rural population for such basic goods as textiles, footwear, soap and other household necessities are met by factory production which displaces small unorganized sector producers.

The vast majority of the population in the unorganized sector is disenfranchised from exerting effective influence on the market since they have very little purchasing power to back their demand. At the same time, they are subjected to the effects of the market economy whose functioning is guided by the savings, investment decisions and pricing policies of the organized sector and, in tandem, of the state.

Upper income households have been able to achieve high rates of saving. In an attempt to direct these savings into modern sector investment, the government has introduced high interest rate savings schemes and tax concessions. These schemes have had the unintended outcome of encouraging a growth of *rentier* capital (see Box 2.4) and financial speculation. As a result, high rates of saving have not necessarily led to high rates of investment and the productivity of investment has fallen.

Since the 1980s, government intervention has thus divided into two main strands: incentives for investment in the organized sector, and poverty alleviation in the unorganized sector. One difficulty of this division arises from the fact that poverty alleviation is not self-sustaining. Policies such as 'food for work' and employment guarantee programmes have been temporary solutions to generate supplementary activities which have had no substantial and permanent effects on the structure of production or income generation. Policies directed towards the creation of small entrepreneurs by giving them subsidized assets or credits have also not had a generalized impact on the economy because the enterprises created have not been viable.

7.5 State intervention in the rural economy

Options for the government

We have seen how problems of poverty and inadequate employment generation are related. We have argued that both poverty and inadequate employment generation form part of an overall process of accumulation and change in the production structure that is closely interlinked with consumption by upper income groups. In this context, what role has been played by the state in relation to the rural poor?

From its inception, the post-independence government of India has had special programmes directed towards alleviating the conditions of the poorer and weaker sections of society, and since the 1970s these have usually been linked with schemes for employment generation. We review the experience of these programmes under three headings which summarize the main options for state action on rural livelihoods:

1   reforming the structure of property relations (usually, land relations) so as to reconstitute production and exchange relations;

2   implementing tax and price-regulating measures to modify incomes, without altering the basic distribution of property;

3   instituting public investment programmes, including poverty alleviation and employment schemes.

These three categories of intervention are approximately coincident with the three phases of state intervention described in the previous section. Attempts to restructure property relations occurred during the phases of systematic state planning in the 1950s. Pricing and fiscal policies were the dominant form of intervention on rural livelihoods during the Green Revolution period in the late 1960s and early 1970s. Targeted programmes have risen to prominence in the period of liberalization from the late 1970s to the present.

The restructuring of property relations

Chapter 3, Section 3.2 described why the restructuring of agrarian property relations, specifically land reform, was a high priority of the Congress Party government after Indian independence, and how the involvement of smaller landlords in the Congress Party, nevertheless, constrained the redistribution of land rights. As a consequence, only a limited land reform was attempted and it had little effect on either the extreme inequality of landholding, or the prevalence of insecure tenancies.

Land reform remains a necessary – though not sufficient – condition to promote productive accumulation, as well as being an important requirement for gaining livelihoods in the rural areas. Indeed, the 'forced commercialization' of the poorer peasants arises from the essential insecurity of their access to land. Rural power structures and the control by landlords over poor peasants' livelihoods could be broken down by tenurial reforms.

The viability of self-cultivation, however, requires more than access to land. Government strategy should also be oriented towards confronting the entire network of market relationships which control credit, labour and commodities, rather than attempting to modify only one aspect.

Pricing and fiscal policies

The second set of government policies attempts to influence existing social relations without challenging agrarian power. The most sustained and substantial example of this set of policies is the water/seed/fertilizer/credit package of the Green Revolution (Figure 7.11), which was launched from the late 1960s onwards in the more 'progressive' regions (as described in Chapter 3). Resources were concentrated on these regions, and credit and price policies were designed to encourage technological changes and productivity growth. The Green Revolution has increased grain output, but significant trends towards either betterment of livelihood conditions or alternative and more secure means of income are not evident. Rather, as we have seen earlier, in many areas livelihoods have become less secure. This is true even of areas of more 'prosperous' agriculture.

Figure 7.11 *'We are all here, Sir – fertilizer supplier, pest controller, seed adviser and soil tester – but I wonder who that man is standing over there!'*

Subsidy, tax and pricing measures which provide incentives and disincentives within existing networks of production and exchange may produce inequities and perverse social and economic results in ways which are not anticipated by policy makers. For example, the fact that landlords and larger farmers dominate in more than one rural market (lending money, say, as well as renting out land) may imply that they are able to appropriate all the benefits of price changes and other government measures, leaving the condition of the poor unaffected. Even where price increases could in principle benefit poor peasants and allow them to break away from the economic domination of large farmers or landlords, the latter will benefit relatively more from price rises for marketed output. In addition, the moneylending and trading activities of large farmers and landlords can undermine any advantage to poor peasants of higher prices.

The failure of pricing and fiscal policies to reduce poverty led policy makers to introduce targeted public investment programmes.

Targeted public spending programmes

Several generations of targeted programmes, such as rural employment-generation schemes, have sought to improve the livelihoods of particular groups of the population. The early thinking on rural works programmes was influenced greatly by arguments that underemployed labour could be mobilized to build public works, such as irrigation canals and roads. Thus from the mid-1950s to the late 1960s there was much discussion of *shramdaan* or gift labour and the creation of a mobile labour army to build much-needed infrastructure. However, little was achieved.

In the early 1970s, rural employment programmes were again considered, but the emphasis on mobilizing labour for asset creation was replaced by the 'food for work' notion. The idea that such work would lead to increased production and would potentially be self-financing was lost.

Schemes of land reclamation, road construction, minor irrigation, civic conservation and village infrastructural facilities (water supply, sanitation, etc.) were identified which would generate work for unskilled labour. These schemes were not, however, designed to create revenue to meet the wages of the workers and the cost of materials for the projects. Without this, it was clear that the financial needs of the programme would be enormous. If the rural poor and unemployed were to be brought to at least the level of the 'poverty line', around Rs 47 billion (US$3.7 billion) in public works programmes would have been required each year (Dandekar, 1988). The sheer size of the finance required was such that it was unlikely to be met, and, in the event, the food for work programme which grew out of this thinking was small in relation to the needs of the labouring classes.

The assets that were created through the food for work programme were of dubious quality. Moreover, the roads, power facilities and irrigation works constructed typically favoured the richer peasants who derived the most benefit from improved infrastructure. Neither the financing of these schemes nor their products were integrated into any planned process.

In the early 1980s, a more systematic plan of intervention provided a package of programmes with a time frame and a separate allocation of resources. Groups of poor households, enterprises and individuals were targeted for income-generation schemes, including the provision of productive assets to poor households, direct wage employment programmes and special integrated area development programmes, especially in drought-prone areas. Evaluations indicate that some programmes have provided only relief while others have had relatively greater success (Planning Commission, 1986, 1987; Hirway, 1986a, 1986b; Dandekar, 1988).

Wage employment schemes have been hampered by bureaucratic and centralized management, inability to mobilize local resources, lack of local decision making and control, corruption and inefficiency, and the poor quality and usability of the products. The greater success of some schemes (for example, the Employment Guarantee Scheme in Maharashtra) indicates that better design and implementation can make rural works programmes more effective. Such programmes have the significant advantage that the poor are self-targeted through being available for such work. The wage rates provided by these schemes can also provide a basis for bargaining with private employers. Rural employment schemes are particularly needed when there are sharp seasonal variations in labour demand or in places where slack seasons continue for long periods (Figure 7.12).

Schemes generating self-employment (notably the Integrated Rural Development Programme, discussed further in Chapter 11) have fallen short of expectations, particularly in view of the resources deployed. They have not been able to create stable and secure incomes for many people, although they have enabled some among the relatively less poor to cross the poverty line.

Conditions required for more effective generation of livelihoods

The problems which have emerged in the conception and implementation of these poverty alleviation schemes are symptomatic of more fundamental problems of promoting small-scale enterprises among the poor in a market system in which the structure of demand is not planned. One of the difficulties many of these enterprises face is the marketability of their output. Indeed, the economic viability of these small household enterprises can be assured only when the entire package – the requisite credit, technical skills, raw materials and the market – is provided. This implies greater planning and co-ordination of the production and distribution networks in the economy.

Small industrial units (in the unorganized sector) attached to large units (in the organized sector) have been relatively more successful than independent small enterprises. The efficient marketing network, the standardization of quality of products and raw materials, and the brand names or product differentiation of organized sector industries have contributed to the viability of these units. The existence of such production does not, however, mean that these wage goods (see Chapter 2) will get to the poor. Production is for a different clientele. Therefore, policies encouraging the production of wage goods will not necessarily achieve a better balance between production, income, employment and demand.

Figure 7.12 Villagers working on a scheme to deepen a large dry pond.

Seeking efficiency of capital use, as is the aim of many liberalization programmes, may not achieve productive use of labour or be compatible with achieving adequate livelihoods. Those who believe that liberalization would lead to a better integration with the world economy and to efficient use for domestic capital mainly stress the constraints over supply and believe that free markets resolve these constraints through efficient resource allocation on the market (see also *Hewitt et al., 1992*, ch.5).

It is here that neo-liberal economists fall into the trap of ignoring the relationship between investment, production, employment and demand. A consistent strategy for providing productive livelihoods for labour cannot rest on the usual and oft-repeated prescriptions of encouraging labour-intensive industries, labour-intensive technologies, infrastructural improvements or market incentives. Whereas all these are necessary and important, employment-generating growth requires a set of mutually supporting activities of small-scale, decentralized production where supplies of credit, capital and raw materials, controls of marketing and quality, and appropriate technology would have to be created to sustain viable reproduction of the activities. This probably requires innovations in organization – for example, new forms of co-operative organization. These would not be the standardized forms for all regions or all activities or for all communities. Perhaps this would be one among the challenges for those who consider that the concept of decentralized planning is not merely a matter of financial and administrative devolution but also one that requires innovative patterns of communitarian institutions and organizations. Decentralized planning could be an important element in a strategy for employment-oriented growth. (For further discussion of public action and development policy, see *Wuyts et al., 1992*.)

7.6 Conclusions

Despite significant changes in the structure of the economy as well as in patterns of production and consumption, India's development strategy has not been able to tackle the problem of pervasive poverty. Indeed, Sections 7.1 and 7.2 argued that many indicators have shown that inequalities have widened between rural and urban populations and also *within* the rural and urban sectors. This suggests that the gains of income growth of the last 30 years have gone largely to a small section of the population, so that the largest proportion – the poorer rural groups in particular – have been excluded from many of the gains of development.

This is not to say that there has been no improvement in material conditions at all; the drop in mortality rates suggests improved conditions of health, nutrition, and medical care, while even for the poor a wider range of goods is now available. Further, the spread of transport and communications networks, the increase in public spending even in the rural areas, and increased integration of economic activities have all been associated with the development of markets in rural and urban areas. This spread of commercialization generally has had important effects not only on material standards of living but also on the labour processes in all sectors.

However, commercialization and the evolution of markets have occurred in a manner which has accentuated differentiation in production and consumption. In turn, differentiation may have reinforced power structures which put severe limits on the bargaining power of the poor and which imply that the dominant groups are able to benefit from control over the labouring groups as well as over the production and distribution of surplus.

One crucial element needed to alter such production and power relations must be the increased availability of income and employment opportunities for the poor so that they can improve their living conditions and bargaining power *vis-à-vis* the dominant groups. Thus the persistence of poverty can be seen as directly related to the inability of the system to provide proper employment or income-generating activities, so that open and disguised unemployment become the primary symptom of the pattern of uneven development in the economy.

Unemployment exists when those without wage

work or other means of income are unable to produce goods currently in demand. Unemployment persists if the type of demand generated with the economy accentuates inequalities between different sectors of the population (for example, differences in demand for luxury or wage goods). In the Indian economy, these inequalities have been reinforced in the recent past by the opening up of the economy to world trade. Patterns of consumption exhibited by the richer countries have been encouraged, and economic imbalances generating unutilized labour have increased, as many workers in rural and urban areas are unable to produce the type of goods currently in demand on their own account.

Therefore, despite the official goals of greater economic equality and the removal of poverty, the Indian state has not been able to resolve the problem. This issue can be dealt with only if the fundamental macro-economic processes which perpetuate poverty and unemployment are confronted – that is, if the state devises policies to change the nature of production, exchange and accumulation relations. Such a set of policies would include, on a macro-economic level, the direction of investment to increase productive employment and generate effective demand among the poorer groups, and, on a local level, the restructuring of market relations so as to allow the poor greater access to all markets, particularly for commodity, land, inputs and credit. The fundamental principle behind all of this should be the effective decentralization of economic decision making. A partial approach, which essentially involves some public spending programmes on targeted groups, may be beneficial in a limited way, but will do little more than mitigate the harsh and polarizing effects of economic forces. Achieving more than this remains the main challenge for state policy.

Summary

1   Household surveys suggest that the extent of poverty has been falling. However, macro-economic indicators show little reduction in poverty and continued material deficiencies of the poor.

2   The gap between agricultural and non-agricultural per capita incomes has been rising. At the same time, in all sectors output has grown faster than employment. Job creation has been inadequate, and is increasingly mainly for work on a casual labour basis.

3   While the country has achieved 'self-sufficiency' in foodgrain production, this has occurred even as per capita foodgrain availability has remained static over the last 30 years. Agricultural growth shows marked regional variations, with dynamism confined to a few states only. In a number of states, per capita agricultural production has been stagnant or falling.

4   There is a strong correlation between low labour productivity and the incidence of poverty. High rates of malnutrition are also associated with low labour productivity.

5   Underemployment in agriculture, industry and services is widespread. The growing importance of casual labour and 'informal' jobs reflects the pressure to find jobs. Employment diversification in the rural areas is also an indication of the desperate need to ensure survival since single occupations often do not provide the minimum necessary income.

6 Poverty cannot be studied in isolation but must be viewed as part of wider macro-economic processes. Targeted anti-poverty programmes therefore have very limited impact.

7 The macro-economic trends of Indian development indicate that while output has registered moderate to good increases, employment generation in all sections has lagged behind, and the benefits of growth have accrued to only a minority of the people. The majority of urban and rural poor have become economically disenfranchised because of the failure of the system to generate sufficient means of productive livelihood.

8 State intervention in the economy has often reinforced those trends, both directly and indirectly, by concentrating on the 'leading' organized sector of the economy and subsequently through a greater reliance on 'market forces'.

8

SOCIAL AND ENVIRONMENTAL CHANGE IN SUB-SAHARAN AFRICA

PHILIP WOODHOUSE

By the 1990s the words 'Africa' and 'crisis' had become inextricably linked in the minds of many people in the industrialized world. Images of starving rural people are used repeatedly by campaigns to raise money for famine relief. Famine in Africa is perceived as the result of farming failure, itself the final manifestation of an ecological catastrophe: 'environmental bankruptcy'.

Such perceptions are reinforced by statistics (like those in Chapter 4) showing falling per capita food production in many African countries, and chronic dependence on food imports or aid to make up for domestic production shortfalls. Chapter 4 suggested that food crisis is one facet of a more general economic crisis in sub-Saharan Africa, which also includes declining production of export crops in some countries. The latter is likewise often taken as evidence of the failure of farming. For many observers, the reasons for this failure are found in the nature of African farming methods themselves:

> "Slash and burn agriculture was appropriate in forest and savannah areas with abundant land and low population density, because it allowed long fallow periods for the land to regenerate. The problem is that, with rapid population growth, the land and resource base has not been adequate in most of Africa to maintain these traditional extensive farming and livestock systems.

> Pressure on the land is resulting in declines in crop yields and in overgrazing. Vegetative cover is weakening, erosion accelerating."
>
> (World Bank, 1989a, p.95)

> "Population pressure has not yet reached the level where intensive farming is unavoidable. But it has reached the level where massive ecological damage can occur if traditional methods go on being used."
>
> (Harrison, 1987, p.41)

Paradoxically, at the same time as these negative verdicts on African farming, a growing body of literature asserts the sophistication and sensitivity of ecological management by African farmers:

> "African farmers and herders have a deep and profound ecological knowledge, which they apply in getting a living from their lands. Virtually nowhere in Africa has their science been sufficiently studied."
>
> (Timberlake, 1985, p.141)

> "Technical characteristics of African land management systems…are inventions appropriate to changing environmental and economic conditions. There is no intrinsic reason why they should be treated as survivals from an era of traditional subsistence production or as evidence of involution in the face of population increase and capitalist exploitation. Recent ecological research

suggests they may be better regarded as advanced rather than backward features."

(Richards, 1983, p.41)

These conflicting views raise a number of questions. If African farmers are good ecological managers, how is it that 'traditional', 'extensive', 'slash and burn' farming methods are so often identified as the cause of environmental degradation? If African farming is to be transformed, what should be the nature of the transformation?

To answer these questions, this chapter explores the nature of 'indigenous' farming in Africa, and the processes of change at work within it.

> **Q** How have changes in farming affected social organization and the environment in sub-Saharan Africa?

The chapter is organized in four sections. The first considers some basic elements of 'agricultural ecology' – the soils, climate, plants and animals that are the raw material of farming – and the general principles which distinguish 'indigenous' African management from those informing European agricultural science. The next section illustrates the nature and extent of change in African farming. The third section examines linkages between farming and environmental degradation in Africa. The final section discusses three different approaches to developing agriculture as a basis for rural livelihoods in Africa.

8.1 Agricultural ecology in Africa

This section sets out the main characteristics which distinguish 'indigenous' African agricultural management from its European counterpart. In the limited space available it is evidently impossible to deal with all the techniques used. Nor is it implied that all the factors discussed here are equally important in all African farming. Readers are referred to more detailed accounts in Richards (1983, 1985).

Tropical rainfall and ecological zones

A fairly typical classification of African farming is given in Table 8.1, and in the accompanying map

(Figure 8.1). If you consider these for a moment, you will observe that the factor which differentiates the zones is rainfall. Although the influence of rainfall is conveniently presented in the form of discrete zones, you should remember that the zone boundaries are rather arbitrary: rainfall decreases along a continuum from the humid rainforests near the equator to the desert areas at the tropics of Cancer (Sahara) and Capricorn (Kalahari and Namib).

In Table 8.1 you can see that the main effect of decreasing rainfall is to reduce the length of the 'growing season', during which there is sufficient rain to support agricultural crops. The small bar charts in Figure 8.1 show that in the equatorial region (Yaoundé and Kinshasa), there is rain almost all year with a slight reduction during two or three months. By contrast, in semi-arid areas like the Sahel (e.g. Kayes and Ouagadougou) all the rainfall is concentrated into less than four months of the year. The bar charts also show that the rain falls at the hottest time of the year: May to November north of the equator, December to April south of the equator. While in most places rainfall is distributed within a single, more or less continuous, wet season each year, in some areas of East Africa, notably Kenya and Tanzania, the rainy season is divided into two distinct parts separated by one or two dry months. In such cases of *bimodal* rainfall distribution, the total annual rainfall will seriously misrepresent agricultural potential, because neither of the two short 'growing seasons' may be long enough for crops to complete their growth cycle.

From the point of view of African farmers, rainfall distribution and intensity are often more important than total annual amount, as illustrated by the Sahelian rainfall pattern in Box 8.1. While the Sahel has become a byword for rainfall uncertainty, the problems identified in Box 8.1 apply to some extent throughout the seasonal rainfall (semi-arid and subhumid) zones, which cover the greater part of sub-Saharan Africa. We can summarize the effects of rainfall in such areas as *seasonal* and *uncertain*. The seasonal rainfall regime limits farmers' productivity by concentrating their work into brief periods of intense activity, causing bottlenecks in labour supply. Rainfall

Table 8.1 Agro-ecological zones in sub-Saharan Africa

Agro-ecological zone	Annual rainfall	Length of growing period (months)	Vegetation and principal agricultural production
Arid	less than 300 mm	less than 2.5	Sparse grassland. Nomadic or transhumant pastoralism (sheep, goats, cattle, camels). Date palm.
Semi-arid	300–1000 mm	2.5 to 4.5	Savannah (corresponds to 'Sahel' and 'Sudan' zones of West Africa). Cereal crops (millet, sorghum), pulses, groundnuts. Pastoralism (sheep, goats, cattle). Many indigenous tree crops ('bush fruit').
Subhumid	1000–1500 mm	4.5 to 8	Moist savannah, dry woodland (corresponds to 'Guinea' savannah in West Africa, 'Miombo' woodland in eastern and southern Africa). Cereals (millet, sorghum, maize), pulses, groundnuts, cotton, root crops (yams, cassava), small livestock (goats, pigs). Coconut, cashew, and indigenous tree crops.
Humid	more than 1500 mm	more than 8	Forest. Food crops (roots, tubers and plantains). Tree crops (cocoa, oil palm, rubber).
Tropical/subtropical highlands	variable	variable	Forest. Tea, coffee, temperate fruit, vegetables, cereals (maize, sorghum, wheat), roots (potatoes, sweet potatoes), bananas, livestock (cattle, goats, sheep).

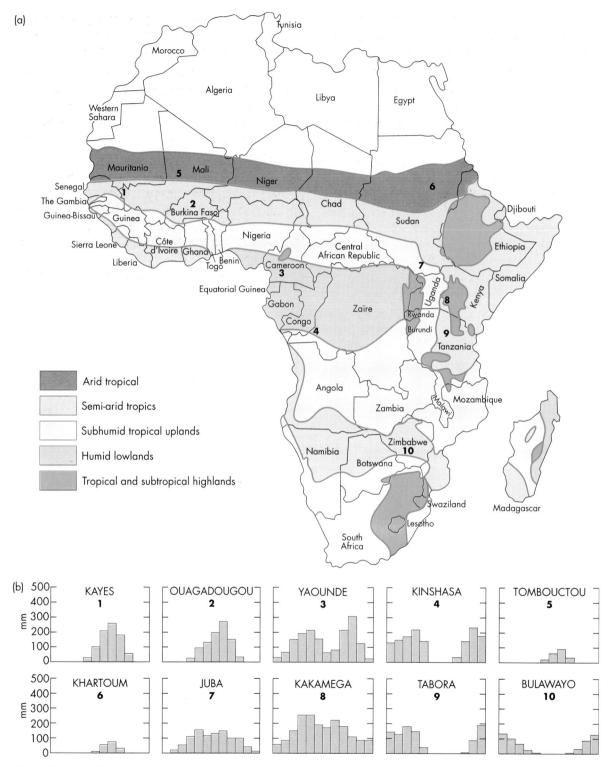

Figure 8.1 (a) *Map of agro-ecological zones in sub-Saharan Africa.* (b) *Bar charts of annual rainfall distribution for selected sites; the bars show monthly totals from January to December.*

uncertainty exposes them to high risks of crop loss. African farming techniques, therefore, aim to reduce or spread the demand for labour (by eliminating bottlenecks) and to minimize risk.

One of the main ways farmers do this is by exploiting local variations in the moisture regime established by the rainfall pattern. There are a number of sources of such variation:

- *Altitude* The highland areas of eastern and southern Africa (e.g. Ethiopia, Rwanda and Burundi, and parts of Tanzania and Kenya) and smaller highland areas distributed more generally through the continent experience higher rainfall and lower temperatures than the surrounding lowlands. In the subhumid and semi-arid areas this translates into a longer growing period for

crops and pasture, and greater diversity of crop choice, often allowing temperate crops (e.g. wheat, apples) to be grown in tropical areas.

- *Relief* The effect of rainfall is strongly conditioned by relief (topography) of the landscape because water accumulates in low-lying areas such as swamps and river floodplains. This 'stored' moisture can be used to grow crops after the end of the rainy season.

- *Soil type* 'Heavy' soils, with a high clay content, retain more moisture than sandy soils. In many cases heavy soils are found in low-lying areas (black clay soils), reinforcing the effect of relief in storing water. However, certain rocks in the underlying geological strata can give rise to higher clay content in upland soils, and the extra

Box 8.1   Rainfall and risk: Sahel and East Anglia

The average annual rainfall at Niamey, the capital of the Sahelian state of Niger, is 562 mm, which is about the same as in East Anglia, UK. Clearly, the higher temperatures in Niger mean that crops will need to absorb (transpire) more water than in East Anglia, but even in the 1984 drought year total annual rainfall at Niamey was 319 mm, above the minimum 300 mm needed by a millet crop. However, while East Anglia averages 175 rainy days in a year, Niamey averages only 44.

In practice, this *concentration* of rainfall is even more acute because a high proportion of rain in Niamey falls in a few violent storms: in the high rainfall (599 mm) year of 1983, 75% of the rain fell in only 13 storms; in the low rainfall (343 mm) year of 1973, two-thirds of the rain fell in only five storms.

This produces a number of management problems, particularly at the start of the rains, when there may be virtually no vegetation covering the soil after the long dry season. The impact of raindrops dislodges particles from the exposed soil surface, and these block the channels through which rainwater enters the soil. As a result, a high proportion of the rain never enters the soil, but runs off the soil surface to be lost, together

with quantities of detached soil particles, as flash floods. Annual variation in both quantity and distribution of rain compounds its concentration in intense storms, and results in *uncertainty* and *risk* for farmers.

First, rainfall uncertainty makes it difficult to know when to plant. Farmers must decide at which point the sporadic thunderstorms that herald the end of the dry season have moistened the soil sufficiently to guarantee germination of their seed. Planting too early risks losing the seed to heat and insects. Planting delayed for more than a few days after the soil becomes fully moist means exposing the germinating crop to competition from rampant weed growth, and losing the benefit of the 'flush' of nutrients that are released by microbial processes reactivated as the soil becomes wet after the long dry season.

Secondly, rainfall uncertainty translates into a high risk of crop loss due to mid-season dry periods. Under the high temperatures of the rainy season, crops quickly suffer if no rain falls for more than about ten days. If a dry period coincides with a critical period of crop growth, such as flowering in the case of cereals, more than half the harvest may be lost.

moisture retained by these 'red clay' soils may enable crops to withstand dry periods better than if grown on neighbouring sandy soils.

The combination of these elements of local variation makes the landscape a patchwork of varying potential productivity; knowledge and exploita-tion of this is the key to many African farming strategies (Box 8.2 and Figure 8.2). The variation may be on a large scale, for example, where there are mountain ranges or large river floodplains. Variations in a smaller area may mean that every small valley presents different production possibilities from those on adjacent hillsides.

Box 8.2   Ecological management in African farming

Transhumance: Exploitation of diversity under-lies the 'transhumant' grazing systems of Afri-can pastoralists, who typically move their herds to graze the pasture that flourishes briefly fol-lowing rain in areas which are too arid to support agricultural crops (Figure 8.3a). As the dry sea-son proceeds and these pastures are used up, the herds are moved to wetter areas: to river valleys, mountain pastures, or to adjacent higher rainfall (semi-arid) areas where the herds graze the stubble on crop fields as well as pasture. Pastoral systems incorporate complex adjustment of herd composition (the relative numbers of cattle, sheep, goats and camels) and division (between fast-moving 'dry' herds and slow-moving milking herds) in order to achieve the appropriate mobil-ity and grazing or browsing 'pressure' for the available resources of grass and trees.

Diversified and dispersed cropping systems: Afri-can farmers typically exploit ecological variation by dividing their total cultivated area between a number of small plots with different soil or land-scape position in order to grow a range of differ-ent crops. A detailed example is given in Figure 8.2.

Minimum cultivation: African farmers frequently plant directly into weed-free soil created by burn-ing or by flooding. Advantages in the humid forest zone are that it causes minimal disruption to the physical structure of the soil, and avoids bringing buried weed seeds to the surface. In subhumid and semi-arid areas the main advan-tage is that it enables areas to be planted very quickly after the first significant rainfall at the beginning of the rainy season (see Box 8.1).

Intercropping: The growing of more than one crop simultaneously in the same field. Crops may be sown at the same time, and even as mixtures in the same planting hole (e.g. sorghum and millet or sorghum and cowpea). Some crops of the mixture may be sown later than others, during weeding operations. Such techniques are also known as relay cropping, and may involve mix-tures of short-duration annual crops (e.g. cere-als) and longer duration crops such as cassava (18–24 months), or tree crops. In typical West African rainforest farming, vegetation is cut and burnt shortly before planting. Upland rice or maize is sown first and later interplanted with cassava and plantains. Once the maize or rice has been harvested, the cassava continues to grow for a further year, and the plantains for a further two years. In some cases the ground between the plantains may be recleared for a further crop of maize or rice in the second year. Intercropping involving tree crops is known as agroforestry. The advantages of intercropping are that the mix of crop plants can be chosen so that the combined leaf canopy is more efficient at intercepting light than a monocrop, and simi-larly the different root systems can exploit the soil in complementary ways. It is often claimed that pests spread more slowly in intercrops than in fields occupied by single crops.

Use of fallows: Often referred to as 'shifting cultivation' (or the more pejorative 'slash and burn'), implying nomadic or semi-permanent settlement (Figure 8.3b). In forest areas, farmers resident in towns or villages commonly build a temporary 'farmstead' next to their cultivated fields. Movement of villages when fields are left fallow is relatively rare, but is reported to occur in the humid forest of the Congo basin. Else-where, the fallowing of fields by allowing the regrowth of either forest or savannah is associ-ated with permanent settlement, and often with continuing rights of access to the fallowed land, particularly where economically important trees remain standing on the site throughout the cultivation and fallow cycle. Fallowing restores the nutrient content of the soil, which is most important on soils of low fertility in humid forest areas. When the wood regrowth of a ten-year fallow is burnt, the nutrients contained in the large amount of ash significantly improve crop productivity. Evidence suggests that suppression of pest and weed build-up is also an important consideration in fallowing land. In drier areas, soils are more fertile and the fallow vegetation is sparser and so produces less ash. If the fallow vegetation is largely grass, then the effect of the fallow may be very small. In some systems farmers compensate for this by collecting wood from a wider area to increase the ash on the cultivated plot (e.g. the *citemene* system in Zambia). Farmers may prevent the regeneration of woodland in fallows by grazing and annual burning. This may be important in areas where trees harbour tsetse fly, the vector of trypanosomiasis (sleeping sickness), a disease which is fatal to both livestock and humans. However, valued tree species are almost always left standing whether land is cultivated or fallowed (Figure 8.3c). Typical trees maintained in farmland in savannah areas are *Acacia albida*, used as a source of animal fodder and believed to enrich the fertility of the soil around it, and *Acacia senegal*, from which gum arabic is obtained.

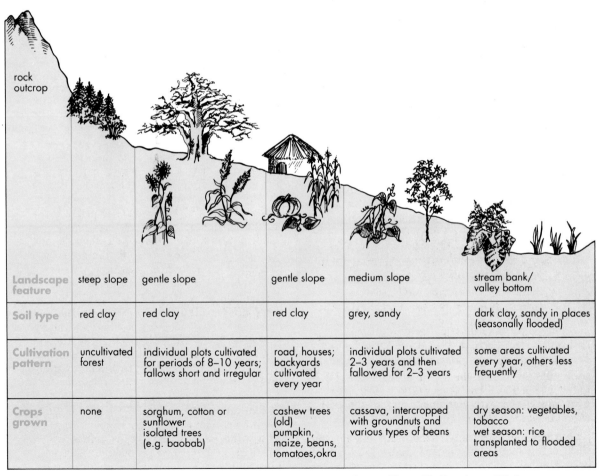

Landscape feature	steep slope	gentle slope	gentle slope	medium slope	stream bank/ valley bottom
Soil type	red clay	red clay	red clay	grey, sandy	dark clay, sandy in places (seasonally flooded)
Cultivation pattern	uncultivated forest	individual plots cultivated for periods of 8–10 years; fallows short and irregular	road, houses; backyards cultivated every year	individual plots cultivated 2–3 years and then fallowed for 2–3 years	some areas cultivated every year, others less frequently
Crops grown	none	sorghum, cotton or sunflower isolated trees (e.g. baobab)	cashew trees (old) pumpkin, maize, beans, tomatoes, okra	cassava, intercropped with groundnuts and various types of beans	dry season: vegetables, tobacco wet season: rice transplanted to flooded areas

Figure 8.2 Farming patterns along a transect in northern Mozambique (Nampula Province, 1984).

(a)

(b)

(c)

Figure 8.3   Ecological management in African farming: (a) transhumance – goats, sheep and cattle being taken to pasture in Burkina Faso; (b) 'shifting cultivation' or 'slash and burn' in Mozambique – after the swamp is dry, trees and bush are cut down and the area is burned; (c) valuable tree species are left standing, such as this baobab tree – its fruit has a tasty pulp, its bark fibre is used for rope and cloth, and the trunks are sometimes used to store water or for shelter.

The combination of intercropping and diversified and dispersed cropping systems illustrated in Figure 8.2 can be found throughout tropical Africa. There are intensively farmed areas fertilized by animal manure or household refuse, complemented by less intensive cultivation involving periodic fallowing. There is also the rain-fed cultivation of upland sites during the wet season, complemented by cultivating low-lying 'wetland' sites either following the peak of the rains when the sites support crops on residual moisture stored in the soil ('flood-retreat' farming), or using crops that can tolerate flooding ('flood-advance' farming). In some cases, supplementary irrigation is practised on a small scale, by using shallow wells (in semi-arid West Africa) or by diverting streams into canal systems (in the Rift valley and the slopes of Kilimanjaro in East Africa).

Access to labour and land

We can note two obvious advantages of the diversified farming patterns outlined. Intercropping, and exploiting different crops on different soils at different times of the year, simultaneously spreads the risk of crop failure and allows farm labour to be used more evenly over the year. Risk and labour productivity are the key issues in understanding African farming systems and their social organization. The unpredictable onset of the seasonal rainfall regime generates a need to complete at short notice the planting, and subsequently the weeding, of crops. This puts a premium on labour-saving soil management, noted earlier, and also on social organization that gives individuals the right to call on the labour of others for planting and weeding.

Chapter 5 showed that elements of the farming system may be under the control of different members of households, with differential access to the labour of members of their own and other households. For example, in millet-growing areas of West Africa, heads of households responsible for securing the staple millet crop can call upon all members of the household to work on the millet fields. By contrast, women growing plots of rice or groundnut are unable to call upon other household members in the same way, although they

may have a reciprocal labour-sharing arrangement with members of their own and other households.

Just as the notion of 'household' needs to accommodate this loose, changing, linkage of individually managed farm activities, so our concept of 'farm' needs to be revised to accommodate different agricultural activities, some more or less constant, others fluctuating, expanding or moving from year to year. This is the basis for understanding customary land rights, which allow great flexibility in land use from one year to the next.

Farming in the rural economy

So far, we have focused on crop production – to outsiders often the most visible activity of rural people. However, cultivation is only one of a number of rural economic activities, which present opportunities for alternative work during the dry season but may also compete for labour during the farming season.

Activities that may be important as alternative sources of income, either seasonally or all year, include fishing and livestock herding. In both cases, considerable expertise is required, and for those with the necessary skill, cultivating crops may be a secondary activity. A number of other activities exploit rural resources through a varying combination of 'gathering' and active management, including the seasonal collection of wild fruit and mushrooms, charcoal production and collection of honey. All are widely used to generate cash income in farming areas in Africa. Many members of farming households are engaged in non-agricultural work: most commonly, carpentry, metal-working, pottery, mat- and basket-making, beer-brewing (Figure 8.4), and transport and trading activities.

The growth of cash crop production in particular areas frequently generates a need for seasonal labour, and migration of labourers from one rural area to another is common throughout Africa; for example, in the development of groundnut farming in Senegal, cocoa and oil palm production in the West African forest zone, tea and sisal plantations in East Africa.

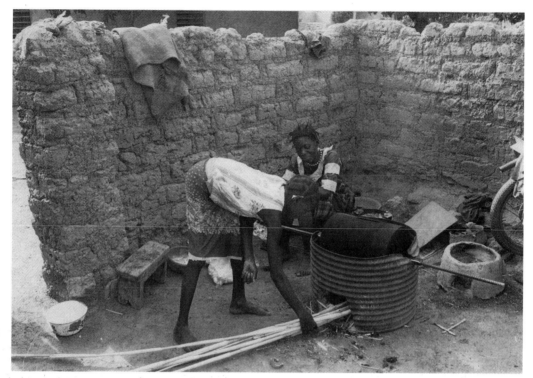

Figure 8.4 Earning 'cash income' in non-agricultural work: brewing beer in Burkina Faso.

Finally, migration to industrial and urban centres has been a growing feature of the rural economy in much of Africa during the past century.

The importance of these alternative sources of income for rural Africans is that they establish the *opportunity costs* of labour in growing crops: the alternative income against which labour productivity in African agriculture must be evaluated (Box 8.3). Clearly, the economic context of African farming has changed greatly during the twentieth century (Box 8.4), and the next section illustrates those changes in a little more detail.

Box 8.3   Opportunity cost of agricultural labour

'Opportunity cost' is a term used in economics to express the cost of any factor in a production process in terms of its maximum value in alternative use. Thus, the opportunity cost of labour spent in farming can be measured as the income which would be generated by the same amount of labour time spent in some other activity, such as trading or working for wages away from the farm. The opportunity cost of labour can only be established, however, where the economic value of alternative activities can be compared, or where individuals allocate their labour according to criteria of the market. Though usually applicable to male labour, these conditions are less often valid for female labour, which is commonly constrained by non-market relationships and commitments, such as responsibility for reproductive activities.

Box 8.4   What is indigenous farming?

The concept of indigenous agriculture in Africa is defined as farming methods which are of local origin and not derived from European (or North American) agricultural science. This definition gains force from, on the one hand, the contempt and hostility with which many European colonial administrators and their technical advisers viewed African farming practices, and on the other hand, the resistance frequently put up by African farmers to the imposition of 'improved' farming methods by the colonial authorities. In the post-colonial era the idea of 'indigenous agricultural technology' has been raised as a response to the failure of many attempts by govern-

ments and development agencies to modernize African farming through the 'transfer of technology' from Europe, Asia, or America. Expressed in this way, however, 'indigenous agriculture' is not a very determinate concept, amounting to little more than 'what African farmers do'. As we shall see in later sections, 'what African farmers do' is to adapt and change as new opportunities arise, or as existing ones are removed. Attempts to fix 'indigenous agriculture' as an unchanging set of practices – whether in order to demonstrate its irrelevance to the future or to claim it as a basis for 'sustainable' development – misrepresent the problem of agricultural development in Africa.

8.2   The changing rural economy

This section illustrates changes in agriculture through three examples drawn from different parts of Africa: the commercial development of tree crops in the forest zone of West Africa; the development of maize production in upland areas of eastern and southern Africa; and the development of irrigated farming in semi-arid West Africa.

Cash crops in the West African forest zone

By the mid-nineteenth century, increasing industrialization in Europe had shifted European trading interest in West Africa away from slaves to trade in raw materials for industry. The equatorial forest region close to the coast was quickly drawn into this growing market. Some of the commodities produced, like timber and wild rubber, were obtained through 'gathering' from the forest, rather than cultivation. In other cases production was developed through cultivation. In particular, the production of palm oil, although obtained from naturally occurring oil palm trees in the forest, was greatly increased through the selective conservation of oil palm during forest clearance for food farms (for yam, cocoyam and cassava). When food plots were left to revert to

fallow, the regenerating forest was dominated by oil palm, and in this way substantial areas of forest were converted to 'oil palm bush' (Richards, 1985, p.107). This process was itself accelerated by the expansion of food farming for sale to the traders, porters and processers using the new trade routes to the coast.

Partly in response to the decline in palm oil prices during the recession in Europe in the 1880s, between the 1890s and the 1920s there was a rapid expansion in cocoa planting in the forest zones, largely financed and controlled by African farmers and traders (Box 8.5).

The onset of the recession in 1929 halted the expansion of cocoa farming as prices fell by 75% in five years, causing the collapse of many of the farmers' marketing organizations. In 1939 an outbreak of swollen shoot virus started and ultimately killed the cocoa trees in most of the older cocoa-growing areas. Many cocoa farms reverted to growing food crops for local sale, or kola nuts for export to the semi-arid areas to the north. In areas where cocoa expansion continued, the colonial authorities banned sales of land to try to contain the pace of rural change and the 'erosion of traditional social control'. Although cocoa farming revived in the 1950s, control had passed from the farmers to the colonial government.

At the outbreak of war in 1939 the colonial government assumed monopoly control of marketing export crops, which enabled it to reduce producer prices in order to achieve wartime goals of reducing import demand while maximizing hard currency earnings. After the war, a boom in commodity prices enabled the state monopoly to raise producer prices while retaining about half the world market value of the crop. In addition, the state increasingly controlled the inputs to cocoa production through investment in research on ways to combat the swollen shoot disease, and the provision of pesticides and fertilizers (Figure 8.5).

Box 8.5   The cocoa boom: 1890s to 1920s

The major characteristics of this cocoa boom were:

- *A market in land* Cocoa farms were established in uncultivated forest land, frequently by migrants or outsiders, who negotiated the acquisition of land with local residents. In Nigeria, traders from Lagos acquired tenancies on land 30–50 km outside the city. In southern Ghana, groups of farmers formed 'companies' to negotiate purchase of large tracts of uninhabited forest, which they then subdivided among individual members. Generally, earnings from cocoa farms were reinvested in further land acquisition, resulting in the accumulation of significant land holdings by individuals such as the Lagos entrepreneur J. K. Coker (1500 acres/600 ha) and the Ghanaian W. A. Q. Soloman, whose estate included 88 cocoa farms by the time he died in 1936 (Hill, 1963, p.238; Hopkins, 1978, pp.87–94).

- *Migrant labour* After cocoa farmers had established temporary 'farmsteads', while retaining their permanent residence in their home town, migrant sharecroppers or wage workers cleared forest and planted cocoa trees, intercropped with food crops (plantains and cocoyams) in the early years. When the cocoa trees began to produce, seasonal labour was required for weeding and harvesting.

- *Wealth and accumulation of cocoa farms* This occurred, particularly in Ghana, where cocoa farmers invested in (a) further speculative land purchases, (b) trade through forming marketing companies, and (c) infrastructure through raising funds for contract-built roads and bridges.

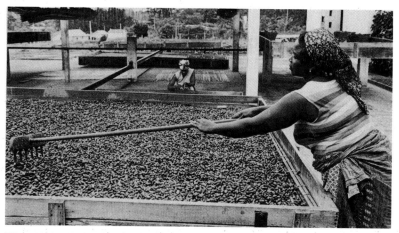

Figure 8.5   Drying cocoa beans in an experimental rack, Tafo Research Centre, Ghana.

After independence, state monopoly marketing to generate revenues continued and farmers never recovered the high profitability that had encouraged cocoa expansion at the beginning of the century. The consequences were:

- Crops were smuggled across borders to obtain better prices.

- The emphasis on food farming continued. According to Richards (1985), farmers' net returns from growing yam and cassava in the Nigerian forest zone were better than those from oil palm production from the 1920s right up to the 1960s. In Ghana in 1981 production of cocoyams and plantains yielded over twice the income of cocoa production (Boateng *et al.*, 1987).

- There was a movement of labour out of farming and into wage labour in urban areas and in mineral extraction.

Households remaining in farming experienced labour shortages with the following consequences:

- Seasonal wage labour, typically from the semi-arid regions to the north, was employed for cocoa harvesting.

- Processing of cash crops like palm oil within households was achieved by saving labour in farming by growing cassava instead of yams.

- Labour shortage was more likely to restrict the clearance of forest for new plots, promoting the use of shorter fallows on existing cultivated areas.

Fallow periods were also likely to be reduced as a result of the shift from a tree crop lasting 30 years or more to food crops like cassava, yams and plantains which occupied the land for not more than two years. This intensification of land use was greatest in the continuously cultivated market gardening areas close to large towns.

By the 1970s, low returns to cocoa farming were reflected in falling output. In Ghana, for example, exports of cocoa halved from the late 1950s to 1979. Efforts to improve producer prices were hampered by a collapse in world commodity prices during the 1980s. In a bid to find an alternative source of foreign currency a number of governments in equatorial Africa contracted with multinational logging companies for timber extraction.

Maize farming in eastern and southern Africa

Chapter 4 described how colonial regimes in Africa established colonies of peasant economy or colonies of settler economy. The case of cocoa production in West Africa is an example of the former, whereas this subsection gives an example of the latter.

Travellers to inland areas of East Africa in the mid-nineteenth century recorded that they crossed large areas of settled agriculture in which permanent cultivation was combined with cattle production. Thus, Speke crossing central Tanganyika in the 1860s wrote that the villages 'followed one on the other, with few intervals of jungle' and that the area 'abounds in flesh, milk, eggs, and vegetables in every variety' (quoted in Kjekshus, 1977). There is evidence from other parts of southern and eastern Africa that the nineteenth century was a period of expansion in African agriculture, particularly in the production of food for sale. In South Africa from the 1840s to the late 1880s, African farmers (often sharecroppers or 'squatters' on land claimed by European settlers) were responsible for most of the marketed grain, wool and cattle products in South Africa, which they sold to European traders in exchange for manufactured goods (tools, weapons, cooking utensils, household goods). During the 1880s the more prosperous African farmers purchased significant areas of land from European 'owners' – some quarter of a million acres in the Transvaal alone (Bundy, 1979, p.205).

In the 1890s an epidemic of the cattle disease rinderpest swept through eastern and southern Africa, originating from cattle introduced by British and Italian armies in the Sudan and Somalia respectively. The disease killed 80–90% of African cattle. This caused immediate famine among pastoralists dependent upon blood and dairy products for their daily diet, and destitution among cultivators, for whom cattle were the source of manure, power for ploughing, bush control by grazing, and, above all, capital. The crisis in

African farming coincided with the seizure of administrative control by European powers and a shift in colonial interest from buying Africans' farm output to employing their labour in European-controlled production. In South Africa, while commercial farming by Europeans was heavily subsidized, restrictions on land purchase or leasing increasingly confined Africans to the 'reserves', where gross overcrowding and the reinvention of traditional land tenure ensured that individual holdings were too small to provide a living, and wage labour in the mines or on European farms became inescapable.

To the north, in Kenya and the Rhodesian territories that are now Zimbabwe and Zambia, the development of African farming reflected that in South Africa, but somewhat later. The pace of European settlement accelerated after the First World War, and by 1930 there were some 7000 settlers, each with 1000 to 2000 ha. African farmers had started growing maize as a cash crop (millet was the staple food in rural areas) in response to the growing market for food due to railway construction and mining development. In Northern Rhodesia (Zambia) in the mid-1920s, Tonga farmers had increased output to the point where they supplied 50% of the cattle and 25% of the grain consumed by the colony (Siddle & Swindell, 1990, p.131). As in South Africa 30 years earlier, the competition from black farmers was removed through legislation: in Southern Rhodesia the 1930 Land Apportionment Act allocated half the agricultural land (disproportionately located in areas of favourable rainfall) to 3000 European settlers, and the other half to an estimated 100 000 African farming households; and the 1931 Maize Control Act levied a tax on maize sales from African areas. As in South Africa, the measures were designed to reduce the viability of African farming in order to force Africans to seek wage labour on European farms.

Male labour migration from the African 'reserves' increased after 1930, but the output of marketed maize was sustained, at least in areas of favourable rainfall, largely due to farmwork carried out by women, coupled with the use of migrants' remitted earnings to purchase ploughs. During the 1940s and 50s rising demand for labour in urban areas, and diversification of European farming towards tobacco and cotton for export, led to attempts to separate migrant workers from their rural homes through permanent settlement in towns, and to improve African farming. Both were reflected in the Native Land Husbandry Act of 1951, which excluded Africans with urban jobs from access to land, and set out to take direct control of African farming. In particular, cultivation of wetlands (*dambos*) was prohibited, and villages were arranged in rows of housing with grazing land allocated on one side and arable land on the other (irrespective of soil type). In addition to the considerable deforestation caused by clearing upland areas to replace previously cultivated *dambo* fields, the Act required the removal of all trees from arable land. Finally, labour-intensive works were mandatory to control the soil erosion that was increasingly evident since the 1930s (Bonnevie, 1987).

The Act provoked strong resistance, particularly from those whose employment in urban areas was neither well-paid nor secure enough for them to relinquish their rights to cultivable land. By 1962, when the Act was repealed, land in the African areas was becoming scarce, a major reason for the war to overthrow settler power. At the independence of Zimbabwe in 1980, the land partitioned in 1930 was occupied by 6000 European farmers and 750 000 African families. When the discrimination against African producers was removed, the maize they marketed increased fifteenfold in five years, and by 1985 accounted for over a third of the national total (Rohrbach, 1987) (Figure 8.6).

This rapid increase was possible because the high-yielding hybrid maize varieties developed for the European farming sector could readily be used by African farmers in the areas of better rainfall. As the technology is heavily dependent on the purchase of inputs (seed, fertilizer), the cash remittances of migrant workers are essential to buy them, and to pay for hiring labour to replace that of the absent worker. Research suggests that households without access to migrants' remittances must seek cash through working on neighbours' farms, thus depriving their own farms of labour (Bonnevie, 1987). This process of differentiation enriches some households while

Figure 8.6 Maize farming in Zimbabwe.

impoverishing others, producing a class of commercial farmers on one hand and a class of rural labourers on the other. Although land allocation in communal areas is nominally 'traditional' (controlled by customary chiefs), the investment by wealthier families in land improvements and rural housing lays the foundation for *de facto* individualized private tenure (see Chapter 9, Box 9.1).

The development of irrigated farming in semi-arid West Africa

European interest in the semi-arid regions of West Africa centred on trade in groundnuts. This developed initially in the Saloum area of Senegal, then spread quickly throughout the 'groundnut basin' of Senegal and The Gambia, and further inland to Mali after the opening of the Dakar–Bamako railway in 1910. In the same year, Kano was linked to the coast by rail and became the centre of groundnut trade in northern Nigeria.

The rapid expansion of groundnuts attracted a large influx of migrant farmers, from as far afield as Burkina Faso. These 'stranger farmers' (The Gambia) or *navetanes* (Senegal) were generally allocated land in exchange for labour on local farmers' fields. In effect, they were incorporated

into the household system of labour obligations, and many married and settled permanently in the area. Others migrated to the groundnut basin as seasonal labourers, returning each year to their own villages; they usually came from areas, such as the Senegal river valley, which were marginal to the colonial economy as suppliers of crops, if not labour.

The areas suitable for groundnuts had sandy, easily worked soils which were also used for the staple food crop, millet. Thus the two crops competed for both labour and land. One response to this, in addition to hiring seasonal labour, was to increase the area farmed through animal-drawn ploughs – horses or donkeys provided adequate draught power on the light soils. However, the basic problem remained that land occupied by millet restricted the output of groundnut, and vice versa. While the effect of this on rural food security is disputed (and by the 1980s the rising price of food staples had resolved the issue in favour of millet), areas growing groundnut could not be relied on to provide food for a growing urban population. From dependence on imported rice, in the post-war period attempts were made to develop irrigated rice production in the great floodplains of the Senegal and Niger rivers.

Early schemes used systems of 'controlled flooding' whereby dykes and sluices controlled the spread of the annual river flood across the floodplain. The schemes did not work as designed, however, due to the large annual variation of the size of the flood of Sahelian rivers. By the beginning of the post-colonial period, irrigation planners favoured infrastructure for 'total water control' to permit crop cultivation throughout the year. Although the dams, canals, and pumping stations required are expensive, international credit for such schemes was readily available during the 1970s credit boom, and large irrigation construction was undertaken throughout the West African semi-arid region, with the objective of supplying rice or wheat for urban consumption (Figure 8.7). Modelled on the colonial Gezira scheme in the Sudan, most of the large schemes were centrally managed by a state agency which provided farmers with water, prescribed what crops could be grown, and provided the inputs (tractors, seed, agrochemicals) required. Equally, most suffered from low productivity, high production costs, rising indebtedness among farmers, and failure to maintain the infrastructure. A central problem was the planners' incorrect assumption that farming households had little remunerative employment during the dry season, or that irrigated crops would automatically be given priority over rain-fed crops (see Box 4.1 in Chapter 4).

Disillusioned with large-scale irrigation, governments and development agencies realized that 'small-scale irrigation' was widespread and growing in the semi-arid region. 'Small-scale' covers many different forms of water management including 'flood-retreat' farming (planting a crop in moist soil following the receding annual wet-season flood in low-lying areas) which had been practised for centuries. Yet two farmers' initiatives had indeed increased irrigation use. The first was the growth of commercial vegetable production in areas such as the *Niayes* close to Dakar and the *fadamas* close to Kano. These are areas with a shallow water-table, and therefore, when provided with drainage ditches to reduce waterlogging in the wet season and shallow wells to provide irrigation in the dry season, allow year-round cultivation. The relatively heavy soils and close attention to water control (using bucket or *shaduf*) make this a labour-intensive system, but a high-value crop could be produced on a relatively small area. A second farmers' initiative was in villages along the middle reaches of the Senegal river valley. To escape the effects of the Sahel drought of the 1970s, village associations were formed to purchase and operate small diesel pumps for irrigating food crops. (For an account of such a project, see Johnson & Bernstein, 1987, pp.67–90.) The objective here was to secure subsistence, and each household was allocated a small (0.2–0.5 ha)

Figure 8.7   Clearing an irrigation canal in West Africa.

plot within a 20–30 ha area irrigated with a single pump. Since the schemes produced little marketable surplus, the cost of inputs (fuel, fertilizer, pump repairs) was usually met from remittances of wages earned by household members working elsewhere, or from sales of livestock.

These developments in 'farmer-managed' irrigation were adopted by development agencies in the 1980s as models for further investment in irrigation: for the privatization of existing state-managed large-scale schemes, and for the wider development of year-round cultivation. A large programme financed by the World Bank promoted the purchase of small petrol-driven pumps by individual farmers to extend the area irrigated in the dry season in the *fadamas* of northern Nigeria. Initially intended to increase vegetable production, the programme rapidly glutted the market, and farmers switched to less perishable crops, and in particular to wheat, which commanded extremely high prices following the banning of wheat imports by the Nigerian government in 1987 (Kimmage, 1991).

The development of irrigated food farming has the following consequences:

- Food production requires a higher level of inputs: for water delivery, fertilizer, and, where the scale of cultivation increases, labour-saving machinery and herbicides.

- Cash requirements for input purchase favours those with a source of cash income, notably those with urban employment.

- For those with fewer resources, crop failure carries the risk of a rapid spiral of indebtedness, terminating in loss of cultivation rights to wealthier neighbours.

- Irrigable land may be scarce, while agricultural land in general is not.

- The establishment of an irrigation scheme frequently means the transfer of land from customary tenure to registered individuals or households, which typically disadvantages women farmers who are rarely registered as landholders, even where they were the previous users of the land.

- Where irrigation is developed through private investment by individuals or groups, the resulting infrastructure (canals, drainage ditches, wells or pump stations) often confers effective freehold ownership of the land, even where, as in many parts of Africa, the land is technically the property of the state and thus cannot be sold. The cost of providing access to irrigation thus translates into the price of the land. Where irrigation has become profitable, an intense struggle for land has developed, pitting local farmers against urban investors, agriculturalists against pastoralists, women against men, and youth against heads of households.

Analysing changes in farming

These three brief case studies show how farming has changed as economic options available to rural people have changed, and in particular how small-scale farmers have responded to market opportunities. They have adopted new cash crops (cocoa, groundnuts, maize, vegetables) and new technologies (ploughs, irrigation, hybrid maize and fertilizer), migrated to new areas, and even switched to new staple foods (maize, cassava and rice) in pursuit of these opportunities. We have emphasized farmers' initiatives to show that what appears 'traditional' may be the product of conscious 'preservation', such as the use by colonial states of customary land tenure to suppress the emergence of land markets, or of deterioration, such as the development of subsistence production as a result of withdrawal of farm labour or collapse of agricultural markets.

Table 8.2 summarizes these changes in terms of the concept of commoditization introduced in Chapter 2, when production, consumption, and reproduction take place through market relations. The table compares the degree of commoditization in three different historical periods for elements of production (land, labour, technology), consumption (agricultural products) and reproduction (intra-household relations). The last two columns assess the extent to which African farmers were able to accumulate wealth and control the management of agricultural ecology.

The three periods in Table 8.2 can be summarized as follows:

1   The 'pre-colonial' period of agricultural change was in response to growth in trade resulting from European expansion in the nineteenth century.

2   The colonial period was one in which commoditization was both promoted and regulated by the colonial government. Cash requirements, to be met from cash cropping or wage labour, were increased by the need to meet colonial taxation; markets in land and agricultural products were controlled by the colonial state in order to withdraw labour from African farms; and in some cases agricultural production was directly controlled by external capital (in the form of settler farmers, plantation owners, operators of forced cultivation schemes).

3   The post-colonial period is one where many of the tendencies of the colonial period are continued. In particular, markets for both inputs and output are controlled by relatively few agencies (including the state), so that the bargaining position of small-scale farmers remains weak and their returns low; hence accumulation tends to be higher in trading than in agriculture. The new technological 'packages' of inputs that have to be bought intensify commoditization. Also, labour and land become increasingly commoditized – the latter often through 'back door' markets for land improvements, such as irrigation, by the development of commercial farming by individuals. Finally, as we have seen in earlier chapters, commoditization tends to lead to, or accentuate, the social differentiation of the countryside.

Table 8.2 suggests an important general observation that commoditization occurs unevenly (see also Chapter 6). Indeed, the commoditization of one factor (e.g. labour) may be promoted by blocking that of another (e.g. land). An important corollary is that the development of agricultural commodity production is not a linear process: as in colonial Rhodesia, the development of cash cropping may be outweighed or reversed by forces promoting a market for wage labour. Development models which define subsistence agriculture as a survival from a pre-market past are

inappropriate to the examples of African farming discussed here.

In contrast to the 'indigenous farming' methods described in Section 8.1, this section has emphasized the adoption of new, and often foreign, techniques by African farmers. Yet, since we have concentrated on examples in which change has largely resulted from farmers' own initiatives, how useful is the distinction? We shall return to this point in Section 8.4.

8.3   Indigenous farming and the environmental crisis

So far, I have considered how ecology, climate, social organization, and political and economic processes have shaped African farming over the past century. Here, I examine the linkage claimed between agriculture and environmental deterioration.

Look back at the passage from a 1989 World Bank report on Africa quoted at the beginning of the chapter. What characteristics does it suggest for African agriculture?

It presents a 'model' with the following features:

- 'Traditional', 'extensive' farming practice has remained unchanged from the (pre-colonial?) past.

- Population growth causes 'pressure on the land' resulting in over-use of land and decline in its productivity.

In the light of our discussion so far, how valid is this model, in your view?

The term 'traditional' is evidently problematic. African farming methods are not static: Zimbabwean women today who plough with oxen and use chemical fertilizer to grow hybrid maize have little in common with their male predecessors who cultivated millet with hoes 80 years ago. According to the 'model', the destructive traditional practices are the 'extensive' ones that use low levels of inputs. Here we will consider an example of relatively low-input farming, such as the millet

Table 8.2 Summary of trends in selected examples of change in African farming

			Commoditization				
	Land	Labour	Technology	Agricultural products	Intra-household relations	Accumulation in farming	Ecology control
Tree crops in West African forest							
pre-colonial	+++	+	+	+++	–	+++	+++
colonial	+	++	++	++	+	–	+
post-colonial	++	+++	++	+	++	–	+
Maize growing in Zimbabwe							
pre-colonial	–	–	–	++	–	++	+++
colonial	–	+++	++	++	++	–	–
post-colonial	++	+++	+++	++	+	+	–
Irrigated farming in semi-arid West Africa							
pre-colonial	–	–	–	++	–	+	++
colonial	–	++	+	++	–	–	++
post-colonial	++	+++	+++	+++	+++	++	++

Key to symbols: (low) – + ++ +++ (high).

growing households in the Senegalese village of Kirene described in Chapter 5. Can such 'traditional' farming (using land, family labour, and possibly animal draught as the only inputs) continue under conditions of commoditization?

Figure 8.8 shows how the commoditization of labour through the growth of wage employment outside the household reduces the pool of unpaid or reciprocal labour that household members can call upon, and creates the need for individuals responsible for production to hire labour. This transformation is often referred to as an *individualization* of production. For high-yielding crops, such as hybrid maize, labour hire may be an economically feasible option. For growing millet in the highly uncertain rainfall of the Sahel, where any additional expenditure may be wasted due to lack of rain, this is much less likely. Commoditization of labour therefore establishes an 'opportunity cost', particularly for male labour, which is too expensive for millet growing to bear. Labour use may be reduced by planting a smaller area, or maintaining the existing area by farming it less productively. A fall in productivity may reflect a reduced ability to carry out tasks such as weeding or sowing at exactly the right time, which can severely reduce the size of the harvest. Late sowing increases the risk that the crop will reach maturity only when the rains have finished. Late weeding allows weeds to remove moisture and mineral nutrients, which are in scarce supply in the sandy soils and erratic rainfall typical of the Sahel. Productivity can also decline when labour-intensive operations, such as clearing fallows, can no longer be undertaken. This is particularly likely where fallow clearance involves tree cutting, previously undertaken by men.

In turn, inability to clear fallows may result in prolonged cultivation of existing farm plots, which requires other labour-intensive tasks to maintain productivity: constructing ridges or terraces to control soil erosion, transporting and spreading manure, or herding animals to graze stubble after harvest. Failure to apply such measures (all of which are found in records of 'traditional' farming in Africa) will lead to soil degradation. Similarly, a reduced ability to travel distances to collect fuelwood intensifies the local exploitation

of trees close to houses, and increases the risk of exposing soil to erosion by rainfall.

Figure 8.8 indicates that the decline in productivity of low-input staple food production tends to reduce the capacity of the household to produce its own food, increasing the cash requirement to cover a food deficit. It also indicates how individuals seek to escape the declining security of the household production system by diversifying income sources. Many, particularly the young, will seek non-farm, often urban, employment. Those that are successful may eventually reinvest in farming through the purchase of inputs (as we saw in the example of Zimbabwean maize production). Those who remain in rural areas may seek to increase their incomes through growing a high-value vegetable crop, or through household-based food processing, activities often taken up by women. Such small-scale producers have proved tenacious and resourceful exploiters of market opportunities, particularly those presented by the peri-urban areas around major cities, where cultivated plots are close to the consumer market.

In more distant rural areas, 'traditional' farming may survive, but in a degraded form: the level of inputs is lower than before, consisting only of the labour of those with least mobility (women with children, the old, and schoolchildren) supplemented by temporary labour during visits of those who have left to earn wages elsewhere. This level of resources is below that required to continue 'traditional' (non-commoditized) farming techniques, let alone that required to apply increased input levels. Such households will tend to have high 'dependency ratios', and will not only tend to overexploit (and thus degrade) soil and vegetation, but will also have few reserves to cope with shocks such as crop failure due to drought or pest damage, which can quickly bring them to the brink of famine.

Is this example a more valid illustration of the 'model' of African farming that is 'traditional', 'extensive', and implicitly for subsistence rather than for sale? The answer is yes and no. First, our example shows how an outwardly 'traditional' agriculture may be little more than a shell, the labour relationships which provided its vigour

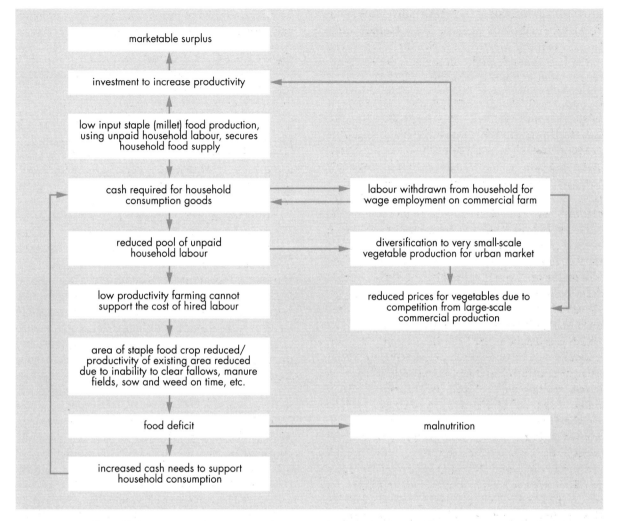

Figure 8.8   Effect of labour commoditization on low-input food production: millet cultivation in the Sahel.

having dissolved in the tide of commoditization. Such a degraded form is no more 'traditional' than high-input maize or irrigated rice in our earlier case studies. The absence of absolute landlessness in many parts of Africa allows the survival of such degraded forms of farming and may mask the deepening poverty of otherwise resource-poor households (see Chapter 9). Secondly, our millet-growing example presents no simple or direct link between population density and environmental damage: indeed, density of settlement may be high and, paradoxically, environmental degradation will result from shortages of labour. Conversely, there is little to suggest that households with high dependency ratios would grow millet more successfully if population density were lower.

The examples in Section 8.2 suggest that 'pressure on the land' may be due to *redistribution* of population as well as to overall growth. We noted the wholesale relocation of Africans as a result of European settlement in southern and eastern Africa. Even in areas with few European settlers, rural people were relocated along roadsides so that the colonial authorities could more effectively enforce the cultivation of certain crops: for example, cotton in Obangui (Central African Republic) and Mozambique. There is also the long history of migrations to areas of 'peasant' cash-crop farming, such as the cocoa- and groundnut-growing areas of West Africa.

In fact, rural populations are *mobile*, and 'population pressure' is created by population redistribution and concentration. This is particularly important in understanding famine in contemporary Africa. In 1991, most of the estimated 27 million people threatened by famine were rural refugees from armed conflict in Mozambique, Sudan, Ethiopia, Somalia and Liberia.

A final aspect raised by our discussion of low-input farming is the ambiguous role of non-farm income. In some cases non-agricultural sources of income are used to invest in raising farm productivity. In other cases, remitted income from absent household members may be insufficient even to hire labour to replace their participation in agriculture. The effect of non-farm income is evidently shaped by class: wealthier individuals able to engage in trading activities are likely to earn, and hence be able to remit, more income than poorer individuals working as labourers. There is also, however, an effect of gender: male household members' non-farm income will be consumed or invested in productive activities according to their priorities. There is no reason to suppose that investment will be made in agriculture largely undertaken by female household members, unless it results in greater control over the product by the male investor. In such circumstances, increasing commoditization of inputs to agricultural production effectively places male household heads in the role of 'capitalist' employer, and other household members – and particularly women – in the role of (unpaid) employees (see Chapters 2, 6). We shall explore further the conflict implicit in this relationship, and its consequences for social development in rural communities, in the final section.

8.4   Options for change

In the previous sections we explored the concept of 'indigenous farming' in Africa: the factors underlying its development and how it may be linked to

the degradation of land and vegetation in rural areas. This section assesses the options for change. In particular, it examines two views of the future role of indigenous technology in African farming. On one hand, a 'modernization' strategy of wholesale substitution of small-scale agriculture by large-scale, high-input 'industrial' farming; on the other hand the rejection of technology 'transfer' from abroad in favour of 'populist' solutions that rediscover and develop indigenous agricultural techniques. The section ends with a discussion of the role of 'social organization' in determining both agricultural and environmental change in rural areas of Africa.

Agricultural modernization

Large-scale commercial farming was well established in many parts of Africa early in the colonial period. Companies that are household names in Britain were directly managing large-scale agriculture in Africa from the 1920s onwards to produce rubber (Firestone), tea and coffee (Brooke Bond), palm oil (Unilever). In the post-colonial era, large commercial agriculture has continued, with multinational companies often participating in joint ventures with post-colonial governments: for example, in sugar (Kenya, Senegal, Swaziland), timber (Swaziland), cotton (Mozambique), wheat and soyabeans (Zambia). The successful application of European agricultural science to African environments in these large-scale mechanized farms has also provided a 'model' of agricultural development adopted by many post-colonial governments, pursued through setting up large-scale state farms or nationalization of existing commercial agriculture (Figure 8.9).

Many of these state initiatives to modernize African farming have been failures, however. Some of this failure has been due to poor knowledge of the local ecology because of inadequate records of rainfall, river flows or soil characteristics. More frequently it was caused by an inability to control labour. We encountered this type of problem in the large-scale irrigated schemes mentioned in Section 8.2 above. Planners underestimated the opportunity cost of the time they expected farmers to spend cultivating irrigated crops or repairing canals. Similar problems are encountered when

Figure 8.9 Large-scale commercial farming: (above) state-run tea estate in southern Rwanda; (right) irrigating tea plant seedlings in Uganda.

assumptions are made about the recruitment of seasonal labour. In this respect post-colonial government agencies are usually (not always) less willing or able than colonial authorities to coerce farmers to provide seasonal labour for large-scale farming. Inadequate control over labour leads to low productivity of other inputs. Since many of these (tractors, fertilizers, pesticides) are imported, low productivity translates into a high cost of production to be paid in convertible currency. This presents particular problems for farms producing food for local markets because they have no convertible currency income. It has frequently been observed of large-scale state farms in Africa that the convertible currency costs would provide more food if used to pay directly for food imports.

Why do foreign-owned and managed commercial farms often prosper in Africa, when state-sponsored attempts to emulate them so often fail? In the late twentieth century, commercial agriculture can offer wages which, although low by international comparisons, compare favourably with what can be earned in local small-scale farming (as in the case of the horticultural company Bud Senegal mentioned in Chapter 5). Coercion is therefore no longer required to obtain an adequate supply of labour. A major advantage of foreign-run farms is that they produce almost

exclusively for export, and hence earn convertible currency. However, even more importantly, they are frequently integrated into the same organization that controls the processing, transport and marketing chains. For a company like Brooke Bond, it matters little how much profit the tea estates make, because it will earn its main profit from selling the packed product in European markets. Even where processing of the crop is minimal, as in the vegetables exported by Bud Senegal, the close contact with the market (in

Europe) was achieved through Bud's own marketing arm in Amsterdam. Indeed, when the vegetable production company in Senegal went bankrupt, the marketing company remained successfully in business, buying vegetables from other producers. Similarly, the capacity of foreign-owned farms to obtain inputs like farm machinery directly from the countries of manufacture, rather than through the locally appointed agent (usually a monopoly supplier), enables them to cut their production costs relative to state farms.

The lesson, therefore, is that the success of capital-intensive production for the export market is determined by the degree of control which can be achieved in the overseas markets. On the whole, African producers, large and small, must rely upon foreign intermediaries in order to gain access to these markets.

Is there any prospect that large-scale, high-input farming will provide an answer to declining per capita food production in Africa? The first important conclusion is that obstacles to such a development are not primarily technical. Former problems of high rates of soil erosion, associated with plough cultivation of tropical soils, have been largely overcome through the development of tractor-drawn implements which loosen soil, remove weeds and plant seeds while allowing the maintenance of a surface mulch on the soil. To some extent, therefore, 'Western' science has learnt the techniques of African farmers. Questions of long-term feasibility remain concerning the intensive use of agrochemical toxins for pest and disease control, but no more than they do throughout the rest of the world. The main obstacles to the 'modernization' of African agriculture are economic and political. Both are connected with the slow growth of industrialization in much of Africa. As shown in Chapter 4, in many African countries during the 1980s there was actually a decline in industrial output – a de-industrialization.

Economically, this means that most of the inputs for mechanized farming on the European pattern must be imported, implying a high cost in foreign exchange. Devaluation and/or foreign exchange shortages seem likely to constrain imports in much of Africa for the foreseeable future. As a result, high-input agriculture will be similarly restricted to export-oriented production, or products commanding high prices on domestic African markets (e.g. wheat in Nigeria). The political obstacle to expansion of large-scale mechanized farming is that, in the absence of an expansion of employment opportunities in industry, it would further reduce the viability of low-input farming and increase the proportion of the rural population engaged in 'sub-subsistence' agriculture.

Indigenous technology

In Section 8.1 we considered a number of farming techniques which distinguish African from European agriculture. We have seen also, however, that new crops and implements have been incorporated into African farming. In considering the future contribution of indigenous technology, therefore, we must consider not only the conspicuous differences of technique, but a framework of agricultural management within which innovation is evaluated, adopted, modified or rejected. We may recall from our reviews of cocoa and maize farming in Section 8.2 that *control* of technology passed from farmers to governments during the 1930s. Thereafter the agenda for technology change was largely determined by colonial, and subsequently post-colonial, government research and extension services. Farmers were not entirely passive in their role as recipients of technology 'transferred' to them by these agencies. In recent years it has become widely acknowledged that failure to adopt new technology is more a reflection of the unsuitability of proposed innovations than any conservatism on the part of farmers.

There has been a parallel growth in willingness to accept farmers' participation in the process of technology development. Under the banner of 'farmer first', advocates of the increased role of farmers' knowledge and priorities in the process of technological change emphasize the role of professionals as 'facilitators', helping farmers to identify options rather than presenting them with solutions. There is undoubtedly reason to believe that closer collaboration between agricultural researchers and farmers in Africa will broaden the scope for agricultural improvement: for example, by ending the arbitrary European distinction between tree growing ('forestry') and agriculture so that trees for fodder and other purposes can be fully incorporated into strategies for growing crops.

The new-found respect among professionals for the ecological knowledge of cultivators and pastoralists opens the way to seek improvements in indigenous practices. For example, Oxfam's widely-acclaimed Projet Agro-Forestier (PAF) in Burkina Faso was able to improve an existing local technique which used stone lines laid across fields to conserve soil and water by introducing a simple system for marking out contours (Figure 8.10). Laying the stone lines along the contour made them more effective in retaining and spreading the surface movement of rainfall, and resulted in an immediate improvement in availability of moisture for crops.

Yet, in considering options for agrarian change in Africa, is it sufficient to rely on the promotion of indigenous technology? There are two reasons to conclude that it is not.

First, it is evident from the account of changes in African farming set out in Section 8.2 that the present pattern of 'indigenous farming' is in fact the product of many adaptations of 'foreign' technologies by African farmers. A development strategy which imposes a blanket rejection of non-indigenous farm technology (as is sometimes suggested in the name of 'sustainability' or self-reliance) is likely to prove as impractical as those which have in the past ignored the importance of non-agricultural or non-rural work in the livelihoods of rural people. In both cases, a stereotype imposes a false division between the urban and rural economy.

Water-lifting (pump) technology provides an example in which externally developed equipment provides the only reliable means to obtain drinking or irrigation water in many rural areas. Precise choice of technology can only be assessed with an eye to existing *linkages* in the wider economy, and few prior assumptions can be made. In many rural areas of Senegal, for example, diesel pumps present a viable option because maintenance facilities are to some extent present in the workshops of motor mechanics who service the diesel-engined minibuses and pickups of the ubiquitous *taxi-brousse* transport system. In more remote areas, such as much of Tanzania or Mozambique, poor roads and lack of vehicles lead to difficulties in obtaining spare parts, fuel and tools, and diesel pump operation can be extremely unreliable. The most graphic evidence which cautions against *a priori* reliance on indigenous

Figure 8.10 Using stone lines to conserve soil and water: the Projet Agro-Forestier in Burkina Faso.

technology is provided by the capacity to adapt unfamiliar technology displayed by individuals in rural areas: for example, a Zimbabwean photographer living in the communal lands near Masvingo prints his own photographs using an enlarger powered by a car battery; a Senegalese villager in Gorou, near Thiès, dismantled the photovoltaic solar panels of a failed irrigation pumping system and used several of the panels to provide lighting for his house and power for his television set.

A second reason for caution in expectations of indigenous technology is that 'indigenous technology' as a set of techniques (intercropping, fallowing, cultivation of wetlands, etc.) no more guarantees the viability of farming by resource-poor households than does 'commoditized' farming using purchased inputs. Overemphasis on indigenous techniques abstracted from their socio-economic context may prove to be no less technocratic – a greener 'technological fix' – than past attempts to resolve the poverty of Africa's farmers through the transfer of technology. This is signalled by Oxfam's assessment of its PAF project in Burkina Faso:

> "There is some concern that the benefits of PAF's interventions have not reached the very poorest farmers, who are unable to provide food for group labour…Improvements to private fields are not enough on their own, and each village must now take responsibility for halting degradation on all its land. Grazing areas pose a particular problem in this respect."
>
> (Critchley, 1991)

In this case, as in many other indigenous techniques that have become 'forgotten', the construction of stone lines requires considerable labour (50–100 person-days/hectare), and feasibility of their use tends to reside in rural households' access to labour. The assessment quoted above also raises the question of *shared* resources. These points lead back to the question of social relationships, access to resources, livelihood opportunities and risks, which emerged from the analysis of the causes of environmental degradation in Section 8.3 of this chapter, and form the subject of the remainder of this section.

Social organization in agrarian change

Employing the linkage between environmental decline and resource-poor farming set out in Section 8.3, we may draw on the observations on the PAF project, quoted above, to set out some key issues which need to be addressed in an agenda for agrarian change.

At the level of the *individual household*, the commoditization of labour means that, even in good years, poor households are unable to afford the cost of labour to improve the productivity of their land (as in the PAF case) by increasing the retention of rainwater on their fields. Inability to take such action increases their vulnerability to crop failure when rainfall is scarce, and thus further undermines their security.

At a wider *social* level, the application of improved stone lines and other measures on individuals' private land holdings is insufficient to manage the environment effectively. This is an observation of general importance: to be effective, environmental management must be undertaken on a sufficiently large scale. This is easily illustrated with respect to water management. Along a continuum from very wet to very dry environments, agricultural activities demand a number of different measures: drainage of waterlogged areas, protection of soil from erosion by excessive surface flow of rainwater, redistribution of water through irrigation, rainwater conservation or 'harvesting' (collection from a larger area onto a smaller, cultivated, area). In each case, it is difficult to achieve success through individual action on small areas because of the influence of neighbouring plots: seepage from surrounding waterlogged soil, or large surface flows of water from upslope areas from which vegetation has been cleared, for example.

At both levels, the implication is that social organization is crucial. We have seen how the process of commoditization has tended to 'individualize' farming by removing the reciprocal obligations of labour, or by creating patterns of unequal exchange (see also Chapter 9) between and within farming households, and that this has increased the vulnerability of poorer households. The erosion of women's independent cultivation rights under irrigation in West Africa is an example mentioned in Section 8.2. From the point of view of the most disadvantaged, therefore, the main objective of a change in rural social organization (and one which commoditized production is unable to make) is the re-establishment of such reciprocal labour relationships: a reassertion of their 'rights'. The examples quoted in Boxes 8.6 and 8.7 indicate that this process of reorganization may already be under way in some places.

Box 8.6 *Foyers* and *femmes*: irrigation groups in the Senegal river valley

Establishment of large-scale irrigation in the delta of the Senegal river in the 1970s was accompanied by land allocation to heads of households at the rate of 0.5 ha of irrigated land for every able-bodied household member. Thus, while household members were expected to work the irrigated fields, the resulting rice harvest was controlled by the land holder – the head of household. This concentration of irrigated land in the hands of older men, in an area where rain-fed farming was impracticable, provided young men with little scope for cultivation except as labourers on their older relatives' fields. This situation prompted the transformation of youth organizations, or *foyers*, from cultural groups into lobbying agencies to gain independent access to irrigated farming for young people. The *foyers* were successful in attracting support from foreign development agencies, particularly non-governmental organizations, to purchase diesel pumps and establish small-scale irrigation. From a single *foyer* in 1972, a movement spread throughout the delta and, under a loose umbrella organization, *l'Amicale des Agriculteurs de Walo*, has become a major social force in the area. Few villages are without an irrigated area divided up into individual plots to which young men and unmarried women are able to gain access through payment of subscriptions to their *foyer*. By contrast, attempts by women's organizations to acquire land to expand irrigated cultivation beyond their current microscale vegetable plots have been consistently blocked by village councils.

Box 8.7 Women's self-help groups in Kenya

From the 1950s, agricultural policy in Kenya sought to 'stabilize' the African peasantry through the development of cash cropping, and 'land consolidation' – the transfer of lineage ownership of land to individual male heads of household. Legally, the output of the land thus became the property of the landowner, despite the fact that in many parts of Kenya men leave their farms to seek wage work elsewhere, and much of the cultivation is therefore carried out by women.

In practice, the distribution of the product of the farm is the result of negotiation between male 'owner' and female 'producer'. Against this background, the strong development of women's *mwethia*, or self-help, groups may be seen as a means of reasserting women's property rights (Figure 8.11). The groups may be savings clubs, with each member in turn receiving a lump sum payment which can be used to purchase 'capital' items, such as livestock or tools, or they may be co-operative business ventures.

'...by channelling cash from crops into self-help organizations, they were preventing the appropriation of their product by their husbands, ... secondly, they were attempting to accumulate capital as a means of protecting and enhancing lost incomes' (Stamp, 1986, p.40).

In Machakos District, the women's *mwethia* have become well known for voluntarily hand-digging terraces on steeply sloping farmland. The terraces are a water-conservation measure which improves crop yields and stabilizes drinking water supply. Their construction requires a large amount of labour (150–350 person-days/hectare) and the magnitude of the *mwethia* achievement is that from the mid-1980s 1000 km of terraces were constructed each year, and by 1990 some 70% of farmland in the district was terraced.

Figure 8.11   A meeting of a Masai women's self-build co-operative in Kenya.

The examples briefly described in Boxes 8.6 and 8.7 are evidence of the emergence of social organizations to undertake agricultural and environmental management. The formation of such groups is relatively recent and their long-term development is unclear. However, we can make the following observations:

• The formation of youth or women's organizations reflects increasing conflict of interest within households brought about by changes in farming.

• The groups are formed to achieve a specific productive goal.

• They represent more homogeneous interest groups than households and hence may be expected to foster greater reciprocity of exchange between members.

• The activities of groups may effectively provide a means for prosecuting at 'community' level conflicts which individuals experience (but feel powerless to resolve) at household level.

• The organizations do not challenge individual cultivation rights, although membership of a group such as *mwethia* might imply some subordination of individual cultivation to an agenda socially determined outside the household.

The implications of these developments are complex but need to be understood if initiatives in 'village-level' environmental management, as proposed for Oxfam's PAF project above, are to have any prospect of success. That such issues will be central to future environmental management in Africa is indicated by the World Bank's recent exhortation:

"Measures are needed to ensure systematic management of forest cover. The lack of financial and other resources places this task beyond the capacity of the forestry services in the sub-Saharan countries. If it is to occur, responsibility will have to be transferred to local communities."

(World Bank, 1989a, p.102)

This clearly raises the question of whether responsibility also implies control. To whom are the 'local communities' to be accountable? And what of the social differentiation by class and gender within 'communities'? There is also an implicit assumption that 'grassroots' organizations will *conserve* the environment. However, there are also prospects that they can actively *manage* and *transform* it.

This is dramatically illustrated by the measures needed to establish agriculture in areas affected by trypanosomiasis, the sleeping sickness disease of cattle and humans carried by the tsetse fly. The entomologist John Ford provided convincing evidence that the vast, lightly populated areas of tsetse-infested Miombo woodland in the centre of present-day Tanzania were densely populated agricultural and grazing lands during the nineteenth century. Ford argues that the area reverted to forest during the early years of this century after the decimation of the cattle herds by the rinderpest epidemic in the 1890s. The ability of African farmers to re-open the forest to farming was blocked by colonial restrictions on burning the forest and on hunting the wild animals which form a reservoir of the trypanosome infection (Kjekshus, 1977). Should 'indigenous technology' be allowed to reclaim such lands lost to farming? This example emphasizes both the nature and scale of environmental management which African farmers may need to undertake if they are to improve significantly not only the productivity but also the long-term sustainability of agriculture. It also should give pause for thought to the many Europeans who consider Africa as a wilderness 'theme park'.

Summary

1 The techniques which distinguish African farming from its European counterpart can be understood as methods of maximizing labour productivity and minimizing risk under tropical rainfall conditions. The success of these techniques rests on an ability to exploit ecological diversity and on social organization which ensures the mobilization of labour for farmwork when needed.

2 By tracing the development of African farming back to the nineteenth century we can see that African agriculture is continually changing in response to the wider economy, and that apparently 'traditional' forms of farming are often manifestations of that response, rather than isolated from it.

3 Much of the change in African farming over the past century can be understood in terms of commoditization. This has proceeded unevenly, however, increasing the use of purchased inputs in some cases, while reducing the availability of farm labour in others.

4 The reduced input of unpaid household labour is an important cause of degradation in 'traditional' farming which may remain outwardly unchanged but continues only at 'sub-subsistence' levels which produce malnutrition and environmental degradation. Although crucial to the link between farming and environmental damage in Africa, this is often ignored by those who emphasize the growth of population as the cause of the failure of African farming.

5 While wholesale 'modernization' of African agriculture is not feasible on economic and political grounds, a reliance on 'indigenous technology' is equally unlikely to resolve the farming crisis in Africa if it fails to recognize the role of social organization in agricultural and environmental management. Agricultural techniques, typically the most visible aspect of farming, can only be understood and evaluated in relation to the economic goals of rural people and the social organization through which they try to achieve them.

6   In the 1990s, as development agencies seek to transfer to 'local communities' the responsibility for managing land, water and trees, it is evident that the outcome of this transfer will reflect not only the degree of control exercised by those communities, but also the social organization, increasingly differentiated by class and gender, within them.

9

CLASS FORMATION AND RURAL LIVELIHOODS: A UGANDAN CASE STUDY

MAHMOOD MAMDANI

This chapter analyses processes of class formation and their effects for rural livelihoods, illustrated by the case of a village in Uganda. The analysis tries to throw light on three key questions:

Q How is the search for a better life by individual peasant households defined by the social relations that account for their poverty in the first place?

Q What are the aggregate or social effects of the search for individual solutions to poverty?

Q Why is the pursuit of individual solutions ultimately futile for most peasant households, worsening their conditions of reproduction and livelihoods?

Here I concentrate on specific processes of commoditization, accumulation and class formation, and their conditions and effects, rather than on the details of other aspects of social organization and culture. Those are discussed from a somewhat different perspective in Chapter 10 which uses material from another village in northern Uganda.

This chapter is based on research I carried out in July of 1984 in the village of Amwoma in Lira District, northern Uganda (Figure 9.1). The village is at least 16 km from the nearest trading centre (Dokolo) and 65 km from the nearest town

(Lira). Most peasants in the village of Amwoma had access to some land which they worked with family labour. This alone distinguishes them from the tenants and landless labourers who form the majority of the rural poor in Asia and Latin America and some parts of Africa. It also makes Amwoma peasants typical of peasants in many villages in the central bulge of Africa, the vast lands that lie between the Sahara and the Limpopo.

Social life in Amwoma shared another attribute with that in much of the central bulge of Africa. This was the persistence of communal forms of labour and property. Recent understanding of these communal forms is subject to considerable controversy. For example, proponents of 'African Socialism' contend that the key to building a just and egalitarian society in Africa, one that would remain true to the 'African genius', lies in the consolidation and expansion of these communal forms.

On the other hand, the host of 'socialist experiments' in Africa has generated a contrary argument, best known through Goran Hyden's *Beyond Ujamaa* (1980). Hyden turns the central tenet of 'African Socialism' on its head, arguing that these communal survivals in rural Africa are the real obstacle to its economic development. The African peasantry is said to retain an independence that its counterparts in Asia and Latin

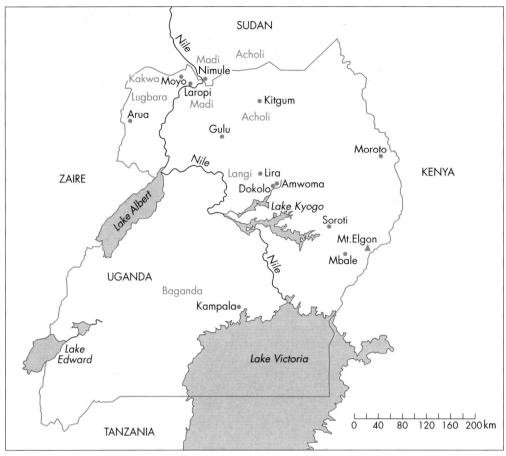

Figure 9.1   Map of Uganda. Names of the ethnic groups in northern Uganda are printed in green.

America have long lost. Consequently, the only way forward for Africa must be to 'capture' this 'uncaptured' peasantry or, to use a less militarist language, to integrate it fully into commodity production and market exchange so that it can serve the interests of capital accumulation.

Both schools of thought begin with the assumption that communal Africa today survives independently of, and as a refuge from, the world of capitalism. The former celebrates this, while the latter decries it, but both share the same conceptual failure: neither is able to grasp processes of differentiation and accumulation taking place behind – or even through – the appearance of 'traditional' communal forms, which this chapter suggests is the case in Amwoma, and by extension elsewhere in Africa (Richards, 1983) (Figure 9.2).

9.1 Organization of land

The main features of the pre-colonial village system survived into the colonial period. Although all land was declared 'Crown Land' and legally belonged to the colonial state, access to it was governed by 'customary tenure'. In Lira District, where Amwoma is situated, prevailing custom was marked by three features.

First, proprietary rights in land, including grazing and water rights, were held by the village. Unoccupied land was free for all to graze; so were swamps which were also considered common land and were grazed in the dry season. Second, the individual household's right to land was a usufruct right – individual households could cultivate the land and pass it on to their heirs or successors, but could not sell or rent it. Finally, there were two

Figure 9.2   A farm work group opening a new field in Moyo District, Uganda. Communal farm work groups are common in rural Africa, but far from being survivals from the past, they may be a response to recent processes of differentiation and accumulation.

types of individual proprietary rights. The right over ant-hills and shea butter-nut trees belonged to the man who first discovered them. It continued to be his personal right even if he migrated, unless he renounced it publicly. There was also a right over certain hunting areas (called *arum*) passed on as hereditary property. This right was not equivalent to absolute ownership, however, as anyone wishing to settle on and cultivate any part of the land only had to request 'permission' from the 'owner'. And requesting 'permission' was considered merely a matter of courtesy; it could not be refused since the 'owner' had no right to the land itself, only to hunting on it. Thus, as population and 'requests' for settlement increased, hunting rights diminished and withered away.

No doubt population increase played an important role in stimulating changes in production practices. As local migrations declined, fallow agriculture replaced shifting cultivation and peoples settled down. Land boundaries began to be recognized in relatively thickly populated areas. As a result, usufruct right of individual families became consolidated, no doubt at first for those influential families whose positions in the colonial administration meant they were most likely to possess larger pieces of land and the necessary instruments (particularly the ox plough) to cultivate it.

The introduction of more individual and private land tenure (Box 9.1) in Lira District was accelerated by a series of laws introduced in response to the anti-colonial movement of the 1940s. These 'reforms' continued after independence in 1962, culminating in the Land Reform Decree of the Amin regime in 1975. The Decree abolished 'absolute ownership of the land' (freehold land, in Uganda held predominantly in Buganda as a result of the 1900 Agreement between the British colonial state and the Baganda chiefs), and also 'the power of the customary tenant to stand in the way of development'. All land was henceforth to be held on a 99-year lease. In legal theory, if not necessarily in practice, the 'customary tenant' thus became a 'tenant-at-will' of the state. The District Land Commission, a body comprising district level state agents and local notables, was empowered to terminate any lease on 'undeveloped' land and grant it to a potential 'developer', who was 'free to evict any tenant occupying any part of the leasehold granted to enable him to develop the land.'

On the face of it, the 1975 Decree seemed designed to clear the mesh of pre-capitalist ('traditional' or communal) relations blocking the path of capitalist development in agriculture. But, as we shall see, its main effect was to accelerate the tendency to land enclosure, without any transformation in the pattern of land use. In fact, it became a weapon in the hands of bureaucrat capitalists seeking to expand their property and activities in the countryside (see Section 9.5).

9.2 The organization of labour

According to elderly villagers – and also British colonial records – agricultural work in the pre-colonial period was organized mainly through co-operative labour teams called *wang tic* (pronounced *wang tich*, meaning 'place of work'). This type of organization is best understood as a response to conditions where land was relatively plentiful, population sparse, the environment

---

Box 9.1   Concepts of land tenure

In Western capitalist societies land is considered property like any other form of property: (a) it is privately owned, (b) its owners have an exclusive right (subject only to legal restriction) to determine what it is used for, and how, (c) its owners have an exclusive right to dispose of it, whether through sale, renting, inheritance, and so on.

As you have seen throughout this book, the question of land is central to rural livelihoods: who has access to land? how much land? what kind of land? on what terms? how do they use it? The distribution and uses of land are a central aspect of agrarian structures and change, and how they affect rural livelihoods, as Chapters 2–4 showed.

This chapter illustrates several dimensions of land tenure and conceptions and practices concerning land which are different from contemporary Western views of land as private property.

First, in the pre-colonial period land tenure was communal: land was held by the village or clan and allocated for use by individual households, on a *usufruct* basis. This means that members of the community had a right to land to farm and feed themselves, but not an individual *proprietary* (or property) right to land that could be transferred through inheritance or sale.

Second, historical change in the colonial period in northern Uganda led to practices of *de facto* individual inheritance, although still on the basis of usufruct: households (especially wealthier ones) started to pass on the land they were

farming to their heirs. On the other hand, in Buganda (southern Uganda) the 1900 Agreement between the British and Baganda chiefs had granted the latter ownership of land as freehold private property, an exceptional case in colonial Africa as far as Africans (rather than whites) were concerned.

Third, there were a series of measures from the later colonial period moving land tenure towards the framework of private property, albeit of a leasehold kind in the Land Reform Decree of 1975. The general direction of change in land tenure, then, was first towards the *individualization* of landholding across generations, and second towards its *privatization*.

The individualization and privatization of land use raise highly topical and controversial issues in Third World countries, not least as they affect *'common property resources'* (CPRs). This refers to areas of common forest, grazing land, water sources for drinking, irrigation, fishing, etc., which are often central to the livelihoods of the rural poor. The productivity of CPRs is affected adversely by overexploitation (undermining the reproduction of nature), whether arising from the pressures of rural poverty or increasing commercial use. Overexploitation and environmental degradation are intensified when the access of the rural poor is reduced by privatization of CPRs, whether *de facto* (through enclosure) or *de jure* (granting of exclusive rights of commercial use to individuals or companies) – see the section below on 'ecological crisis'.

relatively harsh, and productive technology relatively undeveloped. The character of such communal work groups was social as well as economic. Not only was the labour process communal, but so was the drinking that followed: these groups cultivated the land and maintained the cohesion of the labouring community.

The communal groups worked on the farms of their individual members. The groups were more or less permanent, though their numbers and methods of apportioning work varied. The 'host' of the day always provided either 'beer' (*malwa*: a traditional millet brew) or food at the end of the day. Regulated by custom, this was regarded more as a token of appreciation than as a payment for labour performed. The important requisite for labour and its benefits to be equally apportioned was a relative absence of differentiation, and a general equality of conditions between families. The point is that every family could afford to serve the food or beer to the work group when its turn came, and so every family could benefit more or less equally from communal labour.

Colonialism introduced two important changes in this system. The first, though of lesser significance, bent the system sufficiently to serve the immediate interests of colonialism: block farms were created outside the village where every peasant family was allocated an acre (0.4 ha) of land. *Wang tic* labour was then concentrated on these contiguous farms and required to produce a combination of export crops (cotton) and food crops (at first, millet and sesame, later cassava). The areas for these block farms were chosen by colonially appointed village chiefs, but individual plots were allocated by traditional *wang tic* leaders. State-appointed Agricultural Assistants supervised peasants to ensure that the 'right' crops were indeed planted. A blend was created whereby the communal labour system was reorganized to serve the interests of export production, and communal authority was subordinated to state authority.

The second major change was the result of socioeconomic processes. Increasing population, the introduction of commodity production, and state-enforced exactions (taxation), stimulated processes of differentiation among peasant families. Some became too poor to afford the beer necessary to draw on *wang tic* labour. Initially, such a failure must have appeared as an individual and a more-or-less accidental circumstance. And so the response to it was to find compensatory mechanisms characteristic of a classless 'tribal' community to help its individual members in difficulty. But the tendencies to differentiation and impoverishment, though developing gradually over a long period, crystallized at times of crisis. In the drought years of the 1960s, there was a dramatic rise in the number of peasant households unable to brew the beer necessary to have access to *wang tic* labour teams. What had seemed to be exceptional cases before suddenly appeared as a rule.

The impact of the drought years on peasant communities was contradictory. It highlighted the circumstance not only of impoverished households that continued to labour in *wang tic* teams while being unable to 'host' them, but also of a few rich peasant families who had ceased to labour in *wang tic* teams while 'hosting' them with greater frequency. There was an increasing demarcation between those who 'participated' in *wang tic* teams and those who 'hosted' them. The social content of the *wang tic* was changing in these new conditions: it was being transformed from a communal labour team into a *disguised* form of group wage labour, whose payment was fixed by custom. Finally, alongside *wang tic* labour, there also began to develop *leja leja* (retail) labour, in effect individual piece-work in return for cash or cigarettes.

No wonder that the *wang tic* system was reorganized following the drought years of the 1960s. The old *wang tic* now became just an administrative subdivision inside the village, called together only to perform compulsory labour for state authorities (e.g. clearing roads, digging wells). In place of the *wang tic* in farming, a different kind of labour team emerged, called *awak* (or *alulu*). No longer was every family in a *wang tic* compelled to participate in these new teams. In fact, a single *awak* could comprise members from different *wang tics*. The *awak* also did away with the customary appreciation (beer or food). It required that each of its members got a regular turn to benefit from the labour of the team in a sequence determined by lottery, and that each member did a fixed amount of labour every time the *awak* met.

The *awak*, therefore, was not a labouring association of those residing in a common territory, as the *wang tic* was. *Awak* members shared a *common socio-economic position*, not a common locality. The *awak*, with its membership of usually 15 to 20 people, both male and female, was a *labouring association of the rural poor* (of poor and lower middle peasants – see below), consciously defined as a class organization. The literal translation of *awak* is 'team work', but its colloquial meaning is more like 'struggle'. The names of *awak* teams often evoked the common situation of their members, for example, *Ongol-Pyare*, meaning 'broken from the waist' through excessive bending as they dug the land.

Further, labour discipline inside the *awak* was much tighter than in the *wang tic*. Labour required of members was strictly regulated by well-defined rules. Work was done daily, except Sundays, in the first quarter of the day, from 6 to 10 in the morning, when the peasant was at his/her most vigorous, showing that it was given top priority. The work done in the *awak* was also strictly defined. Poor peasants usually carried sticks a metre in length. Field work required of each member in an *awak* was measured as 1 stick × 100. But an *awak* could also take on non-agricultural work like building a hut or digging a pit latrine. Variations on the *awak* included the 'children's *awak*' for those below 12 who took it in turn to cultivate their parents' gardens, and the *awak gedde* ('struggle on a small scale') which was organized between a small group of poor peasants, usually four or five neighbours, for two hours in the evening, and had rules less strict than that of the *awak* proper.

The latter part of the 1970s also saw the emergence of a new labour team, the *akiba*, usually but mistakenly referred to as a 'drinking group'. The *akiba* was really an adulteration of the original *wang tic*. It worked for beer or for money, for

Figure 9.3   Many farm work groups in Africa are not based on kinship relations and reciprocity but on payment. Each man or woman is given a particular section of the field to dig and in return may receive money or food and drink. Here an Acholi work group in Sudan takes a break to drink a bucket of locally made beer provided by the field's owner.

members or for non-members (Figure 9.3). The sharper the social differentiation of a village, the more the *akiba* worked for non-members.

If payment was in beer, it was drunk at the end of the day (a member 'host', though, might be allowed to postpone payment to another day). Non-members, meaning those who did not labour, might also join in the drinking, but not as they did 'traditionally' when this was simply a matter of courtesy and hospitality: now they had to pay. The *akiba* kept its money, either from cash payment by 'hosts', or from payments by non-members who joined in drinking, up to the end of the year (November). Then, a part was usually spent to purchase a bull or a goat, and the meat shared for Christmas celebrations, whereas the rest was divided as cash among members, normally to pay tax and/or to buy clothes.

Like the *awak*, the *akiba* was a class association exclusively of poor and lower middle peasants, with a similar strength of between 15 and 20 members, male and female. Is the formation of such a thinly disguised wage labour team like the *akiba*, so soon after the demise of the *wang tic*, accidental? Or is it a response to the socio-economic conditions in which the rural poor found themselves? A return to the discussion of the *awak* will help answer this question.

The exploitation of labour

The *awak* was an attempt by the labouring poor to put communal labour on a new footing, to mobilize the labour needed to grow food on their own farms for their own consumption. But it failed in this because the rural poor were firmly tied to the wider capitalist-dominated market, and were subject to state demands. Every peasant household was constantly in need of cash, either to purchase consumption needs like matches, salt, paraffin, cloth or medicine, or to buy producer necessities like a hoe or a knife, or to pay compulsory dues like the annual graduated tax or a 'contribution' to a state-sanctioned fund.

Peasants faced with an urgent need for cash would exercise their 'freedom' to take the *awak* team with them on the day of their turn to a capitalist employer, for whom each member performed the required labour (1 × 100 sticks), and then used the group wage for the day to meet some urgent need. It was, of course, a rare poor peasant who would admit to doing this – for it was generally considered a deplorable 'corruption' – but many would point out that others

did resort to this practice, and increasingly so. Similarly, capitalist employers had no qualms in admitting how often they employed *awak* teams. In the drought of 1980, in fact, this practice became so prevalent that the nature of the '*awak*" as a 'disguised' form of group wage labour, rather than a mutual aid team of peasant cultivators, was the more evident (Figure 9.4).

Not all poor peasants (Box 9.2) belonged to an *awak* or *akiba*. In fact, in the neighbourhood where I carried out comprehensive house-to-house interviews, only 58% belonged to an *awak* and 38% to an *akiba*. Those who did not belong were usually the poorest of the poor, who had an almost daily need for cash to buy food as well as other necessities. These peasants gathered at the homestead of capitalist farmers every morning around 6, to labour for cash or salt, millet, cassava, or posho (maize-meal), as might be the need of the day. They were only a step away from being full field labourers in the service of capital.

The rural poor in Amwoma were driven to wage labour, open or disguised – or to other exploitative relations such as renting labour implements – not because they were land poor but because they were implement poor. Because of a lack of tools, many poorer households failed to cultivate effectively the land they did possess, as Table 9.1 shows.

Figure 9.4   Relatively well-off farmers will often employ their poorer neighbours and relatives. Here a woman and her father-in-law stand at the edge of a large field which has just been cleared for them by a farm work group. Food was scarce in northern Uganda at the time (1989), so they were able to attract enough workers by paying them with pumpkins which they had grown in the riverine field behind them.

Box 9.2   Social differentiation: peasant strata in Amwoma

The concept of the differentiation of the peas-antry was introduced in Chapter 2, and applied and illustrated in a number of subsequent chapters. In my research in Amwoma the principal criterion I used in distinguishing strata of the peasantry arising from differentiation was that of 'exploitation'.

My starting point was the 'middle peasantry' (here further narrowed to 'average middle peasantry'), which used family labour with family implements on family land, and was thus not involved in exploitative relations outside the household. The progressive disintegration of this organic unity of property and labour led to the differentiation of the peasantry. Thus, at one end of this spectrum I looked for those peasant households ('poor peasantry') who were compelled to cover a regular deficit from their production by entering into unequal relations outside the household: whether by selling labour power, or by renting in land or implements of labour. Correspondingly, at the other end of the spectrum I found peasant households ('rich peasants') with a *regular surplus* that allowed them to augment family income through exploitative relations outside the household (employment, rent, etc.). In this sense, then, 'lower middle peasant' households experienced an *occasional* as opposed to a regular deficit, whereas 'upper middle peasant' households enjoyed an occasional as opposed to a regular surplus.

Table 9.1   Land owned and cultivated, Amwoma, 1984 (in acres)

	Land owned (per household)	Land cultivated (per household)	Land cultivated as % of owned
Poor and lower middle peasants	4.3	1.9	44
Middle (average and upper middle) peasants	4.2	3.0	71

My estimation of land ownership was necessarily crude. Only capitalist farmers and a few rich peasants possess land titles based on surveys. In all other instances, I had to rely on a combination of peasants' own assessment of their land size and my estimate of it. For purposes of an eye estimate, it may help to keep in mind that an acre is equivalent to the size of a soccer field (0.4 ha).

Poor and lower middle peasants owned as much land as the rest of the middle peasants but cultivated only 65% of that cultivated by the latter, because of their lack of access to the necessary tools. Poor households in Amwoma owned, on average, 1.74 hoes for 2.59 workers, and therefore could not utilize family labour to capacity. Moreover, the second hoe was usually over two years old, with at least a third of the blade worn off. As a result, the hoe was wielded with no more than 50% effectiveness!

Having examined relations *between* labouring households, we can now turn to examine relations *within* households in different peasant strata.

9.3   Relations within households

Middle peasant households

The real repository of 'tradition' in the countryside was the middle peasantry. What was known as the 'traditional' division of labour, the division

Box 9.3   Gender relations and peasant households

Everyone who does research in rural areas shares two experiences: the stimulation of coming across the unexpected and learning from people in the villages, and the frustration of realizing later that there are important things one should have tried to find out but overlooked.

In the case of my research in Amwoma it became clearer to me later that there are a number of issues I should have tried to investigate concerning gender relations and how

households were organized. In this section I describe and analyse what I did learn, but inevitably there are some major gaps.

When you have read this section, it would be useful to make a list identifying what you think the gaps are. In effect, if you were to do research in Amwoma to bring my (incomplete) story of the village up to date, what questions would you ask about gender relations? You can check back with Chapters 5 and 6 to help you with this.

of work along sexual lines into 'male' and 'female' tasks – was more characteristic of middle peasant families than of any other peasant families.

This division could best be characterized as between 'heavy' work and 'light' work (not to be confused with easy work since it could be just as demanding). Tasks such as clearing the land, building shelter, hunting and herding were customarily defined as male. Weeding, fetching water and firewood, drying and threshing grains, cooking, and so on, were consigned to women. Tasks like planting and harvesting where labour needs were dictated by the rhythms of nature, and which had to be done quickly, were often shared by both sexes.

But this 'tradition' had changed over time. In the colonial period, it became customary to speak of 'male crops' as opposed to 'female crops', because taxes were levied exclusively on adult males who were recognized by colonial authorities as the undisputed heads of households. Export crops (cotton, coffee, tobacco), sold for cash to pay tax, were thus classified as 'male' crops, while foods grown for home consumption became 'female' crops.

The development of towns and industry, and the consequent growth of a limited domestic market in food, generated a new situation in which food crops as well as industrial (export) crops became commodities. Often, as in Amwoma, when they had a chance peasants preferred to grow food for sale for immediate payment than export crops

which they had to sell to a co-operative in return for a deferred payment. Now that food crops were potential commodities as well, no longer were certain crops 'male' and others 'female'. Once again, it was more likely that particular *tasks* were defined as 'male' or 'female'.

At the time of my fieldwork in Amwoma, however, even this division was no longer rigid. It was not uncommon to find a woman helping her husband clear the land and a man helping his wife weed, although the woman was more likely to help by weeding than by more difficult clearing work and the man was more likely to assist in weeding beans, planted well spaced, than millet, which was more laborious to weed since it was planted closer together.

What was evident about such 'traditional' gender divisions of labour was their unequal character, especially with changing conditions. As population density increased, with a change from shifting cultivation to fallow agriculture, 'male' tasks like hunting had become obsolete, and those like clearing bush had become rare. As impoverishment became more widespread, farm animals tended to become concentrated in richer strata of the peasantry, reducing the weight of herding as a male activity in the middle peasantry. And yet traditionally 'female' activities, combining farm work, house work and other 'domestic labour' like fetching water and firewood, had put the principal burden of labour in middle peasant households predominantly on women.

This issue has its other side: though unequal, the sexual division of labour in the middle peasant family was a complementary one. While the unequal distribution of the labour burden might give rise to tensions inside a middle peasant family, its complementarity also created cohesion within the family. The middle peasant household was both a unit of consumption and of production (and, like all other families, of the reproduction of labour power), combining labour and property. Common labour and common property were the material bases of the relative internal cohesion of middle peasant households.

It did not take a visitor long to realize that the most stable families in the countryside were found among the middle peasantry. The middle peasant woman was also likely to be more assertive within the family than her rich peasant counterpart. She was more likely to have some say over the disposal of family money and property, to keep a relatively effective check over the husband's drinking habits, and to voice her opinion, at times even in conflict with that of the husband, to an outsider.

Poor peasant households

The social basis and practices of middle peasant households contrasted sharply with internal family relations in other strata of the peasantry. Among the poor peasantry, the burden of abject poverty increasingly tore apart the household as a production unit. First the husband, and later the wife, would move into wage labour. The family farm, progressively neglected, would become the exclusive responsibility of the wife and the younger children.

The family became the focal point where all social contradictions experienced by its individual members became concentrated, as the frustrations of a wretched social existence were brought home. Husbands drunk on cheap but strong country brew would come home at night to find an empty saucepan and would beat up their wives for lack of food. As the woman moved into occasional wage labour, at first to get food and daily necessities like salt, she would also begin to drift away and to drink, although on a smaller scale. The husband, in turn, would use all the power at his disposal, including brute force, to get her to stay home, to work in 'her' garden and not someone else's.

It should not be surprising, then, that the most unstable marriages were found among the poor peasantry. Of the poor peasant homes I visited, a quarter had experienced at least one case of separation. In every instance, the woman had eloped with another man, usually a fellow poor peasant, in order to get away, because to return to her parental home was a slim possibility. It would mean that her father would have to return the brideprice at the time of marriage, an unlikely occurrence as the cattle obtained as brideprice would have been used again as brideprice to obtain a wife for a son, or sold in times of acute distress.

Poor peasant families were volatile and fragmented. Without any appreciable family property under the husband's control, there was only social pressure to tie the wife to the husband. The high separation rate among poor peasant families was not simply evidence of a crisis in the family: that it was nearly always the wife who initiated separation also suggests that poor peasant women possessed greater freedom of action than their middle peasant sisters who enjoyed a greater measure of material security. This recalls the dilemma of women farmers in Africa noted in Chapter 4.

Some of the characteristics of poor peasant households were even more marked among landless workers, but with sometimes different effects. Landless households had little economic autonomy, and were subordinate to the power of capital. They owned no property, except petty consumption goods like a few items of clothing or a pair of sandals carved out of discarded rubber tyres. There was neither a family kitchen nor a family farm. Both the wife and the husband were hired hands, at the command and call of their capitalist employers. The sexual division of labour, too, had only a residual significance. A male might be asked to do 'female' work like weeding millet or fetching firewood and water; similarly, a woman could be commanded to join in clearing the land. Divisions that did persist were confined to activities which called for skills developed over time: e.g. cooking (female) or herding (male).

Landless families differed from those of poor peasants in another important respect. Here, there was no productive property to be passed on: neither land, nor implements, nor cattle. When a man wed, it was the master who paid, not he. And yet, the family of the land labourer was not separated from the rest of society by a Chinese Wall; the ideology of male supremacy as the 'natural' order of things was likely to prevail within landless as well as peasant households.

Rich peasant households

On the other side of the middle peasant we find rich peasants: that minority which was successful in accumulating, whose property was not simply the result of family labour but also of exploitation outside the household. It was only rich peasant men (and, exceptionally, upper middle peasant men) who could afford to marry more than one wife.

A wife might be acquired in one of two ways: by paying a brideprice or inheriting the wife of a deceased relative. The first type of wife was called *dako* (pronounced daho), the second *lako* (pronounced laho). Traditionally, wives (like land) belonged to the clan, not just to the husband. Upon her husband's death, a woman was thus inherited by his closest relative. The only exception was when a woman had no children: if she was too newly wed to have borne any, her family could return the brideprice and recall her; if she was beyond childbearing age, she was free to cohabit with any man of any clan, without any repayment of brideprice. But should she have children, and still be of childbearing age, her family would have to forfeit not only the brideprice but also her children for her to avoid the obligation of remarriage.

The *lako*, however, enjoyed a relative freedom. Whereas the *dako* was bound to the man who paid brideprice, the *lako* was free to reject a clansman cohabiting with her on grounds of

cruelty or negligence, and take another man. It was this relative freedom that rich peasant men had seized on to monopolize control over all the *lako* in the clan, thereby making all wives (*dako*) potentially theirs! Simply put, whereas traditionally the *lako* was inherited by the nearest clansman (according to blood ties), now she was much more likely to be inherited by one of the richer clansman.

The marriage of a second wife by a rich peasant (as will become even clearer in the case of a capitalist) was really the expansion of management capacity as his enterprise expanded. Here, too, 'traditional' sexual divisions of labour broke down as the wife began to take charge of supervising 'her' labourers, at first in the farm, later in the kitchen.

This process was completed with the family's withdrawal from labour. At this point, its income no longer came from two contradictory sources – family labour and exploitation – but from exploitation, the labour of others, alone. Here, then, we have the *capitalist* family. Marriage now was more subordinate to the dictates of capital accumulation than with middle or rich peasants. The husband might marry two or three times, with each wife the focal point of a distinct activity, a distinct investment. One might supervise farm labour in the village, a second might manage her husband's shop in a nearby trading centre, while a third might stay with the husband who might have a lucrative position in the city.

A husband could shed a wife as he would a loss-making activity. In contrast to the poor peasant case, it was usually the husband who divorced the wife because her failure to do her 'duty' – that is, beget children – or her opposition to his marrying yet again. Acquiring wives and begetting children was the basic capitalist strategy for expanding family management, in the short-run and long-run respectively, rather than drafting other relatives into the family enterprise. In fact, there was the story of a rich peasant who made the 'mistake' of relying on a brother to supervise an enterprise, only to be 'cheated' and to find himself impoverished.

Figure 9.5 Nowadays most families in northern Uganda have very few cattle. Many have none. Some rich farmers buy animals as an investment and employ poor relatives to tend them.

It was far more likely that a capitalist would look upon his poor relatives as more a source of cheap, compliant and reliable labour than anything else (Figure 9.5). The history of propertied households in Amwoma suggests that they first exploited the ostensibly 'traditional' ties of family and clan. Their initial sources of full-time labourers were inevitably destitute relatives, invited and attracted to the homestead of their capitalist kinsman. Uncles or cousins, male or female, young or old, were all welcome and, without exception, were integrated into the household as workers in the field or the kitchen. Like landless labourers similarly integrated into the household, they referred to the male employer as 'father' and the female employer as 'mother'. Their real position was, in fact, the same as that of non-related landless labourers, with whom they worked, ate, drank and slept. (Compare this with the example from Peru in Chapter 6.)

Generational differences

As with gender, generational differences were marked by the different class situations of peasant families. True, in peasant culture generally, wisdom is said to be gained more through age and experience than through formal education, and the cultural ideal is for parental authority to reign supreme: the parent commands, the child obeys.

However, in poor peasant households with not much property to inherit, the growing child sees little 'future' with the family. In time, the silent submission of a minor is most likely to be transformed into the open rebellion of a young adult. On the other hand, in a capitalist rural family, a child – and especially a son – learns from the outset that he is the unquestioned master of the world around him. When he cries, everything stops and he is listened to. When he demands, older but poor people (employees of the home) pay heed. The child does not take long to become conscious of the fact that he belongs to the upper class and that this carries definite powers and privileges. In Western terms, he is a 'spoilt' child, unlike the children of poor peasant and landless households who learn that the world is harsh, and that the only way they will gain anything from it is through hard work.

9.4 Accumulation from below

Accumulation from below was the result of competition among petty commodity producers and their class differentiation, whereby rich peasants emerged from the upper ranks of the middle peasantry. This type of accumulation marked the least rupture with existing social relations. The exploitation involved was disguised because of the apparent continuation of customary co-operative practices. But practices that were co-operative in form ceased to be so in content once they linked households in unequal positions.

Consider the example of three households co-operating in herding their cattle; a poor peasant household with two cows, a middle peasant with eight cows, and a rich peasant with 24 cows. They pool the cows and build a common *kraal* (pen) near the home of the rich peasant. They rotate herding, each being responsible for ten days in turn. The labour is shared equally but the ownership of cows is not equal, with the result shown in Table 9.2.

Not only are the benefits of co-operation shared unequally, so are the risks should any cows stray into nearby farms and damage crops. From the

Table 9.2 Ownership of cows and contribution of labour in one herding pool

Peasant household stratum	Cows in pool, number (%)	Labour by contribution per turn, days (%)
Poor peasant	2 (5.9)	10 (33.3)
Middle peasant	8 (23.5)	10 (33.3)
Rich peasant	24 (70.6)	10 (33.3)

point of view of a poor (or middle) peasant, it is of course better to herd 34 cows for ten days in a month than to herd two (or eight cows) every day of the month! But the equal sharing of labour disguises the unequal benefits to each household. In practice, this form of 'co-operation' is really a transfer of unpaid labour from poor and middle peasant households to rich peasant households.

Such unequal relations, whether open or disguised, can develop in relation to each of the major productive forces: implements of labour, land and labour. A major implement of labour like a plough may be rented or 'shared'. When rented, its payment is open for all to see: in Amwoma in 1984 a day's rent was 1500 shillings for a plough, four oxen and two to three labourers. This practice, however, was rare; only an upper middle peasant could afford such a cash payment (Figure 9.6).

More usual was a practice called 'plough sharing' between a capitalist farmer (or rich peasant) and a poor peasant. The capitalist farmer provided a plough, the oxen and at times even a labourer. The poor peasant household provided two members for the ploughing team. Together, they ploughed the lands of the capitalist farmer and the poor peasant in turn. We can note two aspects of this exchange. First, the land of the capitalist farmer would customarily equal six gardens (a garden is one acre, 0.4 ha), but that of the poor peasant would equal only one garden. While the poor peasant household held only 15% of the total land ploughed, it provided as much as 65% of the labour for ploughing! Second, the land of the capitalist farmer was ploughed first, to catch the

Figure 9.6 In much of northern Uganda trained oxen are at a premium. The owners of teams are able to hire them out to neighbours with the money to pay.

rains in time; that of the poor peasant suffered from late ploughing.

The starting point of capital accumulation from below through peasant differentiation was in farming. It was petty accumulation through petty exploitation of the type illustrated. The petty profits so accumulated by rich peasants (combining family labour with exploitation of others through employing their labour, or 'co-operative' herding or ploughing arrangements) were then invested in trade.

The trade a rich peasant could enter with the least difficulty was one that allowed small-scale operation, and was restricted to commodities destined for the local market (e.g. chickens in Amwoma). The outlay required was small, and the licence to trade could be procured from local officials for a small fee.

The next step would be a wholesale operation in the wider internal market. The best example was the cattle trade. Cattle were usually bought one at a time, typically from peasant households desperate for cash, and transported by lorry to urban markets (Busia, Mukono, Kampala). Profits were substantial but were bagged mainly by the transport owner, not the trader. Experience made village capitalists acutely conscious of the need for their own transport, but they also realized that it was not possible to get a big enough bank loan to purchase adequate transport without a state connection. At this stage, their problem was political, a point I return to.

9.5 Accumulation from above

The process of accumulation from above, where a state connection is vital from the outset, was more characteristic of the colonial and post-colonial periods. This type of entrepreneurial activity by bureaucrat capital has a monopoly-like character from the beginning (Box 9.4). Both of the bureaucrat capitalists in Amwoma (which also had one capitalist who emerged from below, from the rich peasantry) were necessarily absentee capitalists; both held important state positions and resided in urban areas. In both cases, their village operations were managed by one of their wives.

In the octopus-like movement of bureaucrat capital into the countryside, we can identify at least four important moments. The *first* involves control over substantial tracts of land, which had a major boost from the Land Reform Decree of 1975. Because clan identities and practices were strong, and the buying and selling of land was a relatively recent phenomenon, the bureaucrat capitalist could not begin by 'buying' land – he had to 'claim' it. This was only possible if the aspiring capitalist returned to 'his own area', to the village of his forefathers, to lay claim to unused land (say, 20 to 40 hectares) as his 'clan right'. The verification of his claim then depended not on the clan, but on preferential treatment by district chiefs and other state officials who comprised the District Land Committee.

Once its physical presence was thus established in the village, bureaucrat capital was well placed to turn to advantage every local development, especially any crises, whether natural or social. The role of crises in facilitating the expansion of capital cannot be overemphasized. The wife of a bureaucrat capitalist who managed the family's village enterprises, when asked to identify their critical period of accumulation in Amwoma, responded: 'What helped us was the famine of 1980. People were hungry and they sold us things cheaply. We could buy land at 250 to 300 shillings an acre and a cow at 2000 shillings. That is when we really started buying.'

Box 9.4   Bureaucrat capital

I use the term 'bureaucrat capital' to refer to capital established and accumulated through political connections with the state bureaucracy, most evidently and immediately by senior bureaucrats (and politicians) themselves. 'Bureaucratic rents' through taxation of the peasantry (see Box 4.2 earlier) and other mechanisms (state enterprises, control of external trade, licensing of internal trade and investment) are the collective basis for bureaucrat capital, which at the same time represents *individual* appropriation of state revenues and opportunities for accumulation. In this sense,

individual bureaucrat capital feeds off 'state capitalism' (state property, contracts, economic controls), at the expense of workers and peasants.

What distinguishes bureaucrat capital from other forms of capital is that it requires a privileged political connection with the state. This implies that other forms of indigenous capital, like the village bourgeoisie of rural capitalists and rich peasants, require similar political connections to expand the scale of their accumulation beyond its local origins.

Every crisis pushed forward processes of class differentiation in the village. Every crisis made clear the distinction between the vast majority who had to pay the price of resolving it, and the tiny minority that had the means to turn it into an opportunity for profit. Land and cattle, neither ordinarily considered by peasants as commodities for sale, were both forcibly converted into commodities and sold in times of crisis. This had advanced so far in Amwoma by 1984 that, in what was considered a cattle-keeping society, 83% of the peasants in the village did not own a single cow!

Though most evident at times of social crisis, this tendency was also manifest in the daily crises of poor households. Hardly a morning would go by without a poor peasant in distress appearing at the compound of a bureaucrat capitalist with a cock, or a goat, or a cow, or even land to sell. The compound of a capitalist was like a magnet. It would attract everyone in the village who came to pawn or sell whatever miserable property he/she might still possess to satisfy an immediate need: for food grains, salt, a hoe, medicine, or cash.

Every peasant intuitively understands the difference between a peasant compound and that of a capitalist. Peasants know that with a capitalist there is never a question of lack of money to buy something, nor the fear that the thing may not be of use to the capitalist. They understand that what matters to the capitalist is not the concrete use value of a thing but its exchange value (Chapter 2). They know that the capitalist does not approach exchange in the same manner: the peasant sells to buy some necessary commodity for its use value in consumption; the capitalist buys to sell, his/her eye fixed on exchange value and profit (Figure 9.7). The poor peasant is compelled by the needs of minimum consumption, the capitalist by the desire for maximum accumulation.

The movement of bureaucrat capital into rural areas to control substantial plots of land was not to develop farming but to get access to legal title which could be presented to a bank – usually, the state bank (Uganda Commercial Bank) – as security for a loan to buy transport, usually a lorry. This is the *second* key moment.

The *third* moment comes with movement into trade, the most lucrative in this area being the grain trade (in millet). The price paid to farmers for a kilogram of fresh millet in weekly markets in July 1984, immediately following the harvest, was 50 shillings a kilogram. The price the same millet fetched in Kampala at wholesale markets was 200 shillings a kilogram. If hired, the means of transport would have cost 12 shillings a kilo. Thus, while transport costs represented 6%, and

Figure 9.7   A new tractor at work in northern Uganda, 1989. It is much easier to mechanize ploughing than weeding. Sometimes a rich farmer will hire a tractor in order to plant a large area, and then employ local people (including his own relatives) as weeders and harvesters.

the peasant got 25%, the bureaucrat capitalist trader retained 69% of the Kampala wholesale purchase price.

Up to this point, the operations of bureaucrat capital are confined to the home market, buying from rural areas to sell in urban centres. Its ultimate goal, however, was to enter the trade in exports and imports, the most lucrative of all (the *fourth* moment). More than any other activity, this required a state connection because it needed substantial bank loans in foreign exchange to buy imports (often necessary consumption goods like second-hand clothing, in great demand in rural areas) and because the export trade (in sesame or maize) required a state licence.

Of the two tendencies to capital accumulation outlined, the stronger and more dominant was that of bureaucrat capital from above, which was closely interlinked with extra-economic (that is, political and physical) coercion in the countryside. No wonder the peasantry, and particularly middle and poor peasants, were especially hostile to it. To them, bureaucrat capital signified the most blatant usurpation of the community's wealth from outside, through practices like the enforced sales of land, crops and cattle that had no historical precedent and were gross violations of local interest. This contrasts with perceptions of rich peasant accumulation through 'traditional' communal practices, bending these to their advantage but over a much longer time-span. Rich peasant property therefore appeared to their neighbours more a result of their own hard work, skill, or just good fortune, than the appropriation of the labour and assets of others.

9.6 Livelihood crises and responses

Peasant production has to be reproduced on a simple basis (see Chapter 2). Of the three elements of the labour process – land, labour and its implements – the implements of labour represent the least dynamic aspect for an individual peasant household. The tendency for more and more peasant households to become implement poor was observed in Amwoma. Although the ox plough was introduced as long ago as the 1920s, its ownership was confined to capitalists, rich and upper middle peasants, who together made up 5.8% of village households. Most peasants cultivated with the hand hoe (Figure 9.8). As we saw, an average poor peasant family with a labouring strength of 2.59 workers could only muster an average of 1.74 hoes (see Table 9.1). Even for those owning ox ploughs, their use was confined to turning over the land. All other farming operations were carried out with the same technology as existed at the turn of the century, the traditional hand hoe for weeding millet and the traditional finger cap or knife for harvesting millet or sesame.

In the conditions indicated, as poor households tried to better their living standards, or simply to defend existing ones, the elements of the labour process that they found relatively dynamic, and that they tried to shape to meet their needs, were land and labour. Yet attempts to maximize the uses of land or labour as immediate solutions to the problems of individual households contributed to a comprehensive crisis of social reproduction. Such is the genesis of both ecological crisis and the crisis of relative overpopulation emerging in countries like Uganda.

Ecological crisis

The ecological crisis is most obvious where landlordism is an immediate barrier to the extensive development of agriculture and the land question is acute, as in southern Uganda. As an expanding peasant population is hemmed in by relatively restricted land frontiers, attempts are made to intensify production, but without a corresponding development of farm technology. Periods of fallow become shorter, as the same land is 'mined' over and again. It becomes tired and yields less and less.

But even when land appears more plentiful, as in Amwoma, similar pressures can operate. As land enclosures gathers momentum, previously communal resources are brought under private control. Swamps are reclaimed as private land, as they have been in western Uganda over the past decade. Sources of water (streams, ponds) or energy (woodland bush) and pastures are privatized. As population increases against a backdrop of diminishing resources, peasant households must search for new sources of energy and new pastures. One inevitable result is deforestation. The process has been observed by numerous writers: for example, in West Africa (Watts, 1983) and Latin America (de Janvry, 1981).

The most acute expression of ecological crisis is where peasant production is pastoral, in those areas where the land question was already most acute in the colonial period. In Uganda, the crisis was most evident in Karamoja whose people lost about 20% of their grazing land through a series of usurpations by colonial and post-colonial authorities over four decades from the 1920s to the 1960s (Mamdani, 1982). The large-scale deforestation that has taken place as a result can only be understood in relation to the search for new pastures in conditions of expanding human and cattle populations.

Figure 9.8   Most peasants cultivated with the hand hoe: preparing for planting in northern Uganda.

A relative overpopulation

The only advantage that peasant farms have over technologically more advanced forms of production is their access to their own labour power. Peasant households are simultaneously units of production and units of reproduction of labour power in conditions where living labour (rather than machinery) continues to be the major input to the labour process.

The tendency of the rural poor in Amwoma and elsewhere in Uganda to have larger families reflected two important aspects of their socio-economic conditions. First, peasant farming allowed for the productive use of child labour. Farm work could be divided into heavy and light tasks, often categorized as 'male' and 'female', as we saw above. The bulk of farm work was tedious, repetitive and time-consuming, comprising tasks that children could help with. The attraction of large families as sources of labour was not lost on female members of peasant households who had the majority responsibility for such 'light' work.

Second, family discipline permitted control over child labour, more so than with other classes. Because the peasant family was at the same time a labouring unit, the discipline of labour was reflected in the internal discipline of the family. Parental authority was supreme. The life cycle of an individual was shaped by demands placed on him/her by the family as a labouring unit. No sooner were they of an age where simple and light tasks could be performed, say five or six years, than children were put to work. Unlike children of the propertied classes, children of the working peasantry did not grow up into adolescents – they became young adults. This generalization remains true for the whole of the labouring peasantry, in spite of other important differences between the strata comprising it, and in spite of a growing rebellious attitude among young adults in homes where there was no productive property to inherit.

As with the search for more land – whether to cultivate, to graze or as sources of energy – the attempt to maximize family labour also had a contradictory consequence. This is because individual and social rationality do not always coincide. What was a solution for an individual family would turn into a long-run social crisis as the contradiction between an expanding population and low-productivity farming surfaced as a crisis of relative overpopulation; relative, that is, to the resources that could be generated in the confines of existing social relations.

Peasant politics: the limits of 'defensive resistance'

The search for individual solutions predominates in situations where radical alternatives do not present themselves. Even so, peasant resistance was never entirely absent. There are two questions that we need to ask:

1 What form did the resistance of the oppressed against the oppressors take?

2 How was the antagonism contained within existing relations – in fact, turned into a force for the reproduction, not the transformation, of these very relations?

The nature of resistance
No matter how peaceful and stable the social order, at no time were the rural poor in Amwoma absolutely reconciled to the existence of oppression and exploitation as if these were natural facts. At no time was their antagonism to the oppressors totally latent. To grasp the nature of peasant action it is necessary to understand how peasants perceived different forces. How did they understand their own social position and how did they view other classes in the countryside? So far as the latter are concerned – the landlords and capitalists at the upper end and landless labourers at the lower end – their conditions of social existence were so clearly demarcated from those of the peasantry (through the absence of any physical labour in the case of the former and of ownership of productive property in the case of the latter) that their existence as distinct social groups was reflected in the consciousness of all villagers.

The same, however, cannot be said of the social recognition of different peasant strata, whether by peasants themselves or by members of other classes. What existed in this case, in fact, was a dual consciousness. If peasants did perceive a differentiation among themselves, it was between two groups: on the one hand the peasant poor (poor and lower middle peasants, but at lean times the rest of the middle peasantry), on the other hand the peasant rich (rich peasants, and during better times upper middle peasants as well). The dominant consciousness, however, was that of peasants as a single community. There was not simply a linear continuity of historical traditions from the past to the present. The community had been shaped and reinforced by resistance to the dominant type of capital accumulation process in the villages concerned. 'Tradition', in this sense, was recreated and born out of confrontation. It was not an historical anachronism. Whether their resistance to capital was individual or collective, the reference point of peasant action was inevitably the community. Because community-centred practices subordinated individual to collective interest, they sought to protect peasant producers from the predatory intrusions of capital, particularly in its bureaucrat form of accumulation from above, and from *outside* the community.

In the name of community and tradition, peasants opposed the fencing of land (except for a *kraal*) and the growing of perennial crops; did not tradition demand that livestock be allowed to graze village land as free range during the dry season? In the same vein, peasants demanded that village land be open to purchase only by members of the community, and that pastures, swamps and forest land remain communal in ownership and control. While peasants appealed to 'tradition', capitalists decried this as 'backward thinking' that blocks 'development and progress in the village'.

There was a daily confrontation between capital and the peasant community. As peasants resisted in the name of the community, they were confronted by the state, the community of capital. As they decried capital accumulation practices as violations of tradition, these same practices were upheld in the name of the law. In this see-saw battle, it was once again extra-economic coercion

that tilted the balance. It was through such experiences that peasant morality developed a character antithetical to that of legal norms. For at such times neither trespassing, nor theft, nor poaching bore any stigma in a peasant community. All were condoned, even heralded by the community, as a defence of its interests against outside usurpation. The active defence of pre-capitalist practices in this context was not an anachronistic hanging on to age-old customs. It made sense only in its modern context: as an active resistance against capitalist accumulation and the pressures it exerted on peasant resources and livelihoods.

But this type of resistance, no matter how pervasive, was at best defensive. It was a rearguard action that was incapable of arresting the process of capital accumulation, let alone transforming the social order. In the dialectic between resistance to the social order and integration into it, its significance was secondary. The primary tendency was the integration of the peasantry into the existing order. In fact, the starting point of the integration of the peasantry was the same as the reference point of its resistance – that is, its constitution as a community – but with one important difference. The community could be organized from above under the leadership of the propertied strata in the village, whether village capitalists or rich peasants, or from below through peasant resistance to exclusively 'external' (bureaucrat) capital.

The outcome of resistance

What are the implications of perceptions and organization of the community from above and from below? I have already suggested that peasants had different perceptions of wealth accumulated by fellow villagers through hard work, skill, or luck, and that acquired by outsiders through a state connection. However, the distinction between the two types of accumulation becomes a relative one after a certain point. As they sought to expand the scope of their accumulations, village capitalists tried to follow the path charted by 'external' capital, i.e. the path of large-scale land enclosures, of legal titles to these, and of securing hefty bank loans on this basis. They clearly understood that none of this was possible without a political connection, and this guided their active involvement in village politics.

The contradiction between the two types of capital hardly ever assumed a violent form, not even in scattered episodes, unlike the contradiction between the labouring peasantry and capital. The process of its resolution, however partial, was essentially a bargaining process. The negotiating strength of village capitalists and rich peasants derived from their ability to enlist peasant support, organized through local-level institutions like the village branch of the party (in this case in a one party state) or the church. In Amwoma, the chairman of the party branch was a rich peasant. He was acutely conscious that this otherwise formal and ceremonial position could become a springboard of important advantage at election time or in a political crisis. For it would allow him the possibility of 'delivering' the village – in terms of votes or other types of support – in exchange for definite gains for himself.

The composition of different types or groups of capitalists was one side of the political process; its other side was the actual disorganization of the peasant community through its formal mobilization from above. To begin with, this was highly personalized. Personalities predominated over issues. One identified with an individual, not his/her stand. The rural poor marched behind local 'opinion-makers' – a school-teacher, a priest, a prominent village capitalist.

Support was given in expectation of, or in response to, the delivery of patronage. Voting was an opportunity to exchange an empty right for a material benefit, such as a few kilos of sugar or a blanket. In practice, the politics of patronage had a disintegrating effect on the peasant community as appointment from above replaced election from below. The lower classes sought out and put their trust in individuals from the upper classes who were said to be 'on our side'. They hoped to maximize their welfare individually through charity from above, not collectively through democratic struggle and movements from below. The poor were atomized

as each sought a personal advantage against another, as each looked for a private solution to a social problem. This type of ruling-class organized politics did not reinforce and strengthen the community, giving it a concrete understanding of wider social forces, whether friend or foe, but rendered it a relatively passive following, and thereby disorganized it and disintegrated it.

This is clear if we shift our focus to the wider political scene. At the local level, I have highlighted differences between the village bourgeoisie and the bureaucrat bourgeoisie as the major contradiction internal to capital. That, however, is not necessarily true from a countrywide perspective. It should be obvious that the village bourgeoisie, by the very fact of its highly localized existence, could not possibly form a cohesive force countrywide. But neither could the bureaucrat bourgeoisie, despite its state connections. The reason lies in the historical background to the capital accumulation process: the moment they turned to the countryside, bureaucrat capitalists necessarily turned like predators to feed on their 'own' nationality or 'tribe' (Box 9.5). That is, they invoked their own ethnic (national or 'tribal') identity to lay claims to land, and to build a political base to defend themselves against both their peasant victims and their bourgeois competitors, by appealing to and manipulating communal traditions and solidarity. Thus, the bureaucrat bourgeoisie had a tendency to organize in factions based on localized nationality, and only then to relate to other similar factions, whether through collaboration or contention.

Organization was the result of a blend of initiative and realism. Initiative involved organizing town-based groups 'from our area', baptised variously as 'tribal' or 'development' or 'cultural' associations. While such associations might include individuals from various social classes – traders, professionals, civil servants, teachers – its real foot-soldiers were students. Without a firm anchor in material interests and with few opportunities at hand, students were forever on the look-out for patrons who would assist them at lean times or provide for jobs upon graduation. At the same time, students had both ample time

Box 9.5 Meanings of nationality

I use the term 'nationality' to refer to constructions of, or claims to, a common political identity and interest based on a geographical region, a shared language and/or other cultural factors. With regard to Africa, this is usually viewed pejoratively as the expression of a primitive 'traditionalism' or 'tribalism'. However, claims to nationality have been advanced and contested throughout the history of the modern world, not least in Europe whose contemporary political boundaries emerged from a long series of wars (and continue to be fought over).

In Africa, 'tribes' have emerged from processes of historical development in which colonial policies of 'divide and rule', and of local administration (the establishment of 'Native' or 'Tribal' Authorities) often played a part. Likewise, 'traditions' are neither timeless nor static, but are actively constructed, manipulated, and invented by different social forces to meet their own interests. This applies as much to the history of Scotland and Wales (Hobsbawm & Ranger, 1983), for example, as to that of northern Uganda.

You may also notice that my analysis of bureaucrat capital integrates elements of all three views on taxation in post-colonial states in Africa indicated in Chapter 4 (Section 4.2). I argue that these states are basically class states of bureaucrat capital (the second view), which appropriate 'bureaucrat rents' (the first view) as a source of individual accumulation. I also suggest why the politics of bureaucrat capital take the form of organizing constituencies based on particular regions, nationalities or ethnic groups (the third view). However, the analysis here does not take regional or ethnic politics at face value – as the expression of a pre-given 'tribalism' – but suggests what produces such politics in this particular instance: the competition and contradictions of bureaucrat capital.

---

at hand (every vacation, they moved from urban to rural areas) and considerable credibility 'back home' – they were the 'children' of the village, its future. It was by patronizing students of 'his' area that a bureaucrat capitalist usually hoped to build his base 'at home'.

Such initiatives notwithstanding, the construction of a nationality following required a strong dose of realism, a process of bargaining with a host of village political bodies, usually led by rich peasants or village-based capitalists. The existence of organizations defined along nationality lines should not be taken to be a negation of class politics. It was in fact an expression of a particular type of class politics, under particular historical circumstance. Where the village community was organized more in such ways from above than through popular resistance from below, each nationality organization was in reality a united front of all classes in that nationality under the leadership of 'its' bourgeoisie.

Neither should this be seen as the result of clever manipulation, of a conscious strategy of divide and rule, or a ruling class conspiracy. It was simply the objective result of a process whereby bourgeois factions, as they struggled against each other, organized popular classes in order to buttress their respective strength and position. It simply reflects the weakness of the peasantry and the initiative of the bourgeoisie in existing social conditions.

Summary

1   Processes of commoditization and social differentiation (class formation) are analysed and illustrated in the case of a village in northern Uganda; a feature of these processes is how they operate through, and hence change the social content of, ostensibly traditional forms of communal labour organization.

2   The basic class categories used are landless labourers, poor peasants, middle peasants, rich peasants and village capitalists (the village bourgeoisie, exemplifying accumulation from below). Bureaucrat capital (exemplifying accumulation from above that requires a political connection with the state) is experienced as an 'external' imposition on the village.

3   The livelihood strategies of the poor are rational for individual survival but have negative social effects, manifested in ecological degradation and relative overpopulation, within existing relations of property and power.

4   These relations are also reproduced by peasant subordination to a politics of 'community' and 'nationality', organized from above by the village bourgeoisie and bureaucrat capital respectively.

10

UPHEAVAL, AFFLICTION AND HEALTH: A UGANDAN CASE STUDY

TIM ALLEN

Chapter 9 looked at some features of life in Amwoma, a Langi village in Uganda's Lira District. This chapter focuses on Laropi, another Ugandan village, located about 150 miles to the north-west of Amwoma, in Moyo District, where the people are Madi (Figures 9.1 and 10.1). The Madi regard the Langi with some antipathy and speak a completely different language as their mother tongue. Nevertheless, the two groups share many cultural traits, and the ways of making a living for peasant farmers are broadly similar. Indeed, some of the processes discussed in Chapter 9 were also occurring in Moyo District during the 1970s, although war and social upheaval in the area during the 1980s undermined the attempts of capitalist farmers to run their farms effectively.

Whereas Chapter 9 looked at commoditization and accumulation, this chapter is concerned with other aspects of rural life. In the 1980s the people of Laropi suffered extreme social upheaval. They had to flee their homes as refugees because of fighting, and then had to flee their place of refuge for the same reason. They had to establish and maintain communal life in a situation where government services were hopelessly inadequate, and where there were very limited markets for agricultural produce. They were thrown back on their own resources, not as 'uncaptured peasants' (Chapter 9), but as a marginalized population with little or no influence in national and

Figure 10.1 Typical homes in Laropi: notice the small gardens near the compounds and larger fields in the distance.

international centres of power. This chapter draws on fieldwork I carried out for a total of 24 months between March 1987 and September 1991. The chapter poses the following question:

Q How did people in Laropi struggle to achieve healthy lives in a marginal, violent and disease-ridden environment?

I begin with observations about the meanings of health, and the kinds of associations that ill health and affliction had for the Madi. Then I review the traumatic events that had happened in this part of Uganda, outline the health situation from the **biomedical** point of view and look at how the people of Laropi themselves dealt with collective and individual suffering. In the final section I comment on some of the implications.

> **Biomedical:** Relating to a body of medical knowledge based on scientific interpretations of illness. Biomedicine recognizes the existence of microbiological phenomena and explains illness symptoms in terms of ideas about germs and/or malfunctions in the patient's body. It is associated with sophisticated technology, a hierarchy of professionals and the curing and control of diseases. It is sometimes referred to as modern, scientific, conventional or cosmopolitan medicine. The term 'Western medicine' is also commonly used, but it is misleading for three reasons. First, there is a great deal of health care in Western countries which is not biomedical (psychoanalysis, acupuncture, homeopathy, faith healing, etc.). Second, biomedicine has historical origins in non-Western as well as Western countries (e.g. Arab countries). Third, biomedicine is established as the basis of formal health care services run under ministry of health auspices throughout the world.

10.1 Perceptions of health and the lack of it

One of the problems with trying to understand health in a different cultural setting is that we have to confront the ambiguities of what we mean by the term 'health'. It is therefore worth pausing to consider what kinds of things it might involve. (More detailed discussion of some of the issues raised in this section can be found in Johnson & Sargent, 1990, and, with respect to control of disease, in *Allen & Thomas, 1992*, ch.2.)

The World Health Organization (WHO, the United Nations body most concerned with health issues) defines health as 'a state of complete physical, mental, and social well-being'. But it is rare for the health of a population actually to be assessed in these terms. More often than not, health statistics refer to mortality rates and morbidity rates, e.g. how many children die per thousand before the age of five? what is the prevalence of HIV (the virus linked with AIDS)? Although it is obvious that health is something positive, it is commonly discussed in terms of its absence. Being healthy means not being sick, and on a day-to-day basis most biomedically trained professionals working for **public health** programmes are concerned with combating ill health rather than promoting well-being.

> **Public health:** Refers to the overall health of populations. It is commonly applied to official government-sponsored strategies for improving health, or to the work of institutions like WHO which operate in collaboration with ministries of health. Although public health policies may include the professionalization of 'traditional healers' (see Box 10.3 below), they are almost invariably based upon biomedical interpretations of illness.

However, the notion of ill health is also problematic. Biomedical practitioners are particularly concerned with disease. They usually see this as the failure of 'normal' physiological activities, and as having a set of objective, clinically or chemically identifiable symptoms. But what is normal, and therefore healthy, from a biomedical point of view may not be normal for someone who feels unwell. Psychosomatic disorders are common, and so are forms of illness that some psychoanalysts have linked to phantasy and the unconscious. Certainly, people often feel unwell as a direct or indirect consequence of interpersonal problems, depression, or some general sense of inadequacy or inner conflict.

Moreover, popular perceptions of 'disease' have much wider connotations than in scientific usage. In Western countries, diseases may be thought of as sinister entitities, always ready to attack their innocent victims, almost like malicious spirits. Alternatively, diseases may be interpreted as punishments. Particular diseases which are, or are believed to be, incurable – notably cancer and AIDS – are imbued with ideas about **moral** probity. It may be thought that sufferers are morally unclean. Along these lines, Susan Sontag persuasively argued that illness can function as a 'symbolic metaphor' for social values. She pointed out that:

> "a surprisingly large number of people with cancer find themselves being shunned by relatives and friends and are the object of practices of decontamination by members of their household, as if cancer were an infectious disease. Contact with someone afflicted with a disease regarded as a mysterious malevolency inevitably feels like a trespass; worse, like the violation of a taboo."
>
> (Sontag, 1979, p.6)

Thus diseases may mean more to the afflicted than they do in biomedical terms. This is why patients will sometimes combine a visit to the hospital with a form of faith healing. In Britain, religious people pray for good health, and many Roman Catholics will consider making a pilgrimage to the miraculous shrine at Lourdes in France if their condition is serious. But, on the other hand, diseases may also mean much less than they do in biomedical terms. People who are biomedically diseased may consider themselves to be healthy. A person who is infected with HIV may be unaware of it, and may feel perfectly well. In Britain certain potentially fatal diseases are commonly thought of as an almost inevitable part of daily life. Influenza, blood pressure problems and disorders associated with the effects of urban pollution may fall into this category. In parts of the world where public health services are less adequate than in Britain and biomedical ideas are less culturally hegemonic, this tendency is more pronounced. For example, in some of the irrigated cotton fields of Sudan, bilharzia is so widespread that it is considered odd not to pass blood in urine. It has been reported that uninfected boys will sometimes cut their penises in order to be the same as their friends. Other potentially serious conditions that have been considered as normal in particular populations include persistent diarrhoea, malaria, and the skin discolorations of pinta.

In response to these kinds of confusions, many health analysts make a distinction between **disease** and **illness**. They employ the term 'disease' in a specifically biomedical sense to refer to objective phenomena of which an infected individual may or may not be aware, and use the term 'illness' to refer to an individual's subjective perception of being unwell. Thus it is possible to be diseased without being ill, and it is possible to be ill without being diseased.

Moral: As used in this chapter, refers to prevalent ideas about the distinction between right and wrong within particular populations. Thus it is not used in a universalist sense, but assumes that moral values are culturally relative.

Disease: A biomedically defined affliction which can be scientifically verified as a malfunction of body organs and/or systems, e.g. measles, cholera, bilharzia, influenza, diabetes.

Illness: An individual's subjective sense of being unwell, e.g. the experiencing of a pain in the head or back, being troubled by strange voices, and feeling ensorcelled (see Box 10.1 below).

Some scholars have questioned the distinction on the grounds that it assumes the biomedical definition of disease is culture free, and that the way diseases are conceptualized by biomedical practitioners is in reality the same for all people, irrespective of whether or not particular individuals or groups recognize it. Critics point out that while there may be 'objective' aspects of biomedicine, its diagnostic procedures are influenced by cultural norms. Cultural differences also explain why clinical practice and biomedical taxonomies vary in detail from place to place.

Nevertheless, the distinction remains useful for some analytical purposes. I found that the people in Laropi were affected by several tropical diseases so frequently that they did not always think of themselves as unwell. It was also common for illness to occur without any obvious biomedical cause. Investigating the struggle for health in the village requires discussion of both these phenomena.

Moreover, distinguishing between disease and illness corresponded with the attitudes to ill health of at least some people living in Laropi. There were certain individuals who almost always asserted a biomedical view of sickness when a relative was unwell. They were usually relatively affluent men who had attended secondary school, and who wanted to promulgate what they regarded as being progressive, scientific attitudes. They may additionally have wished to avoid the economic demands placed upon them by poorer relatives at times when expensive healing rituals were being performed. They would do all they could to persuade family members to take the patient to the dispensary or buy drugs. They were also likely to be unsympathetic when a person claimed to have symptoms which could not be explained in biomedical terms, and might argue that although the afflicted person thought she/he was ill, in fact there was nothing really the matter.

But such an uncompromising view was quite unusual. Most people in Laropi looked upon ill health in several ways simultaneously, and their approach to therapy was correspondingly pluralistic (i.e. different ways of healing were attempted at the same time). It might be recognized that a

Interpersonal: As used in this chapter, this term categorizes interpretations of illness which make reference to relationships between people (or conscious forces), who may be living or dead. Often such interpretations are imbued with moral concerns. In Western countries, interpersonal approaches to illness include many of the interpretations made by psychoanalysts of their patients' behaviour, and may be applied to people suffering from particular diseases, like cancer. However, interpersonal explanations are generally more prevalent in parts of the world where biomedicine has a less dominant position in the ways people conceptualize affliction. In Laropi, interpersonal explanations include the interventions of deceased ancestors, the malign influence of wild spirits, and the activities of sorcerers.

biomedical or herbal remedy was essential, but particularly when the condition of the afflicted person was serious, at least some relatives would also make assessments of what was happening in terms of **interpersonal** relations. To borrow Sontag's expression, illness acted as a metaphor for social values. It became a manifestation of moral concerns. Whether a relative had sleeping sickness or was possessed by ghosts was important in deciding what kinds of therapy specialists were consulted. But why that particular person was afflicted and not someone else was considered to be an equally important question. Commonly, the answer involved the antisocial behaviour of the patient, a relative, or a pernicious neighbour. Sometimes punitive measures were considered to be a necessary part of the cure. There is a Madi proverb, 'suffering does not climb a tree' (laza otu kwe ku), which was used to remind people that things were not coincidental. Disease, illness, misfortune and death did not happen just by accident. In serious cases, some relatives would want to know who was to blame, and medicines might be combined with someone being publicly chastized, or occasionally even killed.

This way of thinking about illness had wide-ranging implications. Arguments about therapy and death frequently necessitated the airing of grievances, the allocation of social accountability, and also supernatural considerations which from a biomedical perspective might seem to have had more to do with religious beliefs than health care. It also meant that healing was frequently seen as a collective enterprise, involving many more people than the patient, close family members and a few therapy specialists. Illness might be viewed as a symptom of shared affliction rather than an individual's problem, and was responded to as such.

It is an attitude reflected in the Madi language. While there were specific words for particular diseases, usually ones that had been the focus of control programmes since colonial times, there was no term which corresponded to the English word 'illness'. In translation, the word *laza* was used. But *laza* had much broader connotations. In the proverb above I have rendered it as 'suffering'. It can also mean crying and pain, both in a collective and individual sense. A distinction between phenomena known to be certain kinds of disease, other ailments and general misfortune was not always made. In fact, it was difficult to separate these things conceptually in the Madi tongue. But a note of caution must be entered here. People in Laropi were multilingual. Most knew how to speak Swahili, and many also knew Arabic and English, as well as the languages of other Ugandan 'tribes' (most could speak Acholi, which was closely related to the Langi language spoken in Amwoma). The Madi consequently had access to a range of linguistic concepts and ideas about illness and affliction which they readily used in their discussions, sometimes grafting foreign concepts into their own language. Those wanting to promote a biomedical interpretation of an illness were likely to use English words to make their point. Paradoxically, those asserting an interpersonal explanation might do the same thing. Instead of using the Madi word *inyinya*, which could most accurately be translated as 'sorcery', they employed the English word 'poison' to give their claims a biomedical gloss (Box 10.1).

> **Box 10.1 Witchcraft and sorcery**
>
> Anthropologists often make a distinction between witchcraft and sorcery. *Witchcraft* is a magical or supernatural means whereby a witch *involuntarily* causes harm to others. *Sorcery* is a magical or supernatural means whereby a sorcerer *deliberately* causes harm to others. Witches may not even be aware that they are responsible for suffering, while sorcerers are evil-doers who have the knowledge of ensorcelling victims. However, in practice this distinction can prove difficult to make. In the Madi language the two notions are not clearly separated, and it is rare for someone to be thought of as harming neighbours unwittingly. The word *lemo*, which is commonly translated as 'witch', is better rendered as 'a very cunning person', and certainly if someone is accused of having *inyinya* it is always clear that he (or more usually she) is wicked and well aware of what is going on. Consequently I generally keep to the term 'sorcery' in this chapter and only use the terms 'witch' or 'witchcraft' when quoting the use of these English words by the actors themselves.

However, even those who wished to promote an exclusively biomedical approach to curing did so in a context in which the broad connotations of the word *laza* reminded everyone of the connections between manifestations of misfortune. Those who denied interpersonal or moral explanations of illness might be accused of not fulfilling their social obligations, and might find that relatives forced them to perform or pay for the very rituals which they had maintained were pointless. There is irony in that, in responding to *laza* and in trying to make things 'good' *(cwui)*, the Madi seemed to come closer to WHO's definition of health as 'a state of complete physical, mental, and social well-being' than many practitioners of 'modern' medicine. Most of them were nevertheless acutely aware that the prospects of attaining a healthy life in the near future were remote.

10.2 Social trauma

I mentioned in the opening paragraph of this chapter that the Madi and the Langi regarded each other with some hostility. The reasons for this relate to the manner in which Ugandan politicians built up followings among their own ethnic groups by appealing to and manipulating communal traditions and solidarity (see Chapter 9).

Following independence in 1962, Uganda's first ruler (initially prime minister, later executive president) was Milton Obote. Obote is a Langi, and came to rely heavily on his own people and linguistically related 'tribes' (particularly the Acholi) for his power base. When Amin seized power in 1971, he immediately set about crushing Obote's supporters. Several prominent Langi 'disappeared' and many of the Langi and Acholi soldiers in the national army were killed. Amin then proceeded to build up his own power base by seeking support from his people, the Kakwa, from other groups living in the north-west part of the country (including the Madi), from Moslems, and from some of the Sudanese refugees then living in Uganda.

After the Tanzanian invasion in 1979 which toppled Amin, and the return to power of Obote in 1980, tables were turned again, and Langi and Acholi soldiers in the so-called 'liberation' army took a terrible revenge on those they held responsible for Amin's regime. Peasant farmers in Moyo District had gained nothing from Amin's period of power. Some of them had even been evicted from their land by Madi capitalists resident in Kampala. But they were now brutally attacked. Appalling atrocities were perpetrated, and Obote himself is said to have claimed that he would clear away the whole population of north-west Uganda, and turn the region into a game park. To escape the killing, some 250 000 Ugandans fled into Sudan (Figure 10.2), where many of them were settled in special settlements under the auspices of the United Nations High Commissioner for Refugees (UNHCR). Almost all the 3500 people of Laropi were located in settlements on the eastern side of the River Nile (see Figure 9.1).

Figure 10.2 'Why we fled our home': a drawing by Thomas Drale (a secondary school student in Moyo District) showing the 'liberation' army killing people in 1980.

Initially, life in Sudan was very difficult. A measles epidemic killed scores of children, and there was little food. But most people said that once they had established farms, things began to improve. In fact many families seemed to have ended up being better off in Sudan than they had been in Uganda. Some would probably have remained in their new homes had they not been attacked again. Unfortunately, a few well-known Madi in Sudan became embroiled in local politics and made an alliance with some of Amin's former Sudanese supporters. This faction was opposed to groups which formed the Sudan Peoples' Liberation Army in the mid-1980s. In April 1986 the refugee camps to the east of the Nile were attacked by the guerrillas. Houses were burned, property looted and some people killed. There was no option but to flee again, this time back to Uganda (Figure 10.3).

Figure 10.3 (above) 'The escape from bandits in the Sudan to Uganda', a drawing by Ceaser Drauku Atori (a secondary school student in Moyo District) showing a Ugandan refugee settlement being attacked – most people assert that the 'bandits' were in fact supporters of the Sudan Peoples' Liberation Army; (left) 'Coming back to Uganda', a drawing by Peter Erika (a secondary school student) showing people leaving refugee settlements in UNHCR trucks; few of the returning refugees were transported as efficiently as this – most had to flee their settlements on foot, carrying what they could of their possessions.

Meanwhile, in Uganda the alliance between the Langi and Acholi soldiers which had bolstered up Obote's regime had fallen apart in 1985. Okello, an Acholi officer, managed to take power for a few months, but was unable to prevent Yuweri Museveni's National Resistance Army (NRA) from capturing Kampala in January 1986 and forming a new government. Museveni drew most of his support from the south and south-west of the country, but adopted a reconciliatory approach to most factions (including many former supporters of Obote and Amin).

A degree of stability returned to much of the country. Even Obote's home district was fairly quiet. But there was almost continuous fighting in Gulu and Kitgum Districts (immediately to the east of Moyo District and north of Lira District). Acholi soldiers from Okello's army retreated to their home area and formed guerrilla units which harried the NRA and occasionally launched offensive campaigns. The most effective of these forces was the Holy Spirit movement, a millennarian battalion, led by a young female spirit medium called Alice Lakwena (Alice the Messenger; Chapter 12), which terrorized northern Uganda during 1986 and 1987. Alice herself has been in Kenya since an unsuccessful campaign into the south of Uganda during late 1987, but other mediums have also managed to secure followings since her departure. At least one group was still active in 1991 and was raiding into eastern Moyo District.

Thus, the area into which the Madi refugees returned in 1986 was still insecure. Poorly maintained dirt roads to the south of the country were frequently closed due to ambushes, and the international relief effort that had been mounted to assist in the repatriation was crippled by transport bottlenecks. According to UNHCR figures, during the first five months of their food distribution operation up to January 1987, only 10% of expected cereals, 14% of beans, 12.5% of cooking oil and 20% of salt arrived in the district and were distributed. Things did not improve in 1987, and most people had to manage as best they could on their own.

In Laropi (located on the Nile), only a few people had fishing nets and hooks, and could catch and sell fish. The rest had to rely on the little they had brought with them from Sudan and on gathered foods: mangoes in December, April and May; wild plants and roots; and seeds of water lilies collected from the Nile (which are dried, ground into powder, and mixed with hot water to make a porridge). Some families harvested a small crop of sorghum before the end of 1986, but yields were low due to poor rains. In 1987 the rains were late, and then were too heavy, destroying any crops planted in the most fertile places close to the Nile. Crops planted later that year and in 1988 did better, but yields were limited by pests and plant diseases. Not enough land had been cleared to prevent forest animals from taking food from the fields, and because people were so short of cuttings and seeds they had planted virtually anything they could find, with hardly any quality control. Three years after returning, many families were still intermittently hungry. Re-opening farms with hand tools is a long and exhausting business. Just removing the secondary forest can take months.

Before fleeing to Sudan, the rural population of Moyo District had been a commoditized peasantry, selling cotton to government buyers in order to raise money for taxes, school fees, radios, bicycles, clothes and better houses. But the war and the continuing fighting eroded government services, and limited the effectiveness of development schemes.

Teachers, medical staff and civil servants often failed to receive their meagre salaries. The ginneries, which separate seeds from raw cotton, were almost always broken down, and farmers who grew cotton were often unable to dispose of it. Markets for other agricultural produce were also very limited. For many families the most important source of income became *waragi*, a distilled alcoholic drink, which women made from cassava flour and sold to those men lucky enough to have some money, such as soldiers and fishermen (Figure 10.4).

Life for the elderly was particularly grim (Figure 10.5). Moses, a man in his late sixties, told of his tragic circumstances, not long before his death:

> "I am suffering so much. My wife and I are old now, but just assisting ourselves. There is no one to help. In the refugee camp, Sudan

Figure 10.4 'Forget your troubles': a woman is shown distilling waragi and a group of customers are drinking it hot from the still. Selling waragi is the main source of income for many families. The drawing is by Maxwell, who is quite a well-known local artist. He ran a small tea shop in eastern Moyo District until it was destroyed by Holy Spirit Movement guerrillas in 1991.

Figure 10.5 Life for the elderly was particularly grim: a blind couple in Laropi.

Council of Churches helped us because I am too frail to work. Only one of my children is here at home with us. He is at Laropi primary school. My wife goes to collect water lilies. There is nothing else to eat. I have daughters who would help, but they are married to men in Metu, and bridewealth has been paid and consumed long ago. One of my daughters is already a grandmother. We have been given some things by the missionaries but it was only enough for a few days. My brothers' sons can do nothing for us. They have their own fathers to look after. In the past they would have done something, but today no one has anything extra. My house was built by my daughters' husbands, who came with all the materials from Metu.

But the children could not stay long, they have their own homes. There is no one to help my wife cultivate. We had a small plot of sorghum, but it was all destroyed by pests."

(tape recorded in the Madi language, July 1987)

In these difficult circumstances, it is hardly surprising that the health situation was so bad. Almost every family in Laropi lost at least one member during the time I was researching in the village, and most people had spells of serious illness. Madi names vividly reflect the situation which the child faced at birth. Box 10.2 lists some of the names parents had given to children since returning, together with their meanings.

Box 10.2 Madi names

The following is a selection of names given to children after 1986. These are the 'African' names which were registered together with a 'Christian' name in the diocese baptismal register at Moyo Catholic mission. I give a literal translation of the name, and brief explanation where this seems necessary.

Kodri: all go to the grave.

Amadrio: without successors, i.e. almost everyone in the family has died.

Ariziyo: unavenged – relatives have been killed without retaliation.

Baku: non-human – the parents are saying that their child is not really a person, and therefore evil-doers and death should ignore him.

Via: in hunger.

Mandera: worn out by family problems.

Dayo: nothing is good.

Drate: awaiting death.

Izama: impoverished.

Anzoako: without joy.

Kareodu: reverted to bush – the family home has been taken over by the forest and there are not enough family members left to rebuild it.

Oliku: short lived – it cannot be expected that the child will live long.

Nyanda: repeated – the last born child died, this one will surely die too.

Vumgbo: empty home – many have died.

Mazapke: without hope.

Dima: you have killed us.

Ambayo: without elders.

Raleo: stop planning – because you will only be disappointed.

Apidra: survived death.

Amaca: dispersed – due to war.

Baatiyo: unfortunate.

Avudraga: went for a funeral.

Adrawa: many of us have died.

Kadabara: it may dawn – things might improve.

Maribio: I am death.

Tabu: all my children have died.

Draa: death.

Drani: for death.

Kitiyo: no peace of mind.

10.3 Infections and diseases: a biomedical perspective

This section looks at some of the different kinds of health problems people in Laropi faced. Following the distinction noted in Section 10.1, I focus first on diseases and other biomedical ailments. Thus I am adopting a narrow perspective, one that relates to 'scientific' or 'clinical' interpretations of symptoms and their causes. It is not simply an outsider's view, because there are some people living in Laropi who promote it, and also because biomedical ideas influence non-biomedical interpretations in both direct and indirect ways. Nevertheless, the discussion of infections and diseases presented in the following paragraphs is not a common kind of approach to things in the village. It summarizes the concerns expressed by senior district medical staff and expatriates working for international aid agencies. The more pluralistic attitudes of local people are addressed in Section 10.4.

Unfortunately, it is impossible to write with much certainty about diseases in Laropi. There were hardly any baseline data for the district. Reporting was very patchy and was based on cases treated at dispensaries or the odd survey aimed at particular diseases. Very little was actually known about what people died from. A great deal, therefore, has to be speculation (as indeed it has to be for much of rural Africa). It can be assumed, for example, that the most serious killer was diarrhoea, even though this was rarely the stated cause of death. Many people suffered from it continuously. The following are notes about diseases and infections derived from the limited information available.

Many people were undernourished and suffered from a variety of nutritional disorders. This is not surprising, given the failure of the food relief programme and poor crop yields due to adverse weather conditions and pests. Children and old people were particularly at risk. A few statistics were collected by a French non-governmental organization working in the area (Médecins Sans Frontières – MSF) and also by the Uganda Red Cross. They carried out nutritional surveys for children under five years. These surveys used small samples of up to 500, took place at long intervals (up to a year), and were not supplemented with mortality surveys. They can therefore only give a crude indication of the overall nutritional situation. The children were weighed and measured, and then compared with standard weight-by-height ratios for healthy children. Children who were less than 80% weight for height were classified as malnourished. Taking an average of all the surveys, 4.8% of children under five years fell into the category. This may not seem a particularly high figure considering the circumstances in which people had been living, but some of the surveys also registered borderline cases. In June 1989, for example, it was found that although only 5.4% of the sample were less than 80% weight for height, a further 14% were between 80% and 85% (Figure 10.6). In fact,

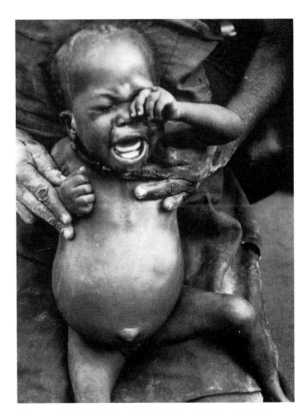

Figure 10.6   While relatively few children in Moyo District were classified as malnourished according to nutritional surveys, many showed signs of an inadequate diet.

probably the majority of children were poorly nourished, at least for the part of the year when there was little food to be harvested. The consequence was that they had a reduced resistance to disease, and children that looked relatively healthy could waste away and die in a matter of days.

Apart from diarrhoeal diseases, the most commonly diagnosed ailments at Laropi dispensary according to the register were: malaria – which affected virtually everyone from time to time, bilharzia (*Schistosoma mansoni*) – which was so

prevalent at Laropi that patients were not treated since they would quickly be reinfected when returning to the river; other intestinal worms – mainly hookworms; eye infections – mostly severe conjunctivitis; infected wounds, often in the feet because of accidents occurring when hoeing without shoes; tropical ulcers; cerebrospinal meningitis; measles; tuberculosis; tetanus; and venereal diseases.

Because it was thought that local cures were better for jaundice than biomedical ones, these cases were often not brought to dispensaries. They were nonetheless numerous. Jaundice can be a consequence of various diseases, including hepatitis (perhaps caught from an unsterilized needle used at home) and alcohol-related liver problems. Alcohol also caused morbidity and mortality in more immediate ways: many men regularly suffered the effects of severe hangovers, and during my stay in Laropi there were several fatalities arising from fights at drinking places or from drunks falling into the river at night.

Other health problems not usually taken to the dispensary were pregnancy and labour complications. Occasionally women died in childbirth. However, antenatal care in Laropi had improved since 1988, when a hard-working and competent nurse started work at the dispensary (Figure 10.7). She gained a reputation for being able to

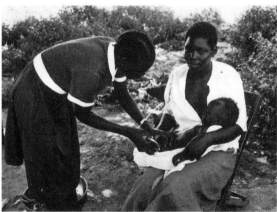

Figure 10.7 (top) Laropi dispensary. A new nurse came to work at Laropi during 1989 and worked hard to improve antenatal care. She became a popular figure, and many people went to consult her who would not normally turn to district medical staff with their troubles. (above) The Laropi nurse administering an immunization vaccine. (right) The nurse with a Traditional Birth Attendant. The TBA's instruments are on the table.

save mothers when everything else had been tried and failed.

The level of infection with the AIDS virus was lower in Moyo District than in other parts of Uganda. A national survey in 1990 indicated that about 3% of the sexually active population of the district were HIV positive, and there were correspondingly few recorded deaths from the disease. Some of those who died were well-known individuals, and this helped spread an awareness of the problem. In 1991 the recently retired head of the civil service in the district died of the disease. Copies of a letter he wrote on his death-bed warning others not to behave as he had done were circulated and read at church services. Such publicity may do something to change sexual practices, but if the area becomes more politically stable and integrated into the national economy, the number of cases can be expected to rise.

More worrying than AIDS was the prevalence of sleeping sickness (trypanosomiasis). Sleeping sickness is a fatal disease which is difficult to treat, particularly when the infection has developed and symptoms have begun to appear (headache, fever, dizziness). It is spread by tsetse flies. By 1960, because of intensive treatment and vector control during the colonial period, the disease was almost eradicated, but since then control measures had become haphazard and many Ugandans were infected with the disease in Sudan. In 1989 the prevalence was estimated at 5% in some parts of West Moyo and Obongi Counties, and as high as 30% in one parish (Lomunga). Tests introduced and implemented by MSF indicated an overall prevalence of 1.5% for the two-year period from October 1987 (in West Moyo and Obongi Counties), and it was feared that this figure would rise as people continued to move across the Sudan border and as tsetse flies continued to infest the secondary forest which grew up over much of the district while the population was in exile. Like tuberculosis, sleeping sickness requires a prolonged course of treatment, which must continue whether or not the patient actually feels ill. Patients with the 'second stage' infection have to be administered with a highly toxic drug, which can itself be dangerous.

MSF responded by collaborating with district medical staff in an attempt to limit the spread of the disease. Screening of the population, combined with hospital care for those infected, kept the overall prevalence under 1.5%. But a continuous effort for the foreseeable future would be needed to maintain this level, tying up a considerable part of the limited resources available for public health activities. It was still hard for people who felt well to understand why they should go to hospital for tests and treatment, particularly since some of those who did go died from the treatment itself. So long as the tsetse flies remained and there were people carrying the disease, reinfection for those who had been treated was always a risk.

Medical facilities in the district were hospital and dispensary based, and were overstretched. Support was provided by MSF, in addition to the work that the organization did for the trypanosomiasis control programme. There were usually about ten French medics based at Moyo town (Figure 10.8), some of them working at the hospital and others providing some assistance to the dispensaries, or in the supervision of trypanosomiasis screening. They certainly helped to save some lives, but short-term contracts, youth and lack of experience tended to mitigate against the effectiveness of many of their activities. Ugandan medical staff were so poorly paid by the government that they had to have an additional income in order to be able to turn up for work. The nurse at Laropi dispensary brewed a special kind of beer to make ends meet. Others resorted to selling medicines to traders for sale in their shops. Immunization coverage against the major preventable diseases was low (in 1989 only about 20% of children had been inoculated for measles, tetanus, whooping cough, polio, diphtheria and tuberculosis), and many people who went to the dispensaries for treatment were told that drugs were unavailable.

Under these conditions, checking the spread of trypanosomiasis was a substantial achievement. But the overall health situation was only likely to improve if food and clean water became more readily available, and general living standards rose. There is evidence that the kind of selective interventions to control particular diseases

Figure 10.8 (left) Médecins Sans Frontières compound in Moyo town – expatriate aid agency workers often live in a fenced world, far removed from the lives of those they are trying to help; (below) members of the MSF team arriving at Moyo landing strip.

adopted by MSF and by other medical aid agencies in Uganda have little impact on overall morbidity and mortality rates. A child cured of sleeping sickness or measles might die soon afterwards of something else. Part of the problem in Laropi, as elsewhere, was that nutritional deficiencies and diarrhoea could not be eradicated by biomedical means alone. They were the effects of poverty.

10.4 The experience of illness

Having made some observations about infections and diseases from the biomedical perspective of district health care workers, I now turn to the issue of illness, as defined in Section 10.1. How did the people at Laropi think about health problems, and what did they do about them?

There are no straightforward answers to these questions. Treatment of illness in Laropi was a matter for negotiation and therapy was pluralistic. Views of illness varied enormously, from those who adopted a biomedical explanation to those who saw the interventions of the dead in the lives of the living. The following case illustrates some of these divergent interpretations and also reveals how in practice they could overlap.

A small boy I knew was very ill. His father arranged for a medical examination, and it was established that the child had sleeping sickness. But the grandmother of the child was convinced that the real problem was that the dead ancestors of the family had not been 'fed' at the ancestral shrine since returning from Sudan. Others suspected that sorcery was at work. Many people were jealous of the boy's father: he had been elected as Resistance Council chairman for the division, an influential administrative position introduced by the National Resistance Movement government. There was considerable opposition from relatives to the idea that the child should go to the hospital in Moyo town because others had been taken there and had died. Pressure was placed on the father to consult various kinds of local healers and to use herbal remedies. He was also persuaded to arrange the sacrifice of a sheep at the ancestral shrine to placate his aged mother. Since he did not have a sheep, a chicken was killed, and a sheep or goat promised to the dead ancestors at a latter date. Eventually the child was taken to the hospital, and his condition subsequently improved.

No one doubted that the drugs given to the child helped cure him, but many nevertheless believed that this alone would not have worked. The boy's grandmother said that the affair revealed how, in spite of all the upheavals of recent years, the ancestors still watched what was going on and played a role in the lives of the living. Certainly, there was no question of the promised sacrifice of a sheep or goat not being performed when one could be afforded. In this way the grandmother asserted an interpersonal or moral explanation of illness. The suffering of her grandchild was a moment to reflect on the ties between members of her family, and between the living and the dead. She did not see this as a contradiction of the claims of 'scientific' medicine, but it explained why her grandchild had become ill at that time, and she had no doubt that if the ceremony had not been performed at the ancestral shrine the child would not have recovered. Incidentally, she also saw no contradiction between this veneration of her ancestors (or, more accurately, the patrilineal ancestors of her deceased husband) and her Roman Catholic faith (Figure 10.9). On another occasion she explained to me that the boy's life was actually saved because *Rubanga* (God) did not want to take him yet.

Thus, alternative explanations of illness might be applied in the same case, and sometimes explanations overlapped in such a way that contradiction did not occur. At other times alternative explanations for an illness might be the cause of debate, and which therapies ended up being adopted tended to reflect the authority and influence of individuals among the patient's relatives. In this

Figure 10.9   Palm Sunday in Laropi chapel, 1988: virtually all the Madi are Roman Catholics, but this is not seen as something that necessarily contradicts beliefs in ancestral ghosts, wild spirits and sorcery. The two white women in the congregation were members of the MSF medical team who had come to Laropi in order to spend a couple of days away from their compound in Moyo town.

instance, the Resistance Council chairman had wanted to demonstrate his 'progressive' attitudes by taking his child to the hospital immediately, but he was forced to accept the demands of his mother, who wished to remind her children of their duties and obligations. He also had to be careful not to be seen as acting in a high-handed manner, because if the child had died he would have been held responsible. If he could not demonstrate that all that should have been done was done, the child's maternal uncles would have demanded compensation for the death of their sister's offspring. If this had happened, he would have had no option other than to pay up (which he could not afford to do), or to have argued that the death of his son was not his fault but due to sorcery (*inyinya*).

In discussing this case I have mentioned a range of different interpretations of illness: sleeping sickness (i.e. a biomedically defined disease generally recognized as such by the population), the intervention of ancestral ghosts in the lives of the living, the will of *Rubanga*, and sorcery. In the following subsections, I comment on these and other local interpretations in more detail, and describe the kinds of therapists consulted in each case. For the sake of clarity I am going to structure my remarks into categories of illness causality. But it needs to be stressed that dividing up personal afflictions in this way does not always correspond with the way Laropi's residents thought about them. Different interpretations of what had caused an illness could be applied simultaneously, and in a serious case, it was common for a variety of therapists to be consulted. It is also important to remember that the Madi term *laza*, commonly used for illness in translation, carries broader connotations of individual and collective affliction.

Illness with impersonal causes

Illness diagnoses which might be classified as **impersonal** or empirical are those interpretations of symptoms which make no reference to interpersonal relationships either with the living or the dead. The cause of the illness is thought to be accidental or arbitrary or a predictable and natural consequence of the patient's own behaviour. Careful attention is paid to the symptoms, and specific therapies are selected to cure them. Speculation about deeper underlying causes is rejected or set aside. There is no generic term in the Madi language for impersonal interpretations, but sometimes people would say that a certain sickness was *laza Rubanga dri*, meaning 'illness/suffering from God' and implying that no one was to blame.

On a day-to-day basis it was generally accepted that some illnesses, particularly minor ones, just happened. It was unlikely that anyone would suggest that what from a biomedical point of view was a common cold was really the consequence of ancestors involving themselves in the lives of the living, and no one would believe a drunkard who claimed that his hangovers were caused by sorcery. Herbal remedies for a variety of complaints had been passed on from generation to generation, and doubtless there had always been occasions when illnesses had been treated without reference to interpersonal problems. It is also apparent that the promotion of empirical causality by many district officials and Christian missionaries since colonial times, the teaching of biology in schools, and the startling efficacy of

Impersonal: As used in this chapter, this term categorizes interpretations of illness which set aside moral or spiritual issues, and concentrate on the specific, physical symptoms manifested in the patients. Biomedical explanations of illness as diseases are an extreme form of impersonal interpretation, but the diagnoses of other healers can also be grouped into this category. Homeopaths, for example, are less prone to represent the human body as an object in which the causes of affliction can actually be isolated and seen, but their remedies are similarly aimed only at relieving the symptoms of a particular illness. (Homeopathy is a form of treatment employing minute amounts of remedies that in massive doses produce similar effects to those of the illness being treated.) In Laropi healers tended to specialize. Obviously district medical staff usually restricted their activities to treating illnesses which they interpreted as having impersonal causes. But so did many local herbalists who did not make biomedical examinations or prescribe biomedical remedies. Other healers specialized in illnesses which they interpreted as having interpersonal causes. However, most patients or their relatives were likely to consult both kinds of healers in serious cases, and impersonal and interpersonal explanations were not necessarily perceived as contradictory.

some imported drugs (like chloroquine) served to broaden this way of approaching personal affliction. Indeed, there were people in Laropi at the time I lived there who invariably began a quest for therapy, even in serious cases, by turning to a healer who specialized in treating illness as impersonal phenomena, and who would only consult interpersonal healers if that did not work.

Two therapeutic pathways were associated with impersonal explanations of illness. One pathway was to go to the dispensary or the hospital in Moyo and consult a nurse (people rarely had the chance to see a doctor). This was done when it was thought that the symptoms were those that could be cured by biomedicine. Of course Madi people did not themselves use the term 'biomedicine'. They called it 'Whites' medicine', or used an expression like 'people of the medical' to refer to district medical staff. The other pathway was to consult a local herbalist who might have alternative remedies to those sometimes available at the dispensary, but who also specialized in the treatment of certain illnesses which were believed to be 'Madi' or 'African' and which 'Whites' did not know about. In practice, patients or their relatives would commonly consult both types of healers, and in fact there was a considerable amount of cross-referral.

'Whites' medicine'

I have already commented on the inadequacies of biomedical facilities in the district in Section 10.3. Here I am concerned with how people living in Laropi thought about biomedical interpretations of illnesses, and how they responded to those biomedical therapies which they were able to obtain.

I have mentioned that some influential individuals, normally men who had attended secondary school, adopted narrowly biomedical interpretations. They did so with more apparent confidence than most people would in Western countries, at least partly for economic reasons. They were often among the more affluent in the village, and promoting biomedicine could be a means of refusing to pay for their relatives' consultations with local healers. Treatment at the dispensary was free or relatively inexpensive, and even biomedicines bought in the market were cheaper than many local remedies obtained from herbalists, let alone the long and complicated rituals performed by interpersonal healers. However, the actual number of individuals who denied the value of non-biomedical healing was very small. Out of a population of 3507 in 1987, only 187 people had attended secondary school (of which 85% were male), and by no means all of these were opposed to local healing. Moreover, even the most vociferous opponents commonly found that they were drawn into pluralistic responses to illness by the social pressures placed upon them, and there were

also several instances during the time I spent in the village when a prominent man's strict views quickly became compromised when he or one of his own children became ill.

For the vast majority of people living in Laropi, attitudes to biomedicine were ambivalent. People may have had some vague notions about the microbiological world and germ theory gleaned from primary school, or from relatives who had greater access to these ideas at secondary school, or from a health education poster that they had seen somewhere. But connections were rarely made between a drug, such as an antibiotic, and the biomedical concept of infection. Nevertheless, it was generally recognized that some illness symptoms could be cured by district medical staff, or by the use of pills and injections. Particular sets of symptoms were associated with words which corresponded to biomedically defined diseases. There were, for example, Madi expressions for sleeping sickness (*mongoto*), yaws (*badra*), leprosy (*dobo*, or *kofo*), meningitis (*lamolamo*), smallpox (*longbongi*), venereal diseases (*kabaruje*, *njuku*, or *vorodra*), and scarlet fever (*yeye*). In all these cases, the diseases had been the focus of control programmes since colonial times (and therefore an awareness of them as distinct phenomena had been advertised), or people in Laropi had discovered that imported medicines could alleviate the particular illness symptoms that they produce. When people had one of these illnesses, there might have been some discussion about why they had become ill rather than a neighbour, but usually they would go in search of a biomedical remedy.

If they lived nearby, people might also go to the dispensary when they had a wound which was deep or had not healed on its own, when they had severe and prolonged headaches (a symptom commonly referred to as 'malaria' although biomedically speaking it might not be this disease), when they had chronic and severe diarrhoea (*oca*), or when they had a persistent runny nose (*ku'duku'du*) (Figure 10.10). They would hope to receive basic first aid, chloroquine and aspirin for 'malaria', and antibiotics for almost anything else. People would be more prone to visit the dispensary if there was an individual working there whom they trusted. They went to consult the conscientious nurse at Laropi about a whole range of things, sometimes unconnected with illness, but attendance dropped dramatically when she was not on duty.

If drugs were unavailable at the dispensary or staff were unsympathetic, treatment might be attempted at home. Chloroquine, antibiotics and aspirin were usually available in the small shops at Laropi trading centre, and sometimes other drugs also found their way into the market. From a biomedical point of view this could be counter-productive. It was common for patients to end up being treated with an inadequate dose, or the wrong drug, while repeated use of syringes at home without proper sterilization caused abscesses and spread certain diseases. The fact

that people resorted to home treatment reflected a common belief that if more 'White' medicine were available, some illnesses would disappear. Sometimes people would say to me things like: 'You Whites are always active and stay young because you have a lot of injections and tablets'. Occasionally there was an implication that 'Whites' medicine' might actually give some protection against interpersonal afflictions, and probably most people suspected that expatriate aid workers, and 'Whites' generally, had more knowledge and resources than they were willing to part with.

Local herbal treatment

Although normally people did not themselves make the link, the re-using of syringes for home treatments could cause jaundice. They called this condition 'yellow fever', although biomedically speaking it was hepatitis and not the yellow fever

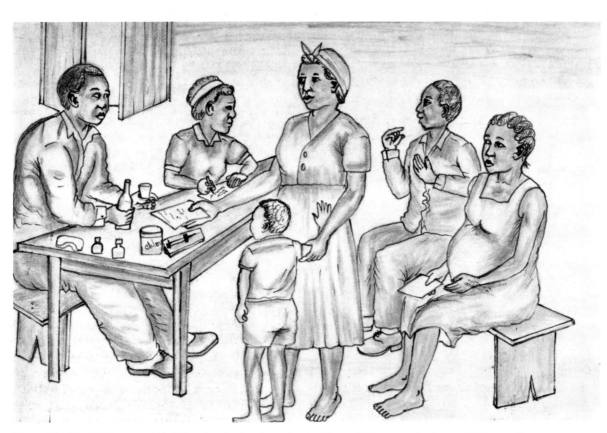

Figure 10.10 'We have no medicine for that': a drawing by Maxwell showing patients visiting a dispensary. Note the tin of chloroquine tablets and the syringe on the table.

disease. This was one of the illnesses which were often thought not to be treatable at the dispensary. It was frequently treated at home or by a herbalist who concentrated on the healing of illnesses thought to have impersonal causes. These specialists called themselves *daktari* (the Swahili word for doctor) or *erowa dipi* (owners of medicine).

All local healers, including healers of interpersonal illnesses, provided herbal treatments of one sort or another, and some simple herbal remedies were known to many people and used at home without consultation. But *daktari* were herbal specialists whose role had become increasingly important as notions of impersonal causality became more current and biomedical facilities did not meet the demand for curative therapy. Like district medical staff, they offered treatment to alleviate particular illness symptoms, and did not usually attempt to deal with any possible underlying interpersonal causes. *Daktari* were usually men, or women who were 'like men' in that they had some sort of status other than that normally available to women (Figure 10.11). The sister of a prominent ex-minister in Obote's government worked as a *daktari* in Moyo town.

Daktari treated patients with medicines made from roots and other parts of plants, which they claimed to have been revealed to them by *Rubanga* in dreams. Sometimes they massaged a patient with an oily substance, and magically removed objects from the body, asserting that these were the source of the problem. More often their remedies were swallowed or rubbed into small cuts made in the skin of the patient. Most concoctions acted as purges or pain killers. From a biomedical point of view, some treatments certainly seemed to be efficacious. A few were dangerous, particularly when unsterilized razor blades were used to make incisions. Occasionally, people treated by a *daktari* ended up in hospital with badly infected wounds.

Apart from jaundice, other illnesses which they had a reputation for being able to cure included the afflictions *jue* and *obui*, 'false teeth', and severe tropical skin ulcers known as *burule*. The terms *jue* and *obui* were used very generally. *Jue*

Figure 10.11   A widely respected daktari, who was 'like a man' because she was the relatively affluent sister of a former government minister; she worked near the market in Moyo town.

refers to any kind of swelling, and *obui* literally means 'worms', but can be applied whenever the patient feels that there are things moving inside his/her body. 'False teeth' are the lower canine milk teeth of small children. At the time of teething many children have diarrhoea and fever and it was believed that these symptoms were caused by worms (*obui*) in the gums. The treatment was to cut out the lower canine milk teeth with a razor blade or a nail. Obviously, from a biomedical point of view this was dangerous. A similar problem arose with some of the treatment *daktari* provided for *burule* ulcers. *Burule* is a virulent tropical skin ulcer which, biomedically speaking, is caused by certain bacteria which live on a grass growing in riverine areas. It often does not respond well to antibiotics, which is the reason why people tended to believe it was a 'Madi' illness needing local treatment. The therapy given by *daktari* usually involved deep cuts being made to drain off pus, in combination with the use of herbal lotions to rub on the wounds. I know of cases where further infections resulted. Some *daktari* said that *burule* were caused by 'eggs', and occasionally they attempted to remove these from the body. These 'eggs' are the swollen lymph nodes of the patient, and I came across instances where patients died as a consequence of this treatment. On the other hand, I knew of two cases where patients were completely cured of *burule* by a *daktari* after having been told by district medical staff that amputation of the affected limb was essential.

Daktari usually denied vehemently any connection with *ojo*, the professional spirit mediums who engaged in healing afflictions thought to have interpersonal causes (see below). They would treat cases of *inyinya* (sorcery/poisoning) when the afflicted person was believed to have drunk or eaten a toxic substance, but they would not attempt to locate the sorcerer or provide vengeance medicine. They asserted that they provided similar services to the district medical staff, and they gained credibility as a consequence of half-hearted attempts to professionalize 'traditional' healers in public health programmes, particularly during the 1970s (Box 10.3). They tended to work close to a dispensary, and model their techniques on hospital-based health care. In effect, they ran their own private clinics, and sometimes they would even have registers of their patients, mentioning the date of the last visit and the medicine given. They charged for their services in instalments: the first after diagnosis and treatment, which might cost 300–400 shillings plus a cockerel, and the second after the treatment had been successful. Charges varied, but one healer I interviewed demanded 5000 shillings for successful treatment of a man, and 6000 for a woman. These were large sums of money. In 1989, 5000 shillings was the cost of a young bull. A large pumpkin cost 100 shillings, a pint-sized mug of maize seeds cost 20 shillings, and a one-litre bottle of locally distilled *waragi* cost 120 shillings.

In addition to *daktari*, there was another group of local healers who sometimes dealt only with impersonal health problems. These were women known as 'traditional birth attendants' (TBAs).

Box 10.3   The professionalization of 'traditional' healers

The expression *'traditional' healers* is frequently used as a catch-all category for all healers who are not practising biomedicine. It includes such healers as herbalists, spirit mediums, exorcists, holders of religious or ritual offices, and individuals trained in the ancient Chinese medical system. Logically it ought to include psychoanalysts and other non-biomedical healers working in Western countries, but in practice it tends to be applied only to non-biomedical healers in non-Western countries. It is an ambiguous and misleading term in that many of the healing practices it refers to are as different from each other as they are from biomedicine, and also because many of them are not actually 'traditional' in that they may have only been introduced recently (for these reasons I prefer to use the term *local healers*).

The idea that 'traditional' healers might be *'professionalized'* has gained currency since the 1970s. Some of those who have promoted this strategy have thought of it as a means whereby the position of non-biomedical healers might be enhanced vis-à-vis qualified biomedical professionals (doctors, nurses, etc.), and thereby ensure that their knowledge is used effectively and handed on to the next generation. However, in practice the 'professionalization' of 'traditional' healers has tended to mean that some kinds of non-biomedical healers have been given a basic grounding in biomedicine. The effect can be to change the role of the healer from a non-biomedical therapist into a biomedically oriented paramedic. The approach has been used as a way of trying to provide biomedical public health care by the cheapest possible means. These issues are discussed in detail in Last & Chavunduka (1986).

The term TBA had no Madi equivalent. It was introduced by district medical staff and expatriates working for aid agencies who were looking for 'traditional' midwives in order to provide them with some basic biomedical training. Women who had experience with difficult deliveries came forward and registered. They attended courses and a few began to operate as paramedics, combining local herbal treatments with what they had learnt, and referring serious cases to the dispensary. But most local midwives did not work in this way. They were healers of interpersonal affliction, and when they registered as TBAs it was partly to gain wider recognition for their activities through their association with outsiders. Some of them were in fact *ojo*, and their role in midwifery is discussed in the following subsection.

Illness with interpersonal causes

There were many different kinds of affliction which could be interpreted as having interpersonal causes. The symptoms of illnesses mentioned above were commonly treated by herbalists or district medical staff, but this did not necessarily mean that an interpersonal healer would not be consulted as well. Going to a *daktari* would not explain why one child was suffering from *burule,* but another was not. Finding this out might be considered essential, particularly if the same symptoms were manifested in several members of the same family. Moreover, some illness symptoms, like repeated vomiting or the hearing of voices in the head, were frequently taken as clear indicators that interpersonal issues were at the root of the problem. Serious illnesses and sudden deaths almost invariably prompted speculation about who was to blame, and an interpersonal healer would usually be consulted by at least some family members to find out what was going on. In addition, when it came to interpersonal suffering a distinction between illness and other afflictions broke down. An interpersonal healer might be expected to provide medicine to prevent the failure of crops, theft, or the capsizing of canoes on the river. He or she might be expected to cleanse the community of evil or amoral elements, and to purify the land of the blood spilt upon it during the killings of the early 1980s. Thus interpersonal healing could be as much about the fabric of social life as about an individual's problem. It might require establishing who was accountable and who was responsible. In an extreme situation, like that in Laropi, this might involve violence.

While particular symptoms in a patient might indicate to people that interpersonal healing was essential, therapy did not so much address these symptoms themselves as confront their deeper causes. An interpersonal healer might give a herbal remedy to help alleviate the symptoms, but the real cure was in effect a by-product of sorting out interpersonal issues. In the past, by far the most important arbitrators of interpersonal causality, and probably in most cases the first resort in a quest for therapy, were the senior male elders.

Ancestor invocation

According to Madi custom, deceased ancestors (particularly patrilineal ones) caused illness and other afflictions to happen to their living relatives. They might do this because their wishes were being ignored, or because one of their descendants was behaving in an outrageous way, or because they had been invoked and asked to intervene. Everyone had the right to invoke ancestors, but male elders had a duty to do so in order to stop wickedness in their homes.

At times of illness, a ritual might be performed by the male head of the home at a shrine located at the edge of his compound where he would communicate with the ghosts of his immediate relatives (including those of his mother's family) (Figure 10.12). In more serious cases, the patient would go with his or her close relatives to see a senior elder who 'owned' the ancestral shrine of the patrilineage. This elder would then call upon

Figure 10.12   A man points to his family's ancestral shrine, which is almost hidden by grass at the edge of his compound in Laropi.

all the patrilineal ancestors to guide him, and would give advice about what needed to be done. Healing rituals would involve a gathering to discuss various interpersonal difficulties (such as jealousy between co-wives), and a special meal was eaten to feed the dead. If possible, unsalted meat was consumed, and a small amount placed under the alter of the shrine. Coercive measures might then be brought to bear upon those whose behaviour was considered to be the root of the problem.

These events used to be the custom, but well before the flight into Sudan the capacity of ritually important elders to regulate social life in this way had been weakened. The effects of colonial administration, commoditization, and the migration of young men out of the area in search of work had undermined their authority. Moreover, in exile, the people of Laropi had lived among families who were not close kin, and who did not share the same ancestors or ancestral shrines. Nevertheless, on returning to Uganda some old people attempted to reassert what they believed to be customary values. Often they made reference to things that were done in their childhood, and asked for ceremonies to be performed by elders who had inherited ritual offices, of which younger relatives professed ignorance. They argued that harvests were poor and people became ill because rites had not been performed to placate the ancestors and to purify the land of the blood spilt upon it. But in most instances these assertions were tactfully ignored. In Dufile, a neighbouring village to Laropi, a ceremony to purify the land and call upon the ancestors to help ensure good rains and fruitful harvests was not performed until 1989, three years after returning. Up until the last time I visited Laropi (1991) it had still not been done. Even the house where the special ritual instruments for ensuring good rainfall should have been kept had not been rebuilt.

In individual cases, however, the elderly might be more successful in promoting their interpretations. For example, when the person who was ill was an ex-soldier, they might argue that his suffering was a punishment inflicted by patrilineal ghosts for having killed innocent civilians and looted their properties. Such interpretations were widely accepted, and stories were told of soldiers dying, or becoming deranged with remorse in the vicinity of their ancestral shrines. All patrilineages in Laropi still had shrines, and numerous homes also had one for the immediate family. Occasionally, therapeutic rituals were performed, as in the instance of the boy who had sleeping sickness. His aged grandmother insisted on a commitment to 'feed' the ancestors being made, before she would agree to his father taking him for treatment at Moyo Hospital. There was, moreover, an event when ancestors were believed by many still to be particularly active in protecting their living relatives – when a woman was suffering from labour pains.

According to custom, a young woman left her father's house at marriage and went to live with her husband's family. In return, her husband's family would pay a first instalment of bridewealth (or brideprice), which might be in the form of animals and hoes or money. This marked the beginning of a process whereby her connections with the patrilineage of her birth were largely replaced by connections with the patrilineage of her husband. But this could take a long time, sometimes several years. It was not really complete until all the agreed bridewealth had been paid, and the woman had given birth to children of her husband's patrilineage (similar customs are described in Chapter 6 for a village in Bangladesh). During the first years of her residence at her husband's home, her mother-in-law and her husband's other female relatives would watch her carefully. She was still an outsider, and her true loyalties might remain with her father and brothers. If things went missing it might be suspected that she was stealing things to send back to her father's home. A jealous co-wife or sister might also suspect that she was a sorcerer, or someone who used magic to make her husband love her more than anyone else.

Childbirth is the moment when these suspicions could be confirmed (Figure 10.13). In the pain of delivery she would be encouraged to reveal her secrets, and the ancestors of the home might make her do so even if she resisted. A woman who gave birth in hospital away from her husband's female relatives would be able to keep her secrets,

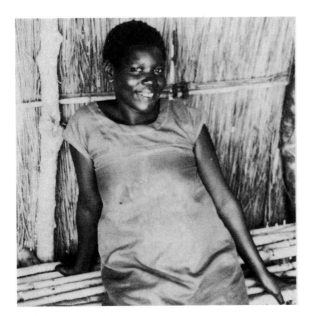

Figure 10.13 New wives do not really become part of the home until they have produced children for the patrilineage.

and might be suspected of escaping from the family because she had something terrible to hide. When there were labour complications an experienced woman came to the home in order to convince her that only by being honest would her husband's ancestors allow her suffering to cease. Several of the women who came forward for TBA training were people who had a reputation for doing this. Some of the best known of them were *ojo*, the female spirit mediums discussed below who acted as diviners when a person was thought to have been ensorcelled or possessed by wild spirits of the bush.

By no means everyone accepted that it was good to put pressure on delivering women in this way, but it only took one or two influential family members to raise questions about a woman's probity for it to become an issue, and in the years following the repatriation from Sudan new wives were probably more suspect than they had been previously. The upheavals had made it impossible for most men to pay the full brideprice. This meant that his wife's association with her husband's patrilineage remained unclear, and so did her children's. Her father could claim that they belonged to his.

Because almost everyone had similar difficulties, a compromise was usually worked out whereby a man would promise to pay brideprice when his family could afford it, and in addition pay a 'fine' for each child that was born before doing so. But this meant that a wife was much less tied to her husband's patrilineage than would be the case if bridewealth had been transferred. If she returned to live with her father and brothers, or moved away to live with another man, her husband could not demand any recompense. In the past, the bridewealth would have had to be returned. On the one hand, this gave some women an independence which they could not have enjoyed in the past, but on the other, it meant that a woman's commitment to her husband's patrilineage was more open to question. As we shall see, the resulting tension could have disturbing consequences.

Communication with wild spirits

Wild spirits were ghosts of individuals or other metaphysical forces which were often not connected with deceased family members. The interventions of ancestral spirits in the lives of the living was conceived as something from within the home or patrilineage, and something morally good, even if it may have adverse short-term consequences. In contrast, the interventions of wild spirits were viewed as something dangerous and unpredictable. In the Madi language, the term *onzi* might be used to describe them – a word which is often translated as 'evil', but might perhaps be better rendered as 'amoral'.

Wild spirits had long been thought of as phenomena affecting women. In the past and still today, infertility was associated with water spirits which could seize a victim if she went to the river alone or late at night. But since returning from Sudan there had been an epidemic of possessions, mainly of women, by ghosts of individuals who roamed the bush, in many cases after having been killed during the war. Symptoms of possession could include those associated with illnesses commonly treated by impersonal healers. If a woman had headaches or swellings which persisted after consultations with herbalists and/or district medical staff, it might be suspected that possession was the real cause. But in other cases the manifestations of possession in the victim were dramatic. She started to do peculiar things, like sleep in a tree, speak in strange voices, or dance as if berserk. So many were affected that to see a woman behaving strangely was a common occurrence. Even scenes like a teenage girl running into the forest screaming and waving her hands in the air, or throwing herself on the ground and covering herself with earth, were hardly noticed. People just remarked that it was *jok jok* and passed on. The term *jok jok* is taken from the Acholi and Langi word for 'spirit' (the Madi word is *ori),* and highlighted the fact that this kind of possession was perceived as something that had spread quite recently from Acholi area.

Cases of *jok jok* were taken to *ojo,* female healers who were themselves possessed. In the past, *ojo* seemed to have been young women who had been seized by an aspect of *Rubanga,* and they would only be consulted as a last resort, usually when sorcery was suspected. At the time of my fieldwork, *ojo* were prominent figures who would be consulted by many families whenever a problem was thought to be serious in much the same way as ritually important elders once were. Some of them became relatively affluent from the payments they received, which certainly rivalled those of *daktari.* In addition to curing people who were ill, they were commonly consulted in matters of theft, or in order to obtain protective medicine when travelling or doing business. They were usually possessed by *jok jok* spirits who spoke through them in strange, high-pitched voices, often in Acholi and other foreign languages. At one time they would have suffered from their possession like their patients, but they had learned to live with their spirits, and now worked as diviners. One of the most successful *ojo* in Laropi was possessed by five named spirits, including an Acholi and a murdered bishop from the south of Uganda. Like most other *ojo* she understood her possession as something that brought her close to *Rubanga,* and the room in which she divined was decorated with pictures of the Virgin Mary. Her spirits told her to pray several times each day, and singing hymns was the way in which she calmed down her spirits when they became too excited. All the *ojo* I met were devout Catholics.

When an *ojo* was employed by relatives to cure someone who was thought to be ill as a consequence of *jok jok* possession, the treatment would normally involve a combination of herbal remedies (which had been shown to the *ojo* by her spirits) and divination séances. The *ojo* would ask her own spirits to assist the patient, and would go into a trance-like state. Sometimes she would dance all night together with the patient and other women who came along for the occasion (Figure 10.14). Young men played drums, and quantities of strong alcoholic drinks were consumed. It could be very exciting. The spirits of the *ojo* would speak through her, calling upon the spirits possessing the patient to show themselves, and explain what they wanted. When this happened, often grievances concerning the home were revealed. At one séance I attended, a spirit explained that it had seized the woman because it felt sorry for her, having heard her crying after being beaten by her husband. On another occasion a spirit surprised everyone by revealing that it was the patient's dead father, and then proceeded to castigate his sons as drunkards who were not looking after their sister properly. The spirits would demand that things be given to the woman they were possessing in order to make them go away.

Not everyone treated in this way was cured. But it was remarkable how their condition, even of people who seemed to be seriously disturbed, tended to improve. Some afflicted women I knew showed no further symptoms, and a few had themselves been initiated as *ojo,* building up their own reputations as healers and diviners (Figure 10.15). Others had at least been calmed by the *ojo* who treated them, and their illnesses held in check.

Observing wild spirit possession as an outsider, I thought that some of those afflicted were individuals who were finding it hard to cope with the appalling traumas they had experienced. The externalizing of their problems in the form of spirits seemed to help. In this respect it is surely significant that the spirits were usually from the bush rather than the home, from the realm outside normal or moral intercourse, and that they often spoke in the language of the soldiers who

Figure 10.14   Jok jok dancing, drawn by Maxwell. A patient dances wildly holding the ojo's spear. The ojo herself is kneeling on a mat in the background. Men play drums and comment on the proceedings.

had committed atrocities. *Jok jok* possession might in this way be understood as a way of making a kind of sense out of what had happened. It was also tempting to see at least some cases of *jok jok* possession as a form of resistance by women. The spirits frequently made economic demands which had to be met by male relatives of the woman possessed, and men were often severely criticized in a manner that would be unacceptable in other circumstances. Possession could therefore be seen as a reflection of the weakened control men had over women as a consequence of their incapacity to pay bride-wealth, and the fact that in many families it was the women who were earning an income by distilling and selling *waragi*. Certainly, some men would talk about *jok jok* possession in these terms. When relatives demanded that they provide money, animals or other things to pay for the treatment, they would complain that their womenfolk were not really ill, and asserted that possession had become an epidemic simply because women enjoyed the séances.

Nevertheless, they were almost invariably forced to contribute in the end.

Sorcery accusation

Men were usually more eager to pay for the services of an *ojo* when affliction was thought to be the consequence of sorcery. While *jok jok* possession was something mainly associated with women, *inyinya* could affect anyone. It was widely held to be the primary cause of serious illness, and to be the major cause of death. From the perspective of most people in Laropi, it was probably this more than anything else which made manifest the upheavals of recent years. Moral behaviour could not be expected even within the home. Social ties were extremely fragile.

Sometimes *inyinya* was referred to by the English word 'poisoning' and vomiting was a symptom occasionally associated with it, but I knew of no case in which it was proved that a chemically toxic substance had been used, and in fact even when people talked of poisoning they commonly indicated the use of magical techniques. The district magistrate in Moyo, for example, told me that she had arranged for her office to be securely bolted in order to prevent anyone from putting poison in it, and a recently ordained Madi priest assured me that he had seen a poisoner harming a victim just by touching him. In August 1989 an edition of a Catholic newsletter reported with approval an incident in Arua District (to the west of Moyo District) in which people had been found guilty by the

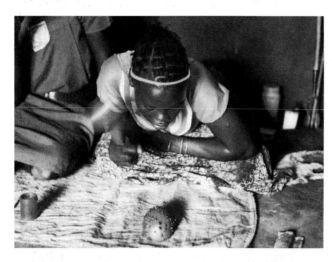

Figure 10.15 (left) Talking to the spirits: an ojo speaking to voices coming from the calabash in front of her. She was training a young woman who had recently become possessed. (below) A new ojo. The daughter of one of the men at Laropi who had been most vocal in promoting biomedical interpretations of illness became possessed in 1991. Relatives persuaded her father to pay for the expensive ceremonies to train her as an ojo. Her spirits had shown her a herbal cure for AIDS, and she already had a large number of people wanting to consult her.

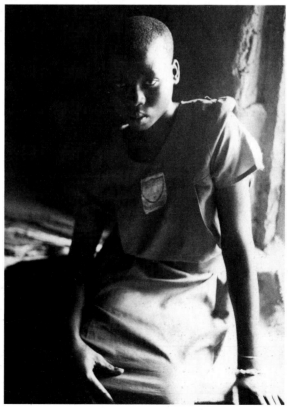

local government chiefs and Resistance Council representatives of having 'witchcraftly' poison. One of the poisons was described as involving the rubbing of a substance on the poisoner's own hand and face and transmitting it by handshake or a hard fixed look at a person he wanted to kill.

Almost any kind of affliction could be seen as a symptom, from crop failure to boils and vomiting. In one instance, a man who had attempted to kill his own mother was believed by his family to have been ensorcelled. The motive was usually *ole* (envy). It was thought of as being practised most commonly by women who had not been fully incorporated into the local patrilineage, or by marginal men, such as those living with in-laws (who were therefore 'like women' because they were not living at their father's home).

When someone was thought to be a victim, an *ojo* was asked to divine the source of the problem. She would throw coins and cowrie shells on a mat and read meanings in how they fell, or put oil on a pebble and see faces in it. All the time she would be talking to her patient and his or her relatives, drawing from them suggestions about who might be to blame. Sometimes she would conclude that it was not a case of *inyinya* but possession or an impersonal illness. She might provide herbal remedies herself, or remove objects from the body of the patient as a *daktari* did, or refer the patient to the dispensary. Usually if she concluded that sorcery was at work, she would provide some kind of magical protective medicine. Occasionally, *ojo* were also employed by a group of families when many people in a particular neighbourhood had fallen ill or other misfortunes were occurring frequently. Her spirits would enable her to find out if harmful magic had been deposited somewhere, perhaps on a path or in a field. All *ojo* whom I interviewed denied ever blaming a particular individual, even when they had discovered who the sorcerer was, but people would often come away with a pretty good idea, probably one which corresponded with their own suspicions. *Ojo* also denied providing clients with their own sorcery in order to take revenge, but it was widely believed that some *ojo* did have the knowledge to do so, and would offer it for a high enough price. This was one

of the reasons why an *ojo* would assert a deep Catholic faith. By emphasizing that she was a good Christian, she would hope to avoid any suggestion that she might be using her skills for bad purposes.

After consultation with an *ojo*, it was unlikely that the afflicted person or any of his or her relatives would publicly accuse anyone. Rumours would circulate about who might be responsible, and individuals could be made to feel uncomfortable, but it was usually only at funerals that people were openly named and confronted. This happened frequently, particularly once the mourners had begun drinking *waragi,* and fierce arguments ensued. As a rule, things eventually calmed down, although people would remember what had been said, and it was difficult to shake off a reputation as a sorcerer once suspicions had been openly expressed. On occasion, however, a consensus built up that a particular person was undoubtedly the culprit. The situation could then become extremely violent. In some instances, the accused ended up being severely tortured and killed. I heard of ten such deaths that occurred in the Laropi area during the course of my fieldwork.

In one of these cases, a man killed a woman whom his father had recently married, together with her daughter from a previous union, at his son's funeral. In other instances, Resistance Council representatives set up a court and killed the accused when they were found guilty. Such courts were not legal, but it was widely believed, even by some magistrates and police officers, that district courts could not act effectively because the laws concerning witchcraft and sorcery accusation required that the prosecution provide biomedical evidence from a post-mortem. This was never forthcoming. The activities of these Resistance Councils were therefore not disapproved of as much as they might be, and some of the Councils in the district had petitioned the president to grant them authority to officially take over the trying of poisoning cases.

It needs to be stressed that there were those who strongly opposed these proceedings, including many Resistance Council representatives. Some people said that the so-called courts were little more than groups of drunk relatives. They cruelly tortured and murdered individuals whom they did not like, and who did not have members of their patrilineage in the vicinity to protect them. Nevertheless, no one involved in these killings was punished. The man who killed the woman and her daughter at his son's funeral was taken to prison for his own protection. Brothers and maternal uncles of those killed came to Laropi demanding an explanation and compensation. But he was later released and was living at home with impunity. Most people seemed to feel that the women must have been guilty, and stories were told that the mother had been accused of sorcery before when living at one of the refugee settlements in Sudan. The mother left behind a baby, but no one would take care of him and he subsequently died.

It may seem odd to end a discussion of healing in Laropi with comments about brutal killings, but the fact is that many people saw these events as part of that process. Communities could not be properly established unless they were cleansed of sorcery, and those responsible for suffering held to account. While elders had not found it easy to convince people that ceremonies had to be performed to bless the land and feed the ancestors in order to allow social healing to occur, most families were much more ready to accept that sorcerers should be eradicated. The evidence of their wickedness was all around them: misfortune, pain, illness and death.

10.5 Wider implications

This has been a detailed and specifically focused case study. What are its wider implications? How should it be put into context? I briefly comment on three issues.

- *The situation in Laropi is not exceptional*

Both this chapter and Chapter 9 have presented case studies of Ugandan villages in which people were living in extreme circumstances. Unfortunately, these circumstances are not exceptional. Mamdani pointed out that the grim processes of class formation he described in Amwoma are occurring in many parts of the world. Similarly, the Madi are not alone in having suffered upheaval. Other populations have had equally traumatic experiences.

Africa has been particularly badly hit. Conflicts in Uganda continue to simmer. Ethiopia, Eritrea, Sudan, Chad, Angola, Mozambique, Liberia, Zimbabwe, Western Sahara and Somalia have all been thrown into turmoil in recent years. South Africa has become increasingly unstable. Many rural populations have been caught up in wars that may have nothing to do with them. Public services have been destroyed and farms abandoned. Thousands have been killed; hundreds of thousands have become refugees. Some regions have become almost synonymous with famine.

In spite of international awareness of what is happening and concern for those caught up in these tragedies, the practical response has been inadequate. There are, of course, relief and development success stories, but time and again aid arrives too late or is inappropriate. Mortality and morbidity rates have sometimes been shockingly high even in places where international aid agencies are operational. In most countries in sub-Saharan Africa the vast majority of the population has little or no access to any kind of public health care. It is a grim situation which is unlikely to change in the immediate future.

It is hardly surprising that some of the most severely afflicted populations are violent. When moral norms and established codes of behaviour are undermined, certain cultural forms may be exaggerated. Inevitably the therapies which help people to cope with suffering vary between individuals and groups. Nevertheless, manifestations of the kinds of phenomena found at Laropi are quite common throughout Africa, and take on particular significance at times of stress. During the independence war in Zimbabwe, for example, the hunting down of sorcerers and the eradication of government informants went hand in hand, while the support given to the guerrillas by spirit mediums was crucial in mobilizing rural people to the cause (Ranger, 1985).

- *Non-biomedical approaches to healing are widespread*

It was pointed out in Section 10.1 that popular conceptions of illness in Western countries may differ from biomedical ones. AIDS and cancer are sometimes imbued with moral significance, and in certain circumstances people may turn with their problems to psychoanalysts, osteopaths, counsellors, aromatherapists, homeopaths, spiritualists, priests, or a host of other specialists. It is also worth remembering that the dominance of a hospital-based biomedical establishment in the West is a relatively recent development. The breakthroughs in medical science which have helped establish its position have largely occurred in the past hundred years or so. Nevertheless, most people will accept that 'doctors know best', and will usually subscribe to biomedical interpretations of illness as a first resort. The ultimate form of diagnosis is generally recognized to be the biomedical post-mortem.

This is not so in many other parts of the world. The majority of people living in poor countries continue to be unconvinced that the human body is a mere system in which malfunctions can be identified, and that it is divorced from the mind, the soul, and moral concerns. There is a vast array of different approaches to healing, which are usually practised from information passed on by word of mouth (although, as in Laropi, practitioners may make reference to spiritual forces). A few have been elaborated into knowledge systems which rival biomedicine in complexity.

Although most spending on public health is aimed at improving biomedical care, in several countries non-biomedical healing receives state recognition or even some state sponsorship. In Zimbabwe, the activities of 'traditional' healers (including both herbalists and spirit mediums) were legalized soon after independence. The Zimbabwe National Traditional Healers Association was formed in July 1980 to regulate their activities, no doubt partly because the government recognized the tremendous sociopolitical influence of local healers, particularly in rural areas. In India, institutions teaching Ayurvedic, homeopathic and other non-biomedical health care systems are partly state funded (Ayurvedic medicine is an ancient body of knowledge associated with certain forms of Hinduism). In China, a remarkably successful public health care programme has used 'bare foot' doctors, most of whom have training in medical skills premised upon conceptions of the human body which relate to philosophical ideas and which make no sense in biomedical terms.

There are many reasons why non-biomedical healing remains so prevalent, some of which are locally specific. But in broad terms two factors are fundamental: first, the limitations of biomedicine and biomedical practice, and second, the strength of alternative shared values about healing.

When WHO was set up after the Second World War, it adopted the far-ranging definition of health mentioned in Section 10.1 ('a state of complete physical, mental and social well-being'). However, in practice it has tended to promote biomedical approaches to the control of diseases, and ministries of health in poor countries were encouraged to develop expensive public health facilities based on Western models. With few exceptions, this had the effect of concentrating limited resources in urban hospitals, and making them available only to a small élite. In 1975 a study undertaken jointly by WHO and UNICEF estimated that only 15% of the rural and non-privileged urban population in developing countries had any access at all to government-run health care services (Djukanovic & Mach, 1975).

Subsequently, attempts have been made to change this situation. Primary health care programmes are meant to be focused on the poorest and most marginalized groups. UNICEF maintains that it is possible to raise levels of health without necessarily raising levels of economic affluence (UNICEF, 1990, p.37). But headway is slow everywhere, and it is hard to see improvement in much of Africa. If anything, the situation has deteriorated, so that even the urban hospitals fail to operate effectively. Sometimes people know that they are suffering from a disease which has a biomedical cure, but it is often impossible to obtain the treatment.

At the same time, doubts have increasingly been raised about the scope of biomedicine both by patients and by practitioners themselves. It is accepted that the application of scientific method has brought dramatic improvements in the diagnosis and treatment of particular health problems, notably nutritional deficiencies, infection and injury, but in other conditions the results have been less striking. As one WHO publication puts it:

"in conditions where behavioural, emotional or spiritual factors play a major role it would be difficult to argue that the scientific method has produced noticeable improvements; some would contend that deterioration is evident. Since psychosomatic disturbance is today one of the commonest of human afflictions, the philosophy and functioning of modern health and medical services is being questioned in many quarters."

(Bannerman, Burton & Wen-Chieh, 1983, p.11)

These limitations partly explain the growing enthusiasm for holistic health care in Western countries. They also underlie what is sometimes thought of as cultural resistance to biomedicine. In a study of pluralistic approaches to healing in the Horn of Africa, it was pointed out that it has:

"gradually become apparent that biomedical programmes, introduced in areas where traditional, indigenous and unorthodox systems are strongly represented, will encounter active opposition and criticism."

(Slikkerveer, 1990, p.23)

Even when some biomedical facilities are available they may be underused, and biomedical interpretations of illness may be rejected.

The values a group of people share are always changing and are always in the process of negotiation. Nevertheless it is through them individuals within the group find meanings in things. Arguments about social responsibility and moral relationships occur everywhere, but it seems to be generally the case that less commoditized cultures do not emphasize individualism to the same degree (or in the same way) as in the West. Indeed, individualism of a certain kind is often seen as the key element which makes Western or Westernized cultures distinct. In other cultures, ideas

about the body and the nature of the person may run counter to the individualistic assumptions of biomedical knowledge. Health and illness may be viewed mainly in collective and metaphysical terms. Biomedical interpretations of illness may be incomprehensible, or when they are understood, they may be ignored or set aside in order to maintain or re-establish the premises of communal life.

But we must be careful here. While recognizing that there may be cultural resistance to biomedicine in some circumstances, we should avoid cultural determinism. Cultures (and especially 'other' cultures) are still commonly discussed as though they are closed and stable systems. They are not. Rather, as is seen in Laropi, they are dynamic, interactive and eclectic.

Where biomedicine seems to be resisted, it is likely that there will be location-specific reasons. Perhaps the state is using the hierarchical organization of biomedical practice as a means of extending its control over the population. Perhaps biomedicine is being promoted by outsiders who have failed to develop a means of communicating with the population. Maybe they do not even speak the language (international aid agency staff rarely do). Perhaps biomedicine has a reputation for not working (there are countless examples where defective or inappropriate drugs have been used to ill effect).

But nowadays a total rejection of biomedicine is very rare. Laropi represents the more usual kind of situation. Here biomedical interpretations were promoted by some individuals, but most people drew upon a range of alternative or overlapping explanations of illness. People were unlikely to reject a particular biomedical cure if it was available and had been seen to work, but this did not mean that they would necessarily adopt a biomedical view of what was wrong with them or a biomedical explanation of what made the cure effective. If public health care facilities improve, it is likely that biomedical interpretations will be more commonly accepted. But this will not displace a deeper questioning of causalities, and is unlikely to have much effect on what from a biomedical point of view are psychosomatic disorders. The killing of sorcerers would probably stop if morbidity rates declined, but they would do so anyway if families managed to re-establish their farms and regain former levels of income.

• *Local knowledge about the human body is different from local knowledge about plants and animals*

It has already been mentioned that the conceptualization of the human body as a system which functions like a mechanical object seems strange to many people. It is worth ending by reflecting on an aspect of this a little more deeply.

Chapter 8 explained how local agricultural knowledge in Africa was largely ignored by 'outsiders' until quite recently. Now it is recognized that the understanding of poor uneducated farmers and pastoralists is often far greater than that of an 'expert'. Agricultural extension tries much more than hitherto to build on what people already know. It is clear that much of the advice given in the past was misconceived or even damaging.

This change in thinking among would-be developers has come about since the 1970s. During the same period there have also been attempts to apply the same principle to 'traditional medicine'. Recognizing that Western-style medical services were not reaching the poor, WHO has encouraged the professionalization of local healers. China is held up as an example of a successful public health programme which employs both 'traditional' and biomedical techniques.

However, elsewhere the dialogue between biomedical and 'traditional' practitioners has been far less fruitful than had been anticipated. Unlike scientifically trained agriculturalists, who have been forced to take on board the local knowledge of farmers, scientifically trained medical staff have successfully resisted any such learning process. A narrowly biomedical approach remains the normative position from which 'traditional' medicine is judged.

In this context, it is important to realize that local agricultural knowledge is more likely to be corroborated by scientific methods than local medical knowledge. This is because a farmer's understanding of crops, soils and weather conditions is commonly based on past experience and experimentation, while the skills of a healer are often based on other criteria. Robert Chambers has elaborated this point in his book on rural development:

"It seems to be more in health and nutrition than in agriculture that harmful local beliefs and practices are found. It might be supposed that the incentive to observe, and to be effective, would be greater with what directly touches human welfare and survival in health and nutrition; but this does not seem to be the case. Several explanations can be advanced. In growing crops, there is a large population of plants from which to learn and a few are expendable; but with people there are fewer and each one is precious. The causes of poor crop performance (drought, flooding, pests, lack of soil moisture) or of good performance (good soils, manure, timely planting, weeding) are often only too obvious, compared with the invisibility of microscopic infections or the spread of disease. Again, learning from agricultural practices can also occur every season, whereas the care of a child through the stages of growth, even in large families, occurs less often, and sickness and malnutrition are spasmodic. Perhaps, also, sickness so engages the emotions that the experimental attitude is driven out. And the coping mechanisms for the awfulness of the illness and death of those who are close are social and spiritual, and so linked with social and spiritual rather than physical explanation."

(Chambers, 1983, p.97)

Chambers overstates his argument. In fact hundreds of biomedical drugs have their origin in local herbal treatments, and some non-biomedical healing techniques (like acupuncture) have proved to be clinically effective therapies even when they cannot be 'scientifically' explained. Furthermore, the actual practice of biomedicine is less directly linked to an understanding of microscopic infections than is often supposed. Most people in Britain never look down a microscope. Rather they believe that diseases exist and that doctors know about them. It is commonly reported that patients' conditions improve even when the drug administered is biomedically ineffective, and it has been asserted that over half of the curative therapies administered by British general practitioners work due to placebo effect.

Nevertheless, Chambers' insight is useful. Most people have less direct or indirect access to the microbiological world than in Western countries. Without clinical tools, even trained medical staff have to interpret symptoms only by looking at the patient, and the diagnosis of local herbalists tends to be by association. That is why *daktari* in Laropi removed 'eggs' for *burule* ulcers. More generally, healing can be part of the very texture of communal life, and cannot be separated out from it in the same way as agricultural knowledge. But it would

be ludicrous to conclude from this that an understanding of local healing is unnecessary. Changing conceptions of the human body, of illness and of affliction are likely to be a crucial part of peoples' experience of social transformation. Particularly at times of upheaval and rapid change, they cannot be ignored.

It is unfortunate that both development analysts and development workers often fail to recognize this. Chambers may be correct in implying that international experts have a great deal more to offer poor rural people in terms of medical knowledge than they do in terms of agricultural knowledge. It makes it all the more sad that so much of the money spent on public health seems to end up turning well-meaning staff into not very effective immunization commandos, who have minimal contact with their target populations other than at the end of a needle.

Summary

1   Perceptions of health cannot be separated from perceptions of affliction, and in many circumstances both are linked to moral concerns.

2   Like many other people in Africa and elsewhere in the world, the Madi of Laropi have experienced social upheaval in recent years. They have struggled to live healthy lives in an extreme situation.

3   An international relief programme launched to assist refugees returning to Uganda from Sudan largely failed, and biomedical facilities in their home district were hopelessly inadequate. They suffered from food shortages and serious diseases.

4   However, their own perception of their health problems was not simply biomedical. They experienced many illnesses as interpersonal phenomena, which made manifest moral and spiritual concerns. Therapeutic pathways were pluralistic, and healing was part of the process of establishing a sense of community.

5   Three sets of implications are discussed: the situation in Laropi is not an exception; non-biomedical approaches to health are widespread; perceptions of the human body play a part in how knowledge is constructed.

TYPES OF ACTION

11

RURAL LIVELIHOODS: ACTION FROM ABOVE

BEN CROW

This is a chapter about the disappointing record of action 'from above' in alleviating poverty. It examines a number of cases, primarily from Asia, including the promotion of household enterprise in India, the reform of property relations in East Asia, and the promotion of collective enterprise among the poor in Bangladesh.

> **Q** Why is action from above rarely successful in significantly reducing rural poverty?

Section 11.1 examines one of the largest attempts to promote income-generating activities among the poor, the Integrated Rural Development Programme (IRDP) in India. This programme, which is characteristic of a range of projects promoting household enterprise, gives poor households access to assets, such as livestock, which can generate saleable goods. From this experience we can identify important constraints to projects which promote household enterprises without altering, or even taking account of, the social relations upon which their viability depends. Failure to challenge rural class relations (specifically those determining access to land and the operation of markets in money and livestock) limits the viability of the household enterprises of the poor.

The second section looks at the case of land reform in South Korea and Taiwan. In these two countries significant change in rural class relations has been associated with rapid economic growth and dramatic improvements in rural incomes. This section suggests that the conditions which made land reform from above possible in East Asia – external pressures and strong states, independent of landed interests – are exceptional rather than generally applicable.

States tend to avoid basic agrarian reform, and their interests and commitment frequently contribute to the failure of other kinds of action. This is the subject of the third section, which also identifies areas in which state intervention is likely to be more or less appropriate and effective.

Section 11.4 asks several questions. Are non-governmental organizations (NGOs) subject to the same constraints as states? Do they have space for action which is denied to states? Using the case of the promotion of irrigation pump groups among the landless in Bangladesh, this section suggests that although NGOs may be subject to some of the same constraints which influence government action, they have greater freedom of manoeuvre and can command space for action in particular processes of agrarian change, with the potential to improve rural livelihoods.

11.1 Promoting household enterprises

One way governments attempt to alleviate poverty is by promoting household enterprises.

Here we examine the case of the Integrated Rural Development Programme (IRDP) in India.

As noted in Chapter 7, India has the largest concentration of poor people anywhere in the world. This programme is also one of the largest anti-poverty programmes of its kind. While the programme was pursued by the Indian Government with considerable commitment, its outcomes indicate some of the limits of this approach within a society which in other respects remains unchanged.

In almost all eras of development thinking, the promotion of some form of commodity production to increase cash incomes has had a central role in government intervention in the rural economy. During the 1960s, intervention tended to focus on agricultural production: examples of this form of intervention include support for the Green Revolution in India (Chapter 3, Section 3.3) and programmes seeking to modernize agricultural production in Africa (Chapter 4 and Box 4.1). During the 1970s, action was frequently directed at small farmer production and income generation for landless households. During the 1980s government intervention to alleviate poverty moved from attempts to provide more widespread income transfers to the poor (for example, through the subsidy of consumer prices) towards targeted policies which selected specific groups. Ideas of growth with equity, agricultural trickledown, 'integrating' women in development, and targeted programmes in the era of structural adjustment have all provided justification for the promotion of commodity production. NGOs as well as government and international aid agencies often place considerable reliance on 'income-generating projects' as a way of tackling poverty.

The promotion of petty commodity production as a response to rural poverty is primarily informed by a 'residual' approach to poverty (Chapter 1). It proceeds on the assumption that the poor (or women) have been marginalized or left out of economic development (see also *Hewitt et al., 1992*, ch.8), and that identifying and removing specific 'obstacles' to their integration will provide them with the benefits of development.

The Integrated Rural Development Programme in India

The name of this programme is misleading. It was initially visualized in 1978 as a comprehensive development plan, generated district by district, to increase rural productivity 'through a strategy of growth with social justice...providing full employment to the rural sector within a ten year time frame' (Draft Sixth Plan cited in Rath, 1985, p.239). But this ambitious plan was not implemented. A more limited programme, targeted on the rural poor, was adopted as an interim measure, and the 'name appropriate for the whole was given to only a part of it' (Rath, 1985, p.239).

In this reduced form, the IRDP set out to provide cheap credit and assets with which poor households could establish small-scale enterprises and generate income. The project has assisted the purchase of irrigation pumps, draught animals, livestock (poultry and cattle) for agricultural production, and the establishment of small craft and trading enterprises (Figure 11.1).

Individuals from 27 million rural households had taken loans from IRDP by 1989 (Pulley, 1989, p.5). A World Bank report judged the project 'among the most ambitious efforts at credit-based poverty alleviation' (ibid., p.iii). Expenditure on the project was US$6 billion between 1981 and 1988, and it is estimated that the programme has increased the asset holdings of a quarter of rural households in India. Let us look more closely at the rationale and effects of this programme.

Why credit?

IRDP identified lack of access to rural credit as the key obstacle to rural self-employment, and state-subsidized provision of credit as the way of removing that obstacle.

In India, rural credit is either very expensive or unavailable at any price for poor households. This is partly because households with little or no land cannot get access to bank credit and may not even be able to obtain more expensive finance from informal sources (moneylenders, traders,

Figure 11.1   The Integrated Rural Development Programme has helped small farmers in India to increase their assets: draught cattle preparing the soil for sowing, Uttar Pradesh.

landlords). Their access to both informal and formal finance is restricted because they have little collateral (such as land) to offer the lender as security against non-payment of loans, and because banks are reluctant to give small loans when the costs of processing small loans are high relative to the returns to the bank. IRDP set out to get round these restrictions by providing subsidized credit and subsidized assets. IRDP planners assumed that proven repayment records would then establish sustained access to future credit for poor households.

For IRDP to improve the livelihoods of poor households, three basic criteria have to be met. Each participating household should be able to:

1   increase its asset holdings;

2   establish a viable production unit to generate sustained income;

3   repay its credit in order to establish future access to banking services.

Whose livelihoods?

Unlike many development programmes characteristic of action from above in the 1960s which aimed to develop cash crop farming, this project targeted *landless* rural people. It does not attempt to give them land, but provides other productive assets to establish people as petty commodity producers. The great majority of loans are given to men. By 1986–87, only 15% of the borrowers were women, compared with the programme's intended 30% (Holt & Ribe, 1991, p.9; see also Box 11.1 below).

The IRDP is one of a number of programmes (outlined in Chapter 7) with which successive Indian governments have sought to reduce rural poverty. After the land reforms of the 1950s (Chapter 3), they have not tried to change property relations in order to challenge the existing agrarian structure and its interests:

"Within the given social, economic, and political parameters of the existing situation, an attempt is being made to enhance income levels and living standards of the poor by self and wage employment programmes and provision of some degree of social services and social consumption through a package of minimum needs."

(Bandhyopadhyay, 1988, p.A-77)

In short, IRDP attempts to promote petty commodity production by landless households without antagonizing landlords, merchants or moneylenders.

Outcomes

The programme has had some success. In the words of a former Indian Government Secretary of Rural Development, it:

"has caught the imagination of the poor because of the high visibility and endowment of the asset, for which there had traditionally been an unsatisfied need among the poor. Moreover, with all the IRDP's weaknesses, the evaluations show that a fairly large volume of assets did reach the very poor."

(Bandhyopadhyay, 1988, p.A-83)

Despite considerable government expenditure, however, the proportion of participating households whose incomes have been raised above the Indian Government's poverty line is relatively small. Although in one or two states around 20% of participating households seem to have been raised above the poverty line, the national average is only 7% (Pulley, 1989, table 5.1), but data enabling a national evaluation of this sort are available only for a two-year period.

A more robust evaluation, by the World Bank, used a four-year survey of one state, Uttar Pradesh (see Figure 7.3 for a map of India). It concluded that IRDP had achieved some success on two of the three criteria listed above – increase in assets, viable production – but none on the third, future credit rating. The programme appears to have 'done nothing to overcome

structural constraints on long-term access to credit' (Pulley, 1989, p.39).

In Uttar Pradesh, 59% of participating households had retained the asset purchased with an IRDP loan after four years, and 58% had only small overdues in their loan repayment. The World Bank took retention of the asset after four years as an indication that a viable enterprise had been established. However, this success rate should be qualified by the fact that some participating households were not poor within the definitions of the scheme. Taking this into account, the survey concluded that 44% of poor households have achieved a viable investment as a result of the IRDP intervention.

The productivity of the investments varied by type of asset, with investments in dairy cattle giving high incomes (Figure 11.2) and investment in irrigation equipment giving lower returns, but almost all types of investment showed declining returns over four years. Although declining returns could be overcome by further investment in assets, the ability of households to do this is restricted by the failure of the scheme to meet its third objective of establishing access to more conventional sources of credit.

The planners' assumption that proven repayment records would establish sustained access to future credit for poor households proved overoptimistic. Nationally owned Indian banks will lend to the limits established by IRDP, but will not provide further finance after that because the fixed interest rate for IRDP loans is below the cost of lending. Thus 'banks choose not to lend additional funds after their obligation to achieve IRDP targets is satisfied' (Pulley, 1989, p.iii). Nor do repayment records establish a basis for access to high-cost 'informal finance' from moneylenders because 'informal markets do not cater to the demand for long-term investment capital. Moneylenders are unwilling to accept such risks except in exchange for long-term labour contracts (i.e. bonded labour) to assure repayment' (ibid.). Moneylenders recognize that the small enterprises established by poor households are risky and that a proven repayment record (at low interest rates) does not provide a secure basis

Figure 11.2 *The IRDP made it easier for landless households to buy dairy cattle, but the livelihoods they made possible were vulnerable to market and household changes.*

for lending at high interest rates over a period of several years.

In fact, the small scale of the enterprises which households are able to establish makes them more risky (i.e. vulnerable to changes in production or marketing conditions) than larger enterprises. Small scale leads to vulnerability for two reasons. First, the household has its funds tied up in few assets (such as a small number of cows) and if one fails, the impact on cash flow is large. Secondly, the reserves of the household are small, so changes in the demand for its product or changes in household expenses (because of sickness, for example), can force the sale of assets, often at 'knock-down' prices (Copestake, 1988, pp.276–7).

Bandhyopadhyay, the former Indian Government Secretary of Rural Development quoted earlier, argues that small-scale production units established on an individual basis by very poor households are inherently vulnerable to the unequal conditions under which the poorest participate in markets of all kinds (see also Chapter 7). This is a general point applicable to many kinds of development through small enterprise promotion:

"What the present IRDP programme [is] trying to do is to make the dependent households economically free agents. But for many, the illusory security of dependent relations is being replaced by the inherent insecurity of impersonal and ruthless market forces. Such a beneficiary is likely to lose both ways as a seller of his [sic] product and as a buyer of inputs. Market forces would toss him about, since he would not be in a position to bargain individually. The market would dominate and defraud such isolated individuals and soon he will go under."

(Bandhyopadhyay, 1988, pp.A-83 to 84)

An analysis of cattle purchase for milk production in south India illustrates one of the ways IRDP-supported households suffer from 'ruthless market forces' when they buy cattle. This study compared livestock purchases by IRDP participants with those of a control group of livestock purchasers not using IRDP loans. The study found:

> "evidence of substantial price discrimination in the market for livestock. Participants in the scheme receive subsidized loans but purchase milch animals [dairy cattle] at inflated prices that are not compensated by higher livestock quality. Such imperfections in the markets for livestock assets have serious adverse consequences for the efficacy of intervention to alleviate the credit market imperfections that are rightly believed to hamper the accumulation of capital by the rural poor."

(Seabright, 1990)

Markets for livestock in this area of south India were small, localized in their operation and often dominated by a small number of cattle traders who were in a position to influence prices. This resulted in IRDP participants paying more for their cattle.

General price discrimination against the poorest participants in these cattle markets was exacerbated by the arrangements for the provision of subsidized credit by IRDP. A group of IRDP participants are given credit to purchase livestock on a particular day and a particular market. In small markets dominated by a small number of traders, the presence of IRDP participants becomes known, and traders are able to take advantage of the low bargaining power and lack of experience of the purchasers.

A further constraint on the operation of micro-enterprises concerns landholding. Many of the poorest households have no land at all; even their homestead plot may be leased to them by a landowner for whom they undertake work or pay rent in cash or kind. Without land, the poor cannot operate enterprises effectively:

> "A man who lives on somebody else's land cannot really operate a mini dairy unit or a small poultry [unit] or even a spinning wheel without the landowner's permission and co-operation. Any sign of the economic betterment of the former which would reduce his dependency on the latter would be frowned upon and prevented unless a significant share of the incremental benefit is passed on to the landowner. In such an event, the poor beneficiary might be used as the front man of the landowner and would act as a convenient conduit for benefits to the ineligible category..."

(Bandhyopadhyay, 1988, p.A-84)

Finally, it is clear that IRDP has been least successful in giving access to credit for women (Box 11.1). Credit is as important to the productive capacities of landless women as it is to landless men, but very few women are able to obtain IRDP loans. In addition, as the box describes, the reorganization of production associated with government support, in the case of dairy production, actually reduces women's control over activities which were theirs under previous social arrangements. The practices surrounding dairy marketing co-operatives give men control over income which would previously have gone to women.

In conclusion

The case of IRDP indicates some of the ways in which action from above, initiated by a government committed enough to devote significant expenditure to poverty reduction, can be limited by the operation of credit, commodity and land relations. In Chapter 7, Ghosh and Bharadwaj argue that the economic viability of these small enterprises can only be guaranteed when government intervention assures access to (continuing) credit, training, raw materials and output markets.

The programme is not a simple 'failure'. Perhaps as many as half the participating households have established enterprises which survived for

---

Box 11.1   Why has IRDP not been able to promote income-generating activities among women?

A World Bank (1991b) study found that women head 30–35% of Indian rural households, and the participation of women in the rural economy is significant: 'the majority of adult women in India are not simply housewives, but in fact *farmers*'. Economic participation is highest among the poorest households (as noted in Chapter 7, Box 7.4).

'Access to credit is the key to almost every form of productive self-employment for poor women because it is the major instrument available to redress, in the short run, the historical imbalance which the tradition of patrilineal land inheritance has caused in the distribution of productive assets directly available to men and women... since land has been the main source of collateral, women's lack of land ownership has barred them from access to the formal system, thus limiting their ability to acquire other productive resources such as cattle, poultry, looms or working capital for trade in farm or forestry produce, food processing, etc. Credit is in a sense the gateway to almost every form of productive employment for poor women.' (p.209)

Nevertheless, few women have been able to use IRDP loans and even in milk production, an economic activity in which the sexual division of labour has given women a longstanding role, the constraints on viable income generation appear to be greater for women than for men:

'In theory, Operation Flood [a co-operative milk marketing scheme] and IRDP should have made it equally possible for poor men and women to gain access to dairy animals and to participate in the co-operative dairy system...IRDP has earmarked 30% of its lending to women. Why has less than 15% of the IRDP credit (for livestock and all other purposes) reached women borrowers even after special targeting for women was introduced into IRDP?

'...The common pattern is for women to handle most of the production aspects and for the men to assume the co-operative membership and control of the cash income. There are a number of problems associated with the traditional 'inside/outside' division of labour. Some affect the overall efficiency of dairy production, others affect the welfare returns to the family and to the women producer herself. Women, for example, do not usually gain access to training in modern livestock management and dairying techniques which is available to men through the co-operative structure. Instead, they must learn second-hand through the men or continue with traditional practices, both of which lower their efficiency and reduce returns to investments in training. Non-member producers also miss the chance to be trained in the responsibilities...

'For women from poor households the greatest disadvantage is that they have no control over dairy income which is collected by the male member. In cases where the women used to deal with traditional milk traders who came to the household compound, they lost what small degree of economic autonomy they had when marketing arrangements were formalized through the co-operative structure.' (p.51)

This study concludes that a 'comprehensive strategy for promoting employment ventures by the rural poor', including training and marketing, would be required before IRDP could effectively support women's enterprises.

---

four years. However, the continued survival of these enterprises is in doubt because future access to credit cannot be assured. In addition, landlessness increases the difficulties of maintaining control over assets obtained through the scheme. The case illustrates the wider context and problems of programmes which 'target' individual households.

The case of IRDP gives grounds for pessimism about actions which do not confront the structural causes of poverty and insecurity in existing relations of property and power. Since the limited land reforms in the wake of Indian independence, the Indian Government has shied away from undertaking change which would confront rural class relations.

11.2 Agrarian reform from above

IRDP is an example of an ambitious poverty-alleviation programme which used income transfers in the form of subsidized credit and assets. A second kind of intervention is action by the state to change property relations. Fundamental change of agrarian relations, notably property relations in land, can have a dramatic impact on rural poverty and on economic growth. The chair of the committee of industrialized country aid donors (the OECD's Development Advisory Committee) recognizes the efficacy of land reform:

> "…where, thanks to historical accident or upheaval (Japan, Korea, China and Taiwan) there has been decisive transformation in the entitlements to land, this has had salutary effects on both agricultural growth and equity."

> (Lewis, 1988, p.15)

This observation has an added impact because it is an exception to the normal reluctance of official aid donors to attribute poverty to rural social relations. (Aid is usually associated with technical conditions of production.) However, as Lewis notes, significant changes in entitlements to land take place only in special circumstances. The purpose of this section is to pick up connections from earlier sections and preceding chapters, and to use the cases of land reform in Korea and Taiwan to explore the question: when can agrarian structures be changed by action from above?

Remaking the agrarian order

Attempts to challenge the causes of rural poverty confront the bases of rural power:

> "…land reform is an intensely political matter, involving substantial conflicts of interest. Indeed, the ownership of land reflects and underpins social power and structure in agrarian economies, so that changes in the pattern of ownership necessarily involve changes in society itself."

> (Bell, 1990, pp.150–1)

This suggests that opportunities to remake the agrarian order usually arise at times when there is a challenge to the whole social order. Most of the important agrarian reforms in the twentieth century have followed some form of social upheaval (defeat in war, socialist revolution or national liberation). Where the ownership of land 'underpins social power and structure', land reform entails transformation of the nature of the state. Agrarian reform requires, in other words, changes in the wider social and political order, including change in the nature of the state itself. Reforms attempted in periods when the social and political order is secure face formidable obstacles because landed interests can successfully resist them (Bell, 1990, p.143).

> "…in countries where no revolutions have happened or are in prospect – especially if these are pluralistic systems given to incremental change – it is apparent that élites have great capacities for delaying and/or evading thoroughgoing tenurial reforms…"

> (Lewis, 1988, p.15)

Types and objectives of agrarian reform

Agrarian reform from above is described in Chapter 7 as state intervention 'reforming the structure of property relations (usually land relations) so as to reconstitute production and exchange relations.' This covers a wide range of possibilities, e.g.:

- redistributing land, which can include:

  (a) giving tenants ownership rights to the land they cultivate;

  (b) setting limits to the area of land to be held by any one landholder, and distributing land above these limits to landless or land-poor households;

- restricting the scale or type of rents which can be charged (tenancy reform);

- establishing the right for women to own land;

- changing collective land use rights into individual, transferable ownership (i.e. privatizing land, which may include elements of enclosure and commoditization).

The *objectives* of agrarian reforms are also varied; not all are intended to benefit the poor. One of the fundamental historical processes bringing about change in agrarian relations, described in several earlier chapters, is the transition to capitalist agrarian relations. This transition may occur without intervention by the state; normal processes of accumulation and differentiation may be sufficient to extend capitalist relations. Nevertheless, many state interventions do seek to initiate, encourage or complete the introduction of capitalist relations (wage labour, private property in land, production motivated by capital accumulation, exchange dominated by markets), as in agrarian policies in colonial India and in twentieth-century Latin America.

In Taiwan and Korea, examined below, agrarian reform was undertaken for several reasons, including the need to accumulate capital and provide food for industrialization, and to secure political support in the countryside. Nevertheless, those reforms from above contributed to a dramatic improvement in rural livelihoods.

There are thus important differences in why agrarian reform is undertaken, what kinds of agrarian reform and in what conditions, all of which will affect its outcomes.

Korea and Taiwan: external pressures and 'high-handed' government

An overview of East Asian economic development by *The Economist* suggests that 'high-handed government' was instrumental in achieving agrarian reform.

"In all three of Asia's biggest successes – Japan, South Korea and Taiwan – the groundwork for both fast growth and the income equality that eased the social strains of development was laid by a radical land reform. It took high-handed government to carry this out: in Taiwan, a dictator (Chiang Kai-shek) imported from mainland China; in Korea, a government carried along by a wave of public anger at collaborators with Japanese colonizers; in Japan, the American occupation army."

(*The Economist*, 29 June 1991, p.16)

Those East Asian cases are arguably the most significant examples of agrarian reform in capitalist economies in the twentieth century. It seems unlikely that a comparable conjuncture of historical circumstances and high-handed, developmental states will enable similarly successful agrarian reforms from above in other capitalist countries.

The success of agrarian reform in Taiwan and Korea relied upon two stages of social upheaval: first, the changes introduced by Japanese colonial rule, then changes introduced after the end of the Second World War. Colonial rule (from 1895 in the case of Taiwan, and from 1910 in Korea) made Korea and Taiwan into suppliers of cheap rice for the industrialization of Japan. In the process, Japanese rule inadvertently laid the foundations for a remaking of the agrarian order in its colonies by subordinating landlords to the state and introducing new agricultural practices. The colonial state transferred new rice production technologies, and in Korea 'landlordism as a means of extracting rural surpluses was encouraged, with many landlords now being Japanese instead of Korean' (Douglass, 1983), and in Taiwan, partial reforms left 'landlordism intact while allowing for new farming practices to be assimilated' (Amsden, 1979).

The defeat of Japan in the Second World War brought the collapse of the colonial state and of landlord power, enabling 'a redistribution of individual rights [in land] imposed on a landed class rendered impotent by the collapse of a state that had reflected its power and interests' (Bell, 1990, p.151). After the Japanese surrender in Korea, popular committees emerged which planned to redistribute land. The interim United States military administration in South Korea suppressed these committees but encouraged land reform and installed a new government. Subsequently, popular backing for land reform was intensified by the example of land reform in North Korea. In the case of Taiwan, the change in the nature of the state was the takeover by the Nationalist party (led by Chiang Kai-shek; Figure 11.3) which had been defeated by revolution on mainland China. The new government came to be identified with a new agrarian order in Taiwan, even though it was associated with landed interests on the mainland:

> "The landlord supporters [of the Nationalist party] who fled with it to Taiwan left their assets behind. [The party] was not beholden to Taiwan's own modest class of landlords. And partly as a pre-emptive strike in case [the landlords] should act as a focus of opposition, the party moved decisively in the early 1950s to expropriate their tenanted-out land above a low ceiling. A low ceiling on agricultural land ownership has remained in force ever since, ruling out – or at least greatly restricting – investment in land as a means of accumulating wealth."
>
> (Wade, 1990, p.263)

Both in Korea and Taiwan the end of colonialism brought to power a state freed from the influence of a powerful landlord class and determined to industrialize in order to build its military strength. Byres notes that the 'passage and sustaining' of agrarian change in Taiwan and South Korea 'have required a very powerful and repressive state' (Byres, 1991, p.58). In Korea land reform was able, according to one study, to 'remove the deadweight of landlordism and install a system of egalitarian peasant farming which ensures growth without mass destitution' (Ghai *et al.*, 1979, p.9; Figure 11.4). This study describes a pre-reform agrarian society far removed from contemporary images of Korea:

> "Prior to land reform, South Korea suffered from the [Indian] sub-continental syndrome of today: endemic rural poverty, a tenancy system characterized by the burdens of high rent and an extreme concentration of agricultural surplus."
>
> (Ghai *et al.*, 1979, p.9)

In Taiwan, too, there was dramatic change:

> "...almost overnight the countryside of Taiwan ceased to be oppressed by a small class of large landowners and became

Figure 11.3   Chiang Kai-shek, Nationalist leader in China, photographed in 1927.

Figure 11.4 *'Egalitarian peasant farming' in South Korea.*

characterized by a large number of owner operators with extremely small holdings."

(Amsden, 1979, p.352)

But the picture of 'egalitarian peasant farming' needs to be strongly qualified. After the land reform, the Japanese colonial practice of maintaining close state control of agriculture was continued. The place of the landlord class 'has been taken by a dominating state' (Byres, 1991, p.58). State intervention enabled new agricultural technologies to be introduced (building on the changes promoted by the colonial state) and the extraction of resources for industrialization. It has been noted that land reform in Korea created the potential for squeezing agriculture without causing famine (*Hewitt et al., 1992*, ch.4). In Taiwan, post-reform agriculture has been described in these terms:

"A self-exploitative peasantry, working long hours to maximize production per hectare, and a super-exploitative state, ticking along effectively to extract the fruits of the peasantry's labours, operated hand-in-hand in Taiwan."

(Amsden, 1979, p.362)

The explanation of improvement in rural livelihoods also has to be widened from egalitarian peasant farming to include the influence of manufacturing industry on rural livelihoods. By creating large numbers of new jobs, industrialization took the pressure off agriculture as the prime source of rural livelihoods. In Taiwan, the dispersed nature of industrialization, spread through much of the countryside, enabled a large proportion of 'peasant' households to earn significant income from industrial employment. In short, industrialization in East Asia contributed to rural as well as urban livelihoods to an extent which Indian industrialization has not (see Chapter 7).

The unique historical and social circumstances of the East Asian experience need to be emphasized. Byres contrasts the conditions under which agrarian change has occurred in Taiwan and South Korea with those elsewhere in capitalist Asia:

"...either landlord classes continue powerful (as, say, in Pakistan), or survive in somewhat weakened form (as, say, in India). Pervasively, unlike South Korea and Taiwan, highly differentiated and significantly differentiating peasantries are to be found. Classes from within such differentiating peasantries – rich peasants and capitalist farmers – have become part of state power and are not separated from it, as has been argued to be the case in South Korea and Taiwan..."

(Byres, 1991, p.58)

Post-war East Asia shows that agrarian reform from above can transform rural livelihoods, but conditions elsewhere in Asia suggest that the example is unlikely to be replicated.

Comparison with Latin America, India and Africa

In this century, land reforms in Latin America, undertaken either in response to popular pressure (from below) or, more usually, to pressure from modernizing coalitions (above), have had mixed consequences for livelihoods. These were frequently land reforms seeking to accelerate the transition to capitalist relations in the countryside. In some cases, a minority of peasants benefited. In most cases, however, the heritage of the *hacienda* is reflected in highly unequal land distribution and the massive scale of many capitalist farms. The influence of a class of large landholders and capitalist farmers is still important in many Latin American states, while many rural people are 'semi-proletarianized' by inadequate access to land and other resources needed for farming to provide adequate livelihoods.

Post-independence land reform in India hastened the end of parasitic landlordism only in some parts of India. This was agrarian reform from above responding to the pressure of unrest from below, but it was constrained by the power and involvement of landowners in the post-independence state. As noted in Chapter 3, 'most landowners...were able to deflect any more thoroughgoing land reform, frequently with official connivance at state and local levels, where reform legislation, compromised as its final drafts usually were, turned into an administrative nightmare.'

In most parts of Africa, reform of property relations in land has not been an issue until recently. (Exceptions are where substantial land was alienated by colonial settlers as in Kenya,

Zimbabwe and South Africa.) This is because the colonization of Africa did not convert land into private property to the extent prevailing in many other areas of the developing world. 'The land used by most African farmers was neither expropriated (as in Latin America) nor incorporated in a system of private property rights (as in India)' (Chapter 4). Even after Independence, land has mostly been plentiful, and its allocation subject to customary practices. Nevertheless, many state-initiated agricultural modernization projects (see Box 4.1) either assume or consciously attempt to create private property in land. And the routes to accumulation from below and from above described in Chapter 9 illustrate the central role of land in alternative forms of differentiation, and how privatization of land is occurring in some parts of Africa, whether *de jure* or *de facto*.

11.3 State intervention: interests, commitments and capacities

East Asian land reform provides examples of intervention by a strong state, independent of landed property interests, and committed to industrialization. Few states in the developing world have these attributes, and the success or failure of their other kinds of rural development policies are closely related to the interests, commitments and capacities of particular states.

Interests

One review of the role of development institutions argues that the relation between states or particular governments, government departments and different social groups may be the prime cause of the failure of development intervention:

"The most telling case against many institutions is not that they are technically inefficient or poorly managed (although they may be) but that in the political and administrative reality in which they operate, they end up pursuing objectives inconsistent with development…In other words, one must go beyond the assumption that government institutions are by definition pursuing the goals of national development to explore what determines the interests organizations work for in practice. This proposition applies as much to international institutions, and to bilateral donors, as to institutions in recipient countries. 'Interests' can relate to the play of foreign policy concerns of states [Figure 11.5], sectional interests brought to bear on aid programs, and the interests of aid officials, departments and agencies in perpetuating their roles [Figure 11.6]."

(van Arkadie, 1990, p.171)

This passage makes implicit use of some common ideas about the state (Box 11.2). It comes from a paper presented to the World Bank's annual conference on development economics in 1989. During the 1980s, the World Bank and the IMF used

Figure 11.6 'All this pressure tactics, blackmail, threats, infighting, is great fun. What spoils it is the bother of running the nation.'

their considerable financial power to promote the idea that the failure of development was primarily the result of excessive, inefficient and self-interested state intervention (the neo-liberal or private interest view of the state summarized in the box). Van Arkadie is arguing that it is not inefficiency which is the principal cause of development failure, but the power relations within which states and government (and other) institutions operate ('the political and administrative reality'), causing them to pursue 'objectives inconsistent with development'. In order to understand the limits to government intervention, he argues, it is necessary to go beyond the idea that states and their governments are neutral agencies of progress (the public interest view) to investigate the 'interests' they represent (this is the structuralist, or class, view of the state).

In other words, to understand why action from above frequently 'fails' with respect to the needs and interests of the rural poor, it is necessary to

Figure 11.5 'No, it is not true that we've lost interest in this project. In fact the Russians showed great interest for three years, the Japanese for two years, later the British, and now again the Russians are showing tremendous interest!'

> **Box 11.2   Alternative notions of the state**
>
> There is ongoing research and debate about how to understand the state and its influence on the policies implemented by government both in industrialized countries and in the very different social, legal, political and economic circumstances of developing countries. Three major positions in this debate can be briefly summarized as follows (*Wuyts et al., 1992*, ch.3):
>
> Liberal theory has portrayed the state as an impersonal structure of power, independent of the power of any social group, and representing the common good or the public interest (the *public interest view* of the state).
>
> Structuralist theory has seen the state as reflecting the interests of dominant social classes, establishing the main parameters of economic activity in their favour, whilst maintaining some degree of independence or autonomy (the *class view* of the state).
>
> Neo-liberal theory, which lies behind structural adjustment policies, has portrayed state intervention as the most serious constraint on economic growth, and its policies as sometimes determined by the private interests of state officials (the *private interest view* of the state).

understand the unequal representation of different social groups in the policy making and implementation of different institutions, including foreign aid donors and international institutions, as well as national governments.

In the case of the IRDP in India, this focus would lead us to ask why a programme first envisaged as a comprehensive 'strategy for growth with social justice' came to be implemented as the promotion of individual household enterprise, in ways which do not threaten the rural power structure. One possible answer would be that the 'interests' represented by the Indian state include those of dominant rural classes. East Asian land reform, on the other hand, was linked with the objective of rapid industrialization, which allied the state with much more powerful interests than (discredited) landlord classes.

Commitment

The World Bank itself recognized 'political and administrative realities', and the importance of problems of government 'commitment', when it launched its new strategy of rural development in the 1970s. The Bank decided that it could not seek the major reforms necessary to *eradicate* poverty, but could realistically try only to *alleviate* poverty:

> "Commitment to solving the poverty problem in the many predominantly rural

economies of the developing world implies major policy, economic and financial changes that amount to restructuring society. Such a change was achieved over generations by a number of highly developed countries and is being attempted more quickly by radical governments elsewhere, but was not expected to happen generally in the developing countries. Hence, the Bank's [rural development] strategy did not concern itself with poverty eradication but with poverty alleviation to reduce the worst attributes of poverty. For example, although the land tenure pattern was identified as a major poverty factor, land reform was not proposed as a project mode. Lack of commitment ruled out widespread land reform as a significant option, although some societies could be expected to address the issue to some degree, out of enlightened self-interest."

(World Bank, 1988, p.42)

This paragraph comes from a major retrospective evaluation of the World Bank's experience of rural development interventions between 1965 and 1986. It reports the Bank's assessment at the beginning of the 1970s that, in many developing economies, poverty eradication could only be achieved through a restructuring of society. Specifically, the Bank recognized patterns of land tenure as a major cause of poverty. Nevertheless, the Bank concluded that the structure of power in many developing countries ruled out a direct challenge to rural class relations through agrarian reform. (In this context, I think, government 'commitment' is being used as a euphemism for the interests represented by the state.) During the 1970s, therefore, the World Bank's rural development interventions ruled out poverty eradication as too ambitious an objective, given the realities of political power, and focused on less challenging initiatives, such as the promotion of income-generating activities.

Lack of government 'commitment' is often identified as an important cause of failure of interventions to improve rural livelihoods. The study cited above took inadequate government finance for the projects evaluated as an indication of lack of commitment, and found this to be the most important adverse factor in 23% of projects studied. There are, however, several possible interpretations of this finding.

First, lack of government 'commitment' may indicate that 'interests' threatened by rural poverty alleviation have exercised some form of veto (in this case, during the process of implementation). A second possible interpretation is that lack of 'commitment' reflects lack of resources. The efforts required for industrialization and agricultural growth make historically unprecedented demands on state resources and capacities, while many Third World states command inadequate resources, and lack the technical and often political capacity (that is, legitimacy with many of their people) to intervene effectively. In other words, many Third World states are caught in a trap between the urgency of major economic and social development in their countries, and their own relative weakness in relation to powerful interest groups, both internal and external.

Thirdly, lack of government 'commitment' may arise from disagreement about what should be done to improve rural livelihoods (Figure 11.7). This is an important explanation to consider in relation to the failure of projects promoted by the major international agencies and aid donors. From the early 1980s, the

Figure 11.7   'Did you hear that? Now he is saying, "We should have a practical common-sense approach..." I bet he is swinging to the right!'

World Bank, the IMF and the major aid donors have promoted neo-liberal policies (including a much reduced role for the state). Opposition to those policies within Third World states could lead to unenthusiastic implementation and the appearance of lack of government 'commitment'.

For whatever reason, it is rare for the government of any capitalist country (South or North) to pursue poverty eradication with the full resources it could mobilize, *unless* there is effective popular pressure.

Capacities

So far, we have touched on how state capabilities (the ability to get things done) as well as development goals (deciding what should be done) are affected by different kinds of interests and 'commitment'. Here we turn to a rather different set of issues about capacity: whether governments are better at doing some things than others (Figure 11.8). For example, Robert Chambers (1988b) has argued that large, centralized and hierarchical organizations, such as governments

Figure 11.8 *'Of course, we have solved the problem – the trouble is about the solution. We are solving that now.'*

or large business corporations, tend to simplify, centralize and standardize. These may be appropriate ways of organizing some kinds of activities, but not others.

Chambers applies this perception in analysing the operation of the 'field bureaucracy' of government in promoting agricultural and health intervention (Figure 11.9). It has successfully implemented two sorts of programme (Box 11.3):

1 simple standardized programmes finding uniform conditions;

2 programmes able to create and sustain uniform conditions.

The suggestion that government institutions are more capable of implementing standardized and simplified programmes may explain the higher rate of failure of intervention in dry-land farming areas of Asia and Africa. The diversity of ecological conditions and of production practices adapted to those conditions means that more sophisticated forms of intervention are required, incorporating flexibility of operation and dialogue with peasant farmers. For diverse, complex and risk-prone agriculture, Chambers concludes, agricultural research and extension for single commodities has a limited role. What is needed is a 'basket of choices', not a 'package of practices'.

A World Bank study of its experience of development projects also supports the conclusion that

---

Box 11.3   Action from above which succeeds

1 'Zipper' programmes. These move from area to area making standard changes which create stable social organizations requiring little maintenance. 'In health, two good examples are smallpox and yaws vaccinations, where a simple and universal intervention including the poorest people, and having to include them, had good effects for all... In agriculture, examples can be found where extension has been able to propagate standard recommendations to many farmers in fairly uniform physical and social conditions in classical green revolution environments... Once zipped up, [these programmes] stay in place or are easy to hold in place' (Chambers, 1988b, p.50).

2 Tight-imperative organizations focusing on a single commodity: projects requiring exacting technical standards, of quality and time, which force high standards of performance. 'The Kenya Tea Development Authority... provides production services, processing and marketing for tea smallholders... in areas of steep terrain, heavy rainfall and difficult road conditions. The tea must be picked carefully... and transported to a factory within six hours of picking... Once started, [such programmes] can remain stably above a threshold for survival by diligently repeating what has been found to work...

'These examples are centralized, standardized and simple...reduced to simple disciplines to be followed regardless of local conditions...[which were either] uniform or were made uniform' (Chambers, 1988b, p.51).

Figure 11.9   Agricultural extension: (top) drenching a cow against worms; (above) vaccinating chickens at Kamurene Farmers' Centre, Kenya.

governments are more capable of implementing standardized programmes:

"...the evidence seems to suggest that institutions that work best more or less approximate to a modern industrial plant. Thus plantation projects in the Bank portfolio have worked better in Africa than projects that support smallholder agriculture."

(van Arkadie, 1990, p.168)

However, as the writer goes on to point out: 'Unfortunately, this does not tell us whether plantations or smallholder agriculture are the more reasonable vehicle for agricultural development in Africa' – i.e. it is a matter of policy choices, rather than methods of implementation.

If governmental institutions tend to lack the capacity for decentralized and interactive innovation, are non-governmental organizations more suited to the latter?

11.4 Are NGOs an alternative?

The scale and scope of activity by NGOs increased markedly during the 1970s and 1980s. Their increased significance, combined with a realization of the limits of state action, prompts some questions. Are NGOs subject to (some of) the same constraints as states? Do they have space for action and capacities that states lack?

In 1970, NGOs based in the North transferred US$0.9 billion for development in Third World economies, and in 1985 they transferred $4 billion. Of this transfer 75% came from private funds raised by NGOs. By 1985, NGO transfers from their own fundraising accounted for 10% of total aid from the OECD countries. In addition, 3–5% of official government aid was channelled through NGOs in the period 1980–85 (Cernea, 1988, p.16).

What are NGOs?

The category 'non-governmental organization' includes a range of very different types of organization. The fact that so embracing a categorization can come into general use may reflect a wide reaction against earlier presumptions that development is above all the business of governments. The category of NGO is a residual one which potentially includes every formal association and informal network which is *not* government. However, economic enterprises from peasant households to transnational corporations, and banks to moneylenders, are generally excluded. So too are formal political parties, and the associations and informal networks which represent religious, ethnic and class interests. In the field of development, the term 'non-governmental organization' has come primarily to mean voluntary groups,

without any particular economic interest or political affiliation, that work for poverty alleviation and economic growth. (The characteristics of NGOs are further discussed in *Wuyts et al., 1992*, ch.5, in which a broader definition is used than in this volume.)

The origins and social character of NGOs vary according to their history. Some emerge from grass roots initiatives, from peasants associations and women's associations, for example. But the more visible NGOs with formal government recognition and links to foreign funding agencies are generally initiated from above.

What sorts of projects do NGOs undertake?

The work of NGOs is diverse, from fundraising and development education to disaster relief and income- and welfare-generating projects, which involve them in negotiations with governments and official aid agencies. A report on how the World Bank can work with NGOs describes one aspect of them in these terms:

"NGOs are germane to the issues involved in local and general development because they articulate and multiply the capacities of mostly powerless rank and file individuals and slowly but gradually increase their bargaining power as groups...They can only lightly, if at all, influence the outcome of current resource allocation and surplus extraction processes, but they may gradually put their membership in a more favourable position to do so."

(Cernea, 1988)

This point focuses on the attempts of some NGOs to empower the social groups with whom they work. Cernea asserts that the capacities and bargaining power of the poor and powerless can be increased, but that existing relations of production and exchange (which he terms 'current resource allocation and surplus extraction processes') that generate poverty cannot be challenged without building the power of the poor.

Many NGO projects provide income transfers. The first international NGOs, such as the Oxford Committee for Famine Relief (now Oxfam), focused almost exclusively on income transfers and welfare activity. More recently, NGOs have focused on a wider range of developmental activity, including promotion of household enterprises. Some of the constraints identified in the case of the IRDP in India apply to similar, if smaller scale, income-generating projects organized by NGOs.

Scepticism about income-generation projects in the context of unchanged rural class and other power relations provides impetus for a third category of NGO projects, which seek to challenge property relations or at least to empower the poor and powerless so that they are better placed to bargain for greater access to economic assets. The idea of NGOs engaging in agrarian reform may seem far-fetched in the context of the significant obstacles to reform emphasized above. NGO work is thus often restricted to bargaining for resources and rights from the state, or the better-off, in the expectation that this process will gradually build collective strengths and skills of the poor to facilitate more fundamental change in the long run.

What is the relation of NGOs to government?

NGOs are not directly bound by the power relations which impinge upon states, nor are they necessarily restricted by the institutional constraints of government hierarchies and procedures. They are, nevertheless, influenced by some of the same social forces and subject to state regulation through laws and financing arrangements. They are not, therefore, entirely independent agents. One study of NGOs in Latin America found that state and NGO activities were interdependent:

> "Grass roots groups in Latin America perceive themselves as in adversarial relation with the state. In practice, they are most numerous and strong where the state is strong, and there is a better organized and prosperous public sector. Organizations develop complex linkages with the public sector."
>
> (Annis, 1987)

Annis concludes that indigenous NGOs (which he calls 'grass roots organizations') and the state come to depend on each other. NGOs depend on the state for resources: significant proportions of NGO funding either come directly from states, or are subject to state regulation. In many countries, organizations receiving money either from overseas or from internal collection have to be approved by the state. In some countries this approval process includes vetting the types of work undertaken by the NGO. On the other hand, the state may also come to depend on the work of NGOs to extend its reach into the countryside. Many states in the developing world have only relatively tenuous links with rural society, particularly where limited development of commodity production means that economic activity is only partially integrated with the national economy. In order to build political support (or at least compliance) and to promote state objectives, many states seek to extend their influence in the countryside, including through supporting NGOs. The interdependence that can result may lead to NGOs experiencing some of the same kinds of constraints that limit government choices and capacities. In particular, dependence on the state for funding or recognition sets up a potential conflict for those groups acting on a 'relational' view of poverty (Chapter 1) not least when a particular state is seen as part of the problem of poverty rather than part of the solution. When the state supports the existing social order in the countryside, it may attempt to limit any challenge by NGOs to existing relations of property and power. Approval or funding may be denied to such organizations, particularly as they become more effective.

Notwithstanding these important relationships between NGOs and states in both the South and the North, NGOs can try to construct some degree of independence to confront relations of unequal power (of classes, gender and states). They also have greater flexibility than states, making them potentially more responsive to social and ecological diversity then state organizations with their tendency to 'simplify, centralize and standardize'. We can examine how a degree of independence and flexibility can inform NGO action through the case of irrigation pump groups among the landless in Bangladesh.

NGOs: innovation and 'collective entrepreneurialism'

NGO promotion of irrigation pump groups in Bangladesh illustrates the potential for non-state organizations to identify and exploit room for manoeuvre within manifestly inegalitarian agrarian structures (Figure 11.10). This example also provides a contrast with IRDP in India because it promotes collective (rather than individual) income generation in an economic activity central to agricultural growth, centred on a resource, water, that in this instance is less affected by property rights than control over land usually is.

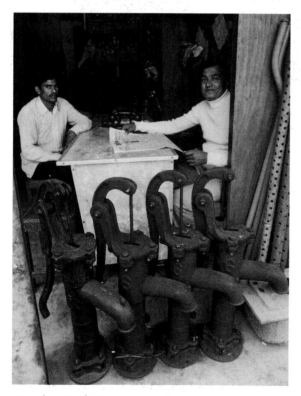

Figure 11.10   Water pumps for sale in Bangladesh: credit can help landless people to buy water pumps from such traders; the landless can then live by providing a service to local farmers who need to irrigate their land.

Over the last ten years in Bangladesh, a new generation of indigenous NGOs has been engaged in the promotion of irrigation pump groups among landless people. They help the formation of groups to install, operate and manage irrigation pumps, by providing credit, training and support. At one level this initiative might be seen as just another example of income-generating activities among the rural poor, notably through the provision of credit and assets. However, the project has some distinctive features.

First, the intervention is based on an analysis of agrarian change which identifies possibilities for modifying agrarian relations by enterprises providing services for agricultural production. The gradual adoption of more productive Green Revolution practices in agriculture means that there is increasing demand for agricultural services to provide water, threshing facilities, seeds, fertilizers and other goods. The second characteristic distinguishing this type of NGO activity from that of promoting individualized petty commodity production is its collective organization based on the shared interests of members of a particular social class. Thirdly, this initiative attempts to challenge newly emerging, and not yet strongly established, property relations associated with the introduction of new technology (the water control required for Green Revolution agriculture).

The background to this form of action from above is high levels of landlessness (Figure 11.11) and low levels of employment generation in rural Bangladesh. Land ownership has become increasingly concentrated, with the majority of rural households effectively landless and dependent on wage labour and petty trade or production for their subsistence. In those conditions 'one way of alleviating their poverty is to place in their hands the control of productive assets not already controlled by landowners' (Wood, 1984, p.55).

Irrigation was chosen as the focus of this project because: (a) it plays a key role in agricultural growth, (b) the social relations controlling water and pump distribution are not yet as strongly established as those controlling land, and

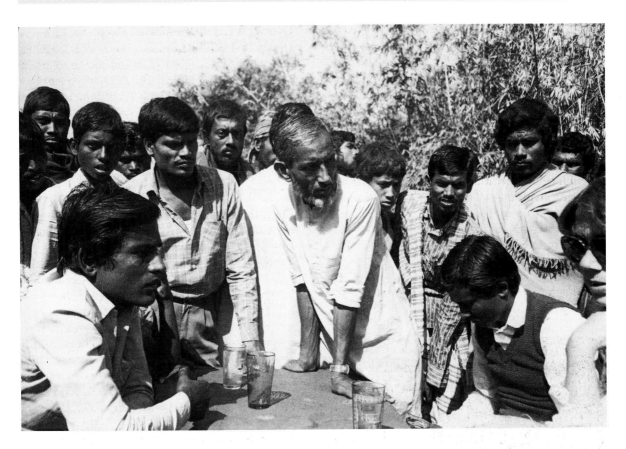

Figure 11.11   A meeting of landless peasants in Bangladesh.

(c) landless people, it is argued, have a collective interest in extensive use of irrigation water, namely that they live by selling their labour power, and the extensive use of irrigation water maximizes the jobs created in agriculture. There is thus a coincidence of pressures for growth and equity; maximized water use increases both agricultural growth and rural employment.

> "Basically the argument consists of appreciating that rural property rights other than land exist and that, since land is the most institutionalized and custom-bound of these rights, its socialization or even modest redistribution is the hardest to achieve...In the rural political economy of Bangladesh...other forms of property, from whose control a rental can be derived, are relatively uninstitutionalized and undeveloped by comparison to land. They are therefore more easily available for capture and control by the landless and near landless marginal peasants, especially through joint group action. This enables the landless to participate in the rural economy not as *supplicants* in the labour market, their value undermined by a surplus of labour or by personal ties of dependency, but as *owners* of commodities and services that have a more buoyant market value. Such a process can lead to a redistribution of assets and incomes, thereby stimulating effective demand and the expansion of production."
>
> (Wood & Palmer-Jones, 1990, p.xxv)

With this analysis, an indigenous NGO, *Proshika*, with backing from some sections of the government and finance from external donors, set out to assist landless people to establish organizations, get credit and build skills. Other NGOs have followed *Proshika*'s example. Flexible and responsive procedures (such as participative action research), and close collaboration with the rural poor, have enabled regional differences and institutional difficulties to be confronted. Landless pump groups are increasing in number, and their outcomes are mixed but promising.

Outcomes

This project has not transformed the rural economy of Bangladesh – it is much too small for that. It has had some promising results though. First, it promotes an economically efficient approach to expanding: '…in certain technical respects (for example, the area irrigated with equipment of a given capacity) the groups are at least as effective as private management…and repayment rates achieved are much higher than is usual for agricultural loans' (World Bank, 1990, p.71). Second, the project has generated incomes for those involved in the pump groups and has increased employment more widely through expanding irrigation. After ten years of this experiment, detailed evaluations provide some support for the claim that:

> "this strategy for agrarian reform, through the provision of irrigation services by the landless, can not only deliver the equity objectives implied by the redistribution of productive assets but can also compete successfully on productivity criteria with the private sector…"
>
> (Wood & Palmer-Jones, 1990, p.197)

There are also indications that this initiative has contributed to reducing aspects of poverty identified in Chapter 1, namely feelings of powerlessness, vulnerability and isolation. Ownership of a productive asset has brought some extension of rights and increased bargaining power. In one case, the irrigation service has been used as leverage to secure employment and increased wages for group members. The social status of group members has also been enhanced. Group members report that:

> "attention was paid to them as actors in the market place with business to transact. They were involved in a wider set of relationships and had a purpose beyond the normal one of buying small amounts of food and other consumption items with minute calculations of how far their daily wages could be stretched…Two of the groups, quite independently of each other and without a leading question, [reported that they] were impressed when their members were offered a chair and even tea by the local bank manager in one case (a welcome exception to the general rule), and the local diesel supplier in another case. To be offered a seat in this society is highly symbolic, a recognition not so much of equality but of worth and value."
>
> (Wood & Palmer-Jones, 1990, p.68)

This case suggests that in some circumstances there may be room for NGOs to undertake action from above which offers hope of improving rural livelihoods. The advantages of NGOs in this situation include: (a) a degree of separation from government (though subject to the potential constraints of dependence on government, noted earlier), (b) more effective dialogue with poor peasants than a government compromised by the involvement of landed classes can establish, (c) potentially greater flexibility of action and responsiveness to rural conditions compared with government. In part, this leads to the conclusion that action from above on rural livelihoods depends on effective representation from below. Action from below is the subject of Chapter 12.

Summary and conclusions

This chapter has argued that there are limits to what can be achieved through action from above on rural livelihoods.

1   One set of constraints arises from unequal relations of production and exchange in the countryside. The case of the Integrated Rural Development Project in India suggests some of the limits on attempts to promote income-generating projects within existing agrarian structures. Conditions of participation in credit and commodity markets may restrict the viability of petty commodity production. Conditions of access to land may also compromise the hard-won achievements of new livelihoods. The small scale of enterprises established by the poor, and their lack of reserves, makes them particularly vulnerable to the risks and fluctuations and markets.

2   These realities of agrarian power draw attention to the importance of agrarian reform, and the striking success in East Asia lends optimism to an otherwise bleak prospect for action from above. The historical circumstances of agrarian reform in East Asia, however, highlight the rarity of the conditions associated with agrarian reform there: 'high-handed government' emerges from conditions of crisis; prior changes in production conditions may be necessary prerequisites for increases in productivity; popular uprising and external pressure may be required for egalitarian reform.

3   A second set of constraints arises from the character of states and their capacities to implement development projects in the rural areas. Externally funded rural development projects are hindered by inadequate government 'commitment'. In some cases, this may be because projects challenging rural power relations are vetoed by a state which supports that power structure. Lack of commitment may also be a symptom of disagreement with the objectives of externally designed programmes, or reflect the lack of resources of state. The institutional structures of government also seem to be more capable of implementing standardized and centralized projects, and less able to promote decentralized initiatives responding to regional variations in social and ecological conditions.

4   Finally, the chapter asks whether non-governmental organizations have an opportunity to act where government intervention may be restricted. This section suggests that many NGOs become dependent upon government for recognition and finance. Comparable constraints to those facing government may therefore limit their ability to intervene in the rural economy. Nevertheless, there may be occasional room for manoeuvre. Where NGOs have an analysis and understanding of agrarian structures and processes of change, their flexibility and independence from landed classes may enable some kinds of effective action on rural livelihoods to be taken.

12

RURAL LIVELIHOODS: ACTION FROM BELOW

HAZEL JOHNSON

Action on rural poverty is not just a question for the state, or for international organizations and non-governmental organizations (NGOs), important though their activities may be. In fact, the lives of poor people consist of innumerable daily actions that attempt to alleviate hardship, from trying to secure ways of growing food or earning income to negotiating the distribution of resources within households. These daily struggles may be accompanied by individual and collective actions to subvert or defeat structures which reinforce poverty. In more collective responses, the rural poor may establish or join peasant organizations or community groups to try and bring about more fundamental change.

The idea that the rural poor can act to improve their own livelihoods is attractive for a number of reasons. States rarely seem to act in the interests of the rural poor, and sometimes act to their detriment: for example, in policies that lead to increasing concentration of resources or environmental degradation (Chapters 2–4, 8 and 11). Furthermore, in spite of many different types of 'actions from above' and different policy orientations (see, for example, *Hewitt et al., 1992*, Part 1; the policy changes in India outlined in Chapter 7; and the discussion of actions 'from above' in Chapter 11), there are still immense problems in creating adequate employment and incomes for all who lack them.

In many Third World economies, there are continuing problems in food production and distribution, combined with low productivity and low remuneration for most people (as well as alternative income sources which do not offer greatly improved means of livelihood). On the other hand, a minority – some groups of producers and land owners – has experienced potential and actual accumulation. If actions from above cannot 'get it right' (or may not necessarily intend to 'get it right' for the rural poor), it is tempting to think that actions 'from below' may have a chance – that the rural poor know what their problems are and seek rational solutions to them. In other words:

1 the rural poor are 'conscious actors' who are constantly adapting to circumstances and actively bringing about change, not just objects or 'targets' of policy;

2 the rural poor have their own kinds of knowledge and skills which have been adapted to local conditions (typically ignored or undermined by agents of 'development' from above – see Chapter 8);

3 the social relations and cultural norms of particular societies or groups have their own validity; the job of 'development agents' is to understand them, and to encourage the improvement of livelihoods of the rural poor in ways that respect local values and draw on local capacities.

Action from below also appeals because it is seen to counteract and change the powerlessness of the rural poor. Because poverty is strongly associated with powerlessness, effective action from below has to involve empowerment (*Allen & Thomas, 1992*, p.91). The poor have to gain a voice, and some control over the social and economic processes that affect them, if they are to improve their productive capacities and quality of life more generally.

Q In what sense do actions from below represent a way forward for improving rural livelihoods in the Third World?

Q Are they an 'alternative' to what states or other policy and planning bodies do?

Q What can be learnt from the decisions and actions of the rural poor?

These questions are pursued in this final chapter by examining how the poor manage in spite of difficulties, how they register protest, and how they try to change structures which create or reinforce poverty. The chapter reflects on evidence presented in this book as well as other examples, and considers the implications of different types of action from below for development.

What is meant by different 'types' of action? This chapter tries to avoid presenting a hard and fast typology of the different kinds of responses made by the rural poor. Action by the same individuals or groups can take many forms depending on circumstances, opportunities, awareness, and courage. At the same time it is useful to indicate some broad categories of actions, and their potential for bringing about change, and this chapter suggests how this can be done.

Section 12.1 shows some of the ways that rural people experience their poverty and how they see it as constraining action. It also illustrates the varied conditions of poverty and the differentiation among the poor themselves. The subjective consciousness and understanding of constraints and possibilities, and of the particular social relations and positions in which poor people are individually and collectively located, affect the types of action they engage in.

Section 12.2 looks at how people cope with poverty. Coping here means the pursuit of everyday survival in the face of more or less difficult conditions and the context of agrarian change in the rural economies of the Third World. This type of action is different from that discussed in Section 12.3 which looks at action as 'resistance'. Often people try to resist changes, or the effects of change, which actually or potentially threaten their livelihoods. But the protection of livelihoods is not the only issue: human dignity, cultural values, customs and social relationships may all be the subject of acts of resistance.

Whereas the types of action in Sections 12.1 to 12.3 may be individual or collective, and their effects short or long term, Section 12.4 discusses another type of action – that of collectively organized groups whose purpose is to bring longer term change in livelihoods, including influencing national strategies for development.

12.1 Experiencing poverty

Statement of Rosendo Huenumán, from the Mapuche people in Chile

"I was born in a *ruca* [hut]. My mother's name is Margarita García Huenumán. She is illiterate. She works on the land and the loom...She is an unmarried mother. I was brought up at my grandparents' house... Well, my life has been pretty hard, particularly my childhood. I started work at an early age. I remember that I first earned money by cutting oats...The next year I travelled further away from my reservation to do the same work, cutting wheat... we also went to pick potatoes...and they paid with ten kilos of potatoes for every sack we picked. We used to do this to take home potatoes for the winter...And so I started learning how to earn a living and how to help at my grannie's home...And there was another way of earning a living...we, the Mapuches of that area live near the sea, so the shellfish is also a product which helped us survive."

(Johnson & Bernstein, 1987, pp.47–8)

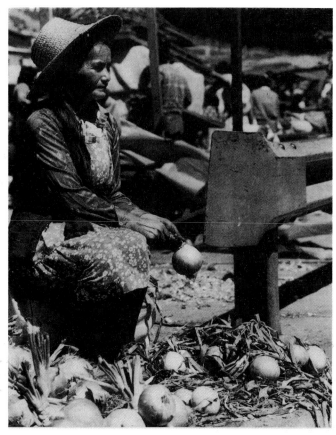

Figure 12.1 Mapuche woman selling onions, Chile.

Interview with women wage labourers, India

" They get up at 5 a.m., get water from the well, collect cow-pats for fuel, cook, clean the floor, take their baths, wash clothes and then go to the fields at 8 or 10 a.m. They work until 6 p.m. – on the days when they can get work – and then return for cooking and household duties until they finally go to sleep at 9 or 10 p.m. Sixteen hours of work a day...

'What do you eat?'

'*Bhakri – jawari bhakri* or *lal bhakri*.' This means a coarse, tortilla-like bread made of millet and sometimes American *milo* (sorghum) which is often imported and sold or given to the poor.

'And vegetables?'

'Vegetables – what shall we tell you?' says Kaminibai. 'If we have vegetables we can have spices but no salt, or salt but no spices, such is our poverty! There is no work. Some collect twigs, some collect wood and sell it, or use it for fuel. What can we do? We are poor...' "

(ibid., p.180)

Interview with a woman farmer, Mozambique

What particular problems do you face when your husband is in Joni [Johannesburg]?

I have no problems. I just work in the fields and stay and wait for his return. He will find me here when he returns!

Do you have lots of work in the fields – do you have big fields?

Well, we do not have seeds. We do have fields but we have no seeds!

Is it not possible to buy seeds from the shops at Homoine with the money from your husband?

There is none in the shops – where shall we find seeds?

Does your husband have cattle?

We have no cattle.

Has he never bought cattle?

No.

So you cultivate with the hoe! Do you have pigs?

I have a young one.

One only?

Yes."

(ibid., p.171)

These descriptions by poor people from different parts of the world show the economic realities of their lives: lack of resources (and difficulties gaining access to them), low wages, long hours, little food, the need to resort to multiple kinds of work to gain an income. While these statements illustrate the *individual* experience of poverty, they also suggest some *general* features: the *amounts of time* that are involved in obtaining livelihoods and in household reproduction, the *low levels of output* and productivity that characterize much economic activity, and the *precariousness* of personal and household survival.

But daily life only partly concerns productive work and making a living. Other aspects of social existence – being a child and growing up, being educated, learning and practising cultural beliefs and values, negotiating kinship and marriage relations, dealing with sexual and generational divisions, dealing with illness and death – are fundamental parts of of rural life (see especially Chapters 5, 6 and 10). They are also experienced in particular ways in conditions of poverty:

Chile

" 'To learn to read and write I had to leave my reservation. Where I lived, a family, not a Mapuche family, opened a school, a fee

paying school…That is how my cousins learnt to read…when one of my cousins learnt to read I always went up to him, to look at his books…And so my cousin started teaching me the letters and I, with that, with that interest in learning, learnt to read without going to school…' "

(ibid., pp.49–50)

India

" 'About divorce. If your husband goes, can you take another husband?'

'Oh yes…'

'Ask Rukmini.'…Yes, she says, her husband left her several months ago and since then she has lived with her sister and worked alone…

'Can't you take another husband?'…

'There is a guest…' The women all laugh. There seems an ambivalence about the high caste standards which define such liaisons as immoral. They recognize the standards, but seem to want me to know that they take them lightly, that they are not really helpless."

(ibid., p.181)

Mozambique

Did your husband [give] lobolo [brideprice] to your family?

No.

Did he not? Has he paid now?

My mother did not accept his *lobolo*. You see, I have no father and who was going to receive the *lobolo*? My mother did not want anyone else to receive my *lobolo* and therefore she did not allow my husband to give *lobolo*.

Your mother took this decision by herself, but what did you feel about it?

What could I have done? There was no one to receive the *lobolo*.

Did you want your husband to pay lobolo?

Yes, he should have given a 'gratification' [a gift in acknowledgement]…did he pick me up from the footpath?

You felt he should have thanked your mother...?

It is okay because he takes care of her now."

(ibid., p.170)

The idea that 'life goes on' under materially hard conditions can be difficult to grasp unless one has experienced it at first hand or been close to it. It almost seems 'immoral' to ask whether poor people enjoy life, or to presume that life can consist of anything but daily grind in the fields, or long hours spent trying to sell a few vegetables or other commodities, or getting up before dawn to prepare food and do domestic work. But the social relations that structure rural life are *multidimensional* and *culturally complex,* and include celebration, ritual and other non-material activities as well as obtaining livelihoods. People's responses to their material conditions and to the effects of change are also multidimensional, and reflect the complexities of cultural values and different aspects of social life. A significant example in this book is the 'story' of responses to illness and disease, both individual and social, told in Chapter 10. As well as particular beliefs about illness and healing, the therapies chosen by the Madi people of Laropi in Uganda reflect conditions of hardship and disruption, and the need to re-establish some sense of community cohesion. For the Madi, experiencing poverty included coping with social upheaval and the potential disintegration of the social relations underlying their moral codes. On the other hand, poverty and powerlessness do diminish the quality of people's lives (also illustrated in Chapter 10) and may actually *exclude* them from participating in or enjoying activities other than the struggle for survival (one defining characteristic of poverty noted in Chapter 1).

The conditions of production and reproduction in rural areas are highly *differentiated* as we have seen in Chapters 2–4, 6 and 9. Not all peasants are 'poor'. In fact, the conclusions to Chapters 2–4 suggested that in rural areas the poorest are 'those for whom the "peasant option" is not viable'. In such cases producers or household members have to resort to other activities as well as – or as a substitute for – their own farming to survive.

This does not necessarily mean that those who *can* choose the 'peasant option' are 'well-off' – on the contrary, most are poor by any standards, with low levels of consumption and limited possibilities to improve their livelihoods, which can be easily undermined by sudden crises and upheavals.

Economic and social differentiation among rural people is a *process* resulting principally from unevenly combined accumulation and impoverishment typically associated with capitalist development (Chapters 2–4 and 9). This means that being poor or wealthy (or somewhere in between) is not just a state – conditions may improve or worsen over time, either as an effect of processes such as commoditization or the uneven development of markets (Chapter 7), or because of actions taken by the poor themselves or by others.

This dynamic aspect of social differentiation reflects the creation of relative wealth, on the one hand, and new forms of economic exploitation and powerlessness on the other. It is illustrated by the following account:

Story of an Indian sharecropper

"[Although Indra Lohar] had been cultivating Plot No. 9...measuring approximately five acres, for more than a couple of decades, he did not get his name recorded during the last revisional settlement operation (1955–62) as that might have been interpreted as an act of disloyalty by his master [Bibhuti]...Indra took orders from Bibhuti and delivered the share of the crop to his...*kharmar* [barn]...After the death of Bibhuti...his son Sachinandan appointed a village quack, Badal Karmakar, known for his shrewdness and ruthlessness, as his estate agent...Indra was summoned by this agent and was told that the new owner would not recognize him as a *bargadar* [sharecropper], [and] he was advised that he should immediately give up possession of the plot of land as well as the produce therefrom to the agent if he valued his life...He did not have a scrap of paper to support his claim..."

(ibid., pp.36–7)

Even people in apparently similar material circumstances do not necessarily experience their living and working conditions in the same way. Access to and control over resources may change over time with *age* and *social status*. For example, it may be very important for people to ensure some form of social security for themselves in old age; on the other hand, old people may be allocated less food because they no longer do productive work. *Gender* is of particular significance in the experience of being poor – gender relations affect access to resources, to labour, output and income, the different responsibilities women and men have in household survival strategies, and the social and cultural content of people's lives (Chapters 2–5; Chapter 6 also discussed the gendered nature of survival strategies). The Chilean and Indian interviewees illustrate some of these problems:

Chile

"...on one of those trips when we are far from home, my aunt gives birth to a baby, delivers a baby without any medical help...at the home of a Mapuche woman who lived alone...very poor, very humble...I remember that one night she became ill in childbirth...as she was not assisted by a doctor nor any specialist person in midwifery, but coped as best she could, she became very ill...On the third day my aunt dies...the child, we gave him to a family who was sort of related to my grandfather. That woman was called Rosalía Huenumán...We left the child in this woman's care..."

(ibid., p.49)

India

"'We have to do the housework and when the housework is finished we have to do the field work and when the field work is finished we have to take care of the children, we have to do all the work!...What do men do? They get up, they take a bath, they eat some bread and go to the fields. But understand what their duty is: they only do the work that is allotted to them in the fields. They only do one sort of work...'"

(ibid., p.182)

Some writers have suggested that poverty engenders its own 'culture' – particular ways of seeing the world and social behaviours. Oscar Lewis, who wrote a vivid and detailed account of the lives of members of a Mexican family in the 1940s and 50s, was a particular proponent of this view (Lewis, 1963, pp.xxiv–xxvii). While poverty and powerlessness undoubtedly affect people's behaviours and responses, as well as limiting the actions they can take, we should beware of stereotypes about the poor. For example, although the poor are often victims of structures and processes beyond their control, it should not be assumed that they will therefore respond with resignation or even fatalism. Although Kaminibai said 'What can we do? We are poor', the actual stories of the people whose accounts are extracted above showed considerable resilience: Indra Lohar never regained his land, but he pursued his claim through the courts; Rosendo Huenumán became a local leader; the Indian women workers were organized in a union; and the Mozambican interviewee participated in the local 'cell' of FRELIMO. These experiences suggest a rather more complex set of processes at work.

In short, being poor does not necessarily prevent individuals and groups or communities from taking action on their own behalf. What affects whether, and how, they might do so? First, there are the *structural constraints* of poverty – the *social relations* which determine people's access to and control over resources and their benefits. Secondly, the *low productivity* typically associated with poverty means that people spend an immense amount of *time* just making a living and maintaining households and families. Thirdly, the rural poor typically have little *political power and representation* and experience great difficulty in getting the state, the judiciary and other institutions to act in their interests (Chapter 11). Fourthly, how poor people perceive and understand their situations – and their possibilities for action – is conditioned in part by their *social position*: age, class, gender, and other determinants of social status such as race or ethnic grouping. Finally, there is the extent to which such understanding is translated into *individual* and *collective* mechanisms for *daily survival*, or organized attempts to *transform* social and economic conditions more fundamentally. These different aspects of 'actions from below' are important threads in the following sections.

12.2   Coping and survival

Different contexts and conditions of poverty affect to what extent action is limited to surviving or goes beyond it, whether it concerns individual or collective ways of coping, and whether coping involves immediate needs or ways of dealing with poverty in the longer term.

Coping with ongoing poverty

Rural poverty can be extremely acute during war, social disturbance and political change, or when environmental degradation or severe weather changes such as drought erode the resource base of the poor (Chapters 6, 8 and 9). Minimal access to resources and alternative means of generating income increases vulnerability in times of crisis (see below). But at the best of times daily survival and making a living mean *ongoing* low income and consumption levels for many rural producers in the Third World.

Coping with ongoing poverty typically involves engaging in many different activities simultaneously, in a constant search for adequate income. Diversification is a necessity as well as a strategy for survival – these are the 'foxes' in Chambers' characterization of the poor's livelihood strategies (Chapter 1) and are exemplified in the Chilean account above. But even the 'hedgehogs', or those who are tied into a particular way of making a living (such as the sharecropping of Indra Lohar), often find ways of supplementing their work with artisanal activities, processing and selling food, other petty commodity production or petty trading, exploiting common property resources and doing casual work for others. A variety of such survival strategies was outlined in Chapter 6 (Section 6.4).

The multiplicity of income streams, small though each may be – and almost invariably connected

with low productivity work (Chapter 7) – may be important for individual as well as household survival. Where men and women have separate budget responsibilities, each may adopt different strategies for maintaining their own as well as their dependants' livelihoods (Chapters 5 and 6). But even where – ideologically and practically – women and men are seen as part of households which pool income (although usually controlled and distributed by the household head), women can try to protect their own separate income sources as part of a personal survival strategy. Kabeer (1990, pp.143–4) lists the following responses of women to the constraints of *purdah* in a Bangladesh village: secret saving; borrowing and lending of rice between women in the village; using relatives to help provide cash-earning activities; selling home-produced goods for lower prices to traders coming to the house (rather than handing them over to male members of the family to take to market – where higher prices can be obtained but women lose control of the proceeds).

Coping with ongoing poverty has a day-to-day character for many households and individuals. Reciprocity, such as loans, gifts and forms of patronage within and between households (discussed in Chapters 5 and 6), is important for survival. These various (low level, small-scale and often unnoticed) actions may have an individual or collective character, even if their existence is not collectively acknowledged. Some such actions may rely on custom (for example, patronage); others may deliberately transgress local norms (such as the secret activities mentioned by Kabeer; see also Figure 12.2). While people have many ingenious ways of coping with ongoing poverty, their vulnerability to unexpected changes or events, however small, can tip them from ongoing poverty into a crisis, as we shall see.

Coping with change

Poverty may be persistent for many in rural areas but it is not unchanging. Several chapters in this book have shown how commoditization can change the conditions of poverty (Chapters 2–5, 7–9) and Chapters 8 and 9 have also discussed the effects of environmental change, resulting from increased or new forms of commoditization and the survival

Figure 12.2   This Sudanese woman manages to maintain her modesty while working in the cotton fields by wearing a large piece of cloth over her head and round her body. She carries the sack for the cotton she is picking under the cloth.

strategies of the rural poor themselves in these changing conditions.

Commoditization affects access to resources and the conditions of exchange of labour and output. The process is typically uneven (Chapter 7): producers may still have usufruct rights to land but may increasingly sell what they produce (or produce more 'cash crops' as opposed to subsistence crops), and may also sell their labour. They may still borrow cash from relatives or neighbours, or receive it as gifts (Chapter 5), but there may also be informal and formal credit markets with varying rates of interest. The extent and conditions of commoditization vary with the type and degree of integration of rural communities into national and international markets, but there are few, if any, communities which are not so integrated in some way (see Allen's comment on the inhabitants of Laropi at the beginning of Chapter 10).

How do rural people respond to the impact of commoditization on their lives, either because of its increasing intensity (for example, with the Green Revolution in India) or changes in its forms (such as the growth of a locally-based wage economy in Kirene)? In the face of inequalities in markets, the types of struggles often engaged in by rural producers are over maintaining access to land and other resources, over prices received for products or wages paid. There may also be market pressures on customary exchanges within and between households that may give rise to conflict between household members.

Intra-household struggles illustrate the pervasive ways in which commoditization affects all dimensions of life – in reproductive as well as productive activities. Some powerful examples of women's responses to the effects of commoditization on rural livelihoods in Tanzania are given by Mbilinyi (1990). As in many Third World economies, Tanzanian agriculture combines peasant farming and large-scale crop production, drawing on family workers, permanent and casual wage labour and migration (seasonal and permanent). The increasing commoditization of labour and crops has affected sexual divisions in household production: men predominate in the wage economy, while women's work has become intensified in both subsistence and cash crop farming, in order to produce enough food or earn enough income for household maintenance. Women have also had to move into non-agricultural activities: Mbilinyi lists 'beer-brewing, food processing, petty trade and prostitution' (ibid., p.117).

Intensified by structural adjustment policies applied in the 1980s, these economic pressures on rural women in Tanzania have generated different responses. First, women have refused to do unpaid work for male household heads. Second, they have protested against the general intensification of their labour distributed between domestic work, farming on their own account, or working on village farms, and working in women's cooperatives. Women have spoken out against existing sexual divisions of labour in the household and the appropriation of their labour by husbands (see Box 4.3 in Chapter 4) and have also refused to provide casual labour for the large estates because of the low wage levels.

These responses express a more general and complex set of issues arising from commoditization. First, there are apparently external and 'objective' effects on rural people: for example, the pressures to earn cash incomes through petty commodity production or wage work, and the intensification of work (particularly for women because of their reproductive roles) that often accompanies commoditization (especially, too, when there are technical changes – note the effects of the Green Revolution analysed in Chapters 3 and 7). Second, there is the relationship – often a clash – between existing social relations within and between households and the new sets of social relations engendered by commoditization. The combination of these two dimensions means that commoditization has an important social and ideological content. In the Tanzanian case, for example, changes in market conditions and in the demands being placed on the rural population (to raise both food and export crop output) are absorbed into patriarchal relations in households that in turn generate particular pressures on women (see Chapter 4, Section 4.5).

The pressures on livelihood struggles created by environmental changes are also often gendered. Common property resources may provide an important contribution to the food provisioning strategies of poor women, and Chapters 6 and 7 mentioned the problems of declining access to such resources. The following account of one woman's way of coping with environmental degradation resulting from deforestation shows ingenuity and knowledge of regenerative techniques of resource use, but also shows how increasing constraints on access to resources make it impossible to sustain 'sound' practices.

Afi is a Ghanaian woman who farms under precarious conditions and who sought to diversify her income after drought devastated her crops. She became a charcoal burner. This is highly labour-intensive work but there is a ready market for charcoal in nearby Accra.

How Afi copes

"The constant exploitation of trees has put so much pressure on the vegetation that Afi's choice of wood species for fuel and charcoal has shrunk. Now she has to choose from only three different species. The Nim..., a drought-resistant tree, is her favourite source because it is widely available, produces excellent fuel and makes good charcoal. Haatso..., though not as popular as the Nim, is also highly combustible and good for charcoal and so is Nokotso...But Afi scarcely uses these two species because they are not available on her fallow land. The increasing scarcity of wood compels her to produce fuelwood only from her farm, where she has cutting rights.

The problem of fuelwood resources encourages Afi, just like other women in the village, to manage the wood carefully. She leaves the farm fallow to rest for a long period of time to regenerate. To assist regrowth and prevent soil erosion after felling, tree stumps of about two feet are left standing. This is her major method of sustaining the vegetation. But, because of the constant pressure for wood and the need for agricultural, commercial and residential lands, the fallow has been drastically shortened. The result is that the trees are not given enough time to recuperate fully before they are cut. Occasionally she used to plant a shade or fruit tree, but she has not done so for a long time. The number of trees on her farm is dwindling."

(Dankelman & Davidson, 1988, p.82)

This example shows some of the contradictions of an individual response to social and ecological changes. Some sort of coping strategy was necessary for family survival but the means chosen would only work in the relatively short term because of the limits on land and trees available to Afi – an example of powerlessness in the face of the patterns and impact of economic change.

Coping with crises

The discussion of responses to ongoing poverty and change – particularly changes in the forms and types of commoditization – shows how vulnerable the poor are and how easily crises may develop. Crisis may take many forms – from sudden calamities, such as illness, shortage of water, a rise in prices or losing land to general social upheaval or war.

To analyse crises, we cannot just look at events from the outside and assess their impact on rural people. It is also necessary to see *why* particular groups of people are vulnerable and why, for them, certain events or processes become crises. Crises do not necessarily have to be dramatic or large scale. My own interviews with maize producers in the Central American country of Honduras in 1988 showed that crises arose when someone fell ill and medicines had to be paid for, or when crops failed and debts could not be repaid. The margins of survival were often so narrow that relatively small events could spell economic disaster.

The example of Kirene villagers (Chapter 5) illustrates some of these points. The basis of livelihood for many villagers was a delicately balanced combination of access to different types of land and labour. The growing squeeze on land was reinforced by the insertion of a transnational company into the local economy, which further threatened the local farming system and undermined food production, especially for those who lost their land to Bud. Why was this so critical if Bud offered alternative sources of employment and income? In fact, the alternative sources of income could not provide the economic security that villagers had tried to ensure through careful management of their ecology and the redistributive mechanisms operating within and between households.

Coping with drought, for example, had been relatively manageable through the diverse cropping systems and use of different kinds of ground (see also Chapter 8 for additional examples). Although the existing squeeze on land was making this more difficult, such crisis avoidance became even more precarious for those households that lost land and had only temporary wage work in a company that, ultimately, proved not to be a commercial success.

Coping with crisis (as well as ongoing poverty) often results in immediate and (objectively) short-term responses rather than actions which can be sustained as part of a longer term strategy. In the Ugandan village of Amwoma (Chapter 9), Mamdani identifies two main crisis areas affecting the villagers: environmental degradation and population pressure on the land. The first arises from peasant responses to the privatization of landholdings and reduced access to communal land resources. To maintain access to land, poor producers have been deforesting new areas as well as using existing areas more intensively – but with technologies inappropriate to more intensive land use. The second results from the childbearing strategies of poor families who have large numbers of children to carry out high labour input farm work. But, as children grow up, there is increasing pressure on the land available to the poor. In this example, one can see how crisis constrains the ability of poor farmers to respond to the dilemmas they face, and how their responses then intensify those dilemmas for the future. Land privatization has been reinforced by internal processes of accumulation in the community, and poor farmers are not in a position to improve their livelihoods by changing their technologies and techniques of production. Thus peasants' coping strategies (deforestation and having many children) are rational responses to immediate crises of poverty, but do not provide a strategy for longer term survival.

Dealing with crisis also involves the reproduction of knowledge and cultural norms. In Laropi (Chapter 10), dealing with the crisis of social disruption and war involves rebuilding moral codes and social cohesion, in this instance expressed through approaches to illness and healing. Although these responses centre on individual actions, they are part of a collective survival strategy based on commonly held beliefs and values.

One of the most frequent ways of coping with rural crisis is migration, in the most dire cases moving as refugees to camps and centres where aid is disbursed (see Chapter 10). Seasonal migration commonly occurs to supplement income sources, but many poor farmers abandon the land and seek waged work or self-employment because the 'peasant option' is no longer viable for them. While this can happen on a large scale in cases of drought or other forms of devastation, migration in cases of economic crisis is an ongoing phenomenon involving many individual or household life decisions.

12.3 Resistance

Most of the cases described so far concern immediate and individual responses to difficult situations. Moreover, the borderlines between coping with ongoing poverty, change and crisis are often extremely fine. If the rural poor engage in many different small actions as part of survival strategies, to what extent do they manage to *protect* or *advance* their interests? Their actions take place in conditions of severe constraint but are often informed by considerable knowledge and ingenuity, suggesting an important (and necessary) capacity to resist. Is resistance just a defensive response to meet immediate needs, or does it suggest possibilities of more transformative or strategic change?

A recent approach to analysing forms of resistance (Scott, 1985, 1989) suggests that rural people do often lodge clear protests and even bring retribution to those considered the source of their problems, but that these actions do not usually take the form of open rebellion and are not necessarily directed towards longer term change. Such behaviour has been called 'weapons of the weak' (Scott, 1985) or 'everyday forms of resistance' (Scott, 1989). An example would be the covert actions of the Bangladeshi women cited by Kabeer; an example of actions involving retribution would be

the vengeance exacted against supposed poisoners or odd, non-conforming people suspected of bringing illness or transgressing moral codes in the Ugandan community of Laropi (Chapter 10).

Scott (1989, p.5) lists the following 'everyday forms of resistance': 'foot dragging, dissimulation, false compliance, feigned ignorance, desertion, pilfering, smuggling, poaching, arson, slander, sabotage, surreptitious assault and murder, anonymous threats' – all characterized by their covert nature, and avoidance of open confrontation. Although they may comprise many individual small actions, they require collective complicity so that silence and secrecy is maintained *vis à vis* the people the actions are aimed at. While these actions in themselves are not strategic or consciously directed to fundamental change, Scott suggests that they may have cumulative effects and achieve changes that open or organized protest may not.

Scott argues for a particular view of the political, in which peasants and rural workers act in the relative safety of covert action, to pursue or protect their interests, register protest and try to undermine the positions of the more powerful (see Table 12.1). Point 1 in the table gives examples of protest against material conditions, already mentioned above (and see Figure 12.3). Some types of action may be as much symbolic as material, such as those in point 2. Poverty and powerlessness can also involve lack of dignity and forms of humiliation by the more wealthy or powerful. While resistance may take the forms of hidden aggression, it can also be expressed in drama, dance and music where the humiliated symbolically reestablish their social position or take revenge. Other apparently symbolic or non-material actions (for example, those in point 3 of Table 12.1) are related to underlying protest about economic or political conditions. The movement led by the spirit medium, Alice Lakwena, mentioned in Chapter 10, would be such an example (Figure 12.4). In disguising what the protest 'agenda' is really about, those resisting may be able to preserve their anonymity (at least from their targets) to protect themselves from retribution or retaliation – though such actions may also, at times, build up to form part of an open resistance movement.

Figure 12.3 *Heavy drinking may also be viewed as a form of resistance.*

Figure 12.4 *Alice Lakwena (Alice the Messenger), the celebrated Acholi medium, was able to use her capacity to communicate with the spirit world to replace male holders of ritual authority. In 1986 she secured a large following of some 6000 people, many of whom were ex-soldiers. She led what became known as the Holy Spirit Movement in a 'Peasant's Revolt' against the Ugandan government. Thousands died in her uprising before she escaped from Uganda into neighbouring Kenya in October 1987. This photograph shows her with some of her supporters.*

Table 12.1   Domination and disguised resistance

Form of domination	Forms of disguised resistance
1 Material domination: appropriation of grain, taxes, labour, etc.	Everyday forms of resistance, e.g. poaching, squatting, desertion, evasion, foot dragging. Direct opposition by disguised resisters, e.g. masked appropriations, carnival.
2 Denial of status: humiliation, disprivilege, assaults on dignity.	Hidden transcript of anger, aggression, and a discourse of dignity, e.g. rituals of aggression, tales of revenge, creation of autonomous social space for assertion of dignity.
3 Ideological domination: justification by ruling groups for slavery, serfdom, caste, privilege.	Development of dissident subculture, e.g. millennial religion, slave 'hush arbors', folk religion, myths of social banditry, and class heroes.

Source: Scott, J. C. (1989) 'Everyday forms of peasant resistance' in Colburn, F. D. (ed.) *Everyday Forms of Peasant Resistance*, M. E. Sharpe, Inc., New York, p.27.

One of Scott's concerns is the relationship between action from below and 'development'. In particular, he feels that while radical critiques of top-down development and the 'liberal democrat' participation approaches are justified, structural change brought about by revolutionary action from below does not necessarily bring about development. He cites the 'tragedies of the Great Leap Forward in China, collectivization in the USSR, and state-directed agriculture in Vietnam' as examples which 'are hardly encouraging...of participatory development. The problem with both the liberal-democratic and the radical view of development is...that neither is sufficiently radical. What they miss is the nearly continuous, informal, undeclared, disguised forms of autonomous resistance by lower classes' (ibid., p.4).

A broader definition of the 'political' is welcome and helpful in understanding resistance from below that might otherwise pass unrecognized, but one of the paradoxes of such covert or indirect actions is that, while they may individually or cumulatively bring change, they may also reinforce conditions of oppression they seek to resist. While the 'weapons' that Scott lists suggest how the poor and weak protect their interests and dignity, it is worth asking whether such actions can transform the conditions that produce poverty and powerlessness, rather than just make them more bearable.

To explore this issue, I look at two examples of 'actions from below'. Both illustrate types of resistance to changes imposed 'from above'. The first discusses resistance by peasants and rural workers to state policies which were, in principle, directed to their interests as well as those of the national economy. The second looks at the reaction of rural workers to decisions imposed on them by their employer.

The first example is from Nicaragua under the Sandinistas, who overthrew dictator Anastasio Somoza in 1979 and established a socialist-inspired government with a mixed economy development strategy which lasted until the national elections of 1990. The example is based on a study by Colburn (1989) who examined the responses of peasants and rural workers to

Sandinista agrarian policy. It should be noted that the majority of rural workers and peasants in Nicaragua welcomed the Revolution and had strong expectations of its new social and economic policies. Distribution of land had been highly unequal, and rural wages inadequate. So how did rural workers and peasants respond to the radical changes introduced by the Sandinistas? Colburn looks first at rural workers, then at peasants.

The main foci of Sandinista agrarian policy were land distribution, the improvement of peasant agriculture, and increasing productivity and output in the export crop and livestock sectors. After 1979, the Sandinistas appropriated all of ex-President Somoza's land and other assets. In addition, there were many spontaneous land seizures by tenants and rural workers from other large landowners, especially absentee landlords. Spontaneous takeovers of large farms often ran into problems because of lack of technical assistance, infrastructure and organizational experience, and the state frequently intervened to take over the management, employing former tenants or labourers as workers on these new state farms. Rural wages were raised, benefiting permanent and seasonal workers on both private and state farms. However, the Revolution of 1979 was accompanied by aspirations on the part of rural workers that were difficult for the state (or even private enterprise) to meet. Colburn suggests that rural workers hoped for less arduous working conditions as well as higher wages and, in general, a much 'easier life' than previously, and this was manifested in labour indiscipline: 'Poor rural Nicaraguans, frustrated in their desire for an immediate improvement in their standard of living, simply took their "historical vacation"' (ibid., p.181). In addition, land seizures continued and many rural workers went on strike for further wage increases. The difficult economic situation, reinforced by the economic boycott from the United States, meant that the Sandinistas resorted to a policy of austerity, freezing wage increases, banning strikes and calling for labour discipline. Appealing to workers' sense of duty in defence of the national economy was not successful: 'An administrator of a state tobacco farm summarized the evolution in peasant [sic] response to exhortations to increase productivity: "After the revolution they said, 'No, we are free.' Now they say, 'You are too demanding; the salary is very low'"' (ibid., p.183). It was only when the state introduced output-related wage rates that rural workers began to respond positively to increasing output.

What about the peasants? Nicaragua's peasants grow basic staples (maize, beans, rice) consumed in both urban and rural areas. The Sandinista strategy to increase simultaneously peasant welfare and national food output was based on a programme of land reform providing cheap access to land, as well as co-operative organization of production and access to credit. Peasants (especially those gaining land after the revolution) were encouraged to form co-operatives and there was a massive credit injection into the countryside. To ensure cheap and adequate food supplies to the towns, peasants were also to sell their 'surpluses' to a new state marketing board at fixed prices.

How did peasants respond? Many were guided by individual interests rather than 'national' needs. A particularly difficult area was access to and use of credit. According to Colburn, while peasants could gain access to credit relatively easily, they tended to use it for personal consumption rather than buying farming inputs (although inputs were, in practice, difficult to get hold of). This use of credit was reinforced by lax debt collection. Furthermore, farmers' marketed output of basic foods remained low partly because of inadequate communications and because government-controlled prices could not compete with the higher prices being offered for foodstuffs on the black market. It became cheaper for peasants to buy grains from the state than to produce them for sale at state prices.

Colburn (ibid., pp.188–9) concludes that there was a range of individual peasant responses to government policies, which were to:

"1   increase production using their own resources;

2   increase production using state provided resources, principally land and credit;

3   use part of the credit or inputs as consumption income;

4   switch crops (especially to those without controlled prices);

5   avoid selling to the government, and seek higher prices in the private market;

6   produce only for consumption, and seek to earn needed cash elsewhere;

7   switch occupations altogether."

While many peasants who belonged to peasant organizations actively supporting Sandinista policies adopted the first two options, many individual producers exhibited responses 3 to 7. Thus, although peasant resistance to state policies did not represent a political threat (at least until the 1990 elections), it did help to undermine Sandinista attempts to run a regulated war economy and to provide cheap food for the towns.

How does this account relate to the discussion of resistance by the rural poor at the start of this section? On the one hand, there was a Revolution which had the popular support of rural and urban workers and peasants. On the other hand, there were different types of resistance to policies which in part were directed to the welfare of the rural poor and in part to the needs of urban consumers and the national economy. So, even though the state was apparently acting in the interests of the poor, many peasants and rural workers chose not to co-operate fully with particular aspects of agrarian policy, but to pursue individualized needs and interests.

Was the problem one of 'lack of consciousness' among certain rural workers and peasants, or that the state did not perceive these groups' real needs and problems? Both and neither. Peasants and rural workers *did* have collective organizations that worked in support of the government. The state also provided for many peasants' needs in land and credit (although the high costs of a credit programme with a low repayment rate, in a generalized economic crisis, limited credit availability later). The situation reveals complex social dynamics: in the past, rural workers and peasants had faced multiple hardships and many took advantage of new possibilities to improve their lot, even though poor urban consumers would face great difficulties in obtaining food at low prices; the government faced multiple problems devising an economic strategy based on 'redistribution with growth' as well as trying to cope with a war; political tensions over democratic processes and the promotion of 'social responsibility' created further contradictions.

Thus, forms of resistance – or 'weapons of the weak' – can be complex in their origins, their rationale and their outcomes. The fact that a state has a progressive agrarian policy does not necessarily mean that the rural poor will individually (or collectively) always choose to collaborate with it. In particular, the interests of individual peasants and the state (or urban consumers of staple foods) do not necessarily coincide, especially in such difficult conditions as those of the Nicaraguan war economy. This example shows that resistance may have contradictory outcomes and it leaves some important questions: what part do *interests* play in 'actions from below'? whose interests are they? how do those interests (multiple and conflictive as they are – in this case, for example, between the rural and urban poor) bear on 'development'? to what extent do the interests behind 'actions from above' and 'actions from below' converge or diverge? (See *Wuyts et al., 1992*, ch.6, for example, for a discussion of state action in relation to the empowerment of women.)

The second example is based on the case study of the Senegalese village of Kirene (Chapter 5). While the Nicaraguan example exemplified 'sectional' and individual interests among the poor, this one focuses on collective action (albeit on a small scale) against individualism promoted by those with economic power. The instance concerns how villagers who worked for Bud tried to maintain forms of solidarity that were part of social relations in Kirene and that were threatened by very different types of work organization involved in Bud (Mackintosh, 1989, pp.103–6).

When Bud set up operations, it appropriated about 50 hectares of village land in Kirene. Demands for compensation were not met by Bud, and villagers came to see their jobs in the business as a form of recompense – the jobs became *their* jobs, and in their minds only available to people from Kirene. Villagers wanted priority in terms of hiring, and control over the hiring process, whereas

the management wanted to control hiring and firing and to employ workers from other villages, often to undermine workers' bargaining power in cases of conflict.

A conflict arose when a new packing shed was opened and Bud hired workers (women) from Bandia, a nearby village, as well as from Kirene. The Kirene women refused to let the Bandia women work and demanded to speak to the management. The management did not arrive and all workers, men and women, from Kirene and elsewhere, went on strike. After negotiations, it was agreed that *Kirenoises* would have priority in the new packing shed.

How was this conflict interpreted? Kirene women saw management's actions as subverting social relations both in the village and also between villagers and their nearby kin in other villages. They saw their own actions as trying to maintain solidarity between the different groups.

Statement of a Kirene woman with a married sister in Bandia

"If we had fought with the women of Bandia, there would have been women here who would have beaten their mother's child...and the children might fight without knowing they were related through their mothers... There have been women from Bandia who picked on the harvests...we did not forbid them to come, we could not...it was the method of doing it that caused the trouble. They should have asked us, and we would have said, go and get certain people."

(ibid., p.106)

Mackintosh's interpretation

"Bud's very existence was necessarily a source of upheaval and conflict within the village. In a society which put a high value on agreement and consensus, and on mutual assistance between kin, the act of publicly closing ranks had the effect of denying the importance of such conflict, and blaming dissension upon specific machinations of the Bud management. The villagers were in effect asserting that the wage work experience could be inserted into the older social structure of the village."

(ibid.)

This example gives an additional dimension to the issue of *interests*. In this case, the villagers, male and female, were strongly motivated to maintain some sense of social cohesion in the face of upheaval (see also Chapter 10). But that sense of social cohesion involved a complex view of 'insiders' and 'outsiders': Bandia workers were 'insiders' as long as Kirene villagers had some control over their employment in Bud. (The story and statements indicate that the *Kirenois* preferred their *kin* in Bandia to be employed.) So while villagers were acting in what were perceived as communal (as opposed to household, or male or female worker) interests, the community could be extended to others. The *real* 'outsiders' were Bud and its threat to an 'older social structure'.

Another aspect of this action was its collective nature, arising not so much from wage work and co-operation at the workplace as from the 'older social structure' from which the workers had come. These responses also had *social* benefits – reproducing community solidarity in

the increasingly vulnerable position in which the community found itself. Jobs at Bud were not a secure alternative to land, subsistence food provision and local crop markets – in that sense, Bud was not entirely beneficial nor an 'answer' to the problems of economic development in the area.

Do both these examples (from Nicaragua and from Senegal) indicate actions that are 'weapons of the weak'? Yes, in the sense that in neither case did the people taking action have economic power, although in the Nicaraguan case, the Sandinistas were eventually voted out of political power (Figure 12.5). Nicaraguan peasants and rural workers were able to make life difficult for the Sandinistas; the Kirene villagers only marginally so for Bud, although they formed a union to carry out collective negotiations and actions, in the course of which individuals, particularly young people, and especially young women, gained an informal education in organization and mobilization. What did such actions achieve in both cases? In Nicaragua, there were short-term gains for small farmers, but longer term improvements in basic food production and conditions for rural workers were made difficult by the complex interplay of actions from above and below, and the war. In Kirene, in one

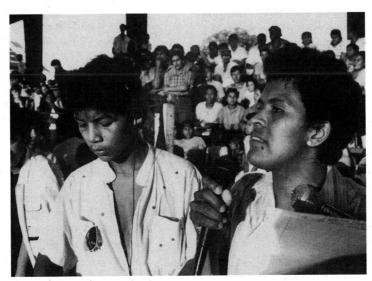

Figure 12.5 Voting the Sandinistas out did not necessarily serve the interests of the rural poor: here Miskito people are meeting to protest against the threatened loss of collective land following the change of government in 1990.

sense the gains were only short term because the Bud 'experiment' did not succeed (see Chapter 5), but the actions of the *Kirenois* suggested a longer term intention of protecting livelihoods in a new context of combined wage employment and farming. Some of these points are summarized in Table 12.2.

Table 12.2 Types of action

Type of action	Nicaragua	Senegal
Individual	yes	no
Collective	no	yes
Short-term goals	yes	?
Long-term goals	no	?
Survival	yes	yes
Transformation	no	no

It is becoming clear that poor rural people do not *necessarily* have a commonality of interests. What they have in common does not always outweigh their differences, nor do they necessarily express their interests and demands in the same way (see also the Colombian example below). It is obviously important to try to understand action from the actors' perspectives and to recognize that their interests might well differ radically from those of development agencies such as the state or aid organizations. But it is also important to try to distinguish the meaning and significance of different types of action: whose interests are involved (in terms not only of class and gender, but also of communities, ethnic groupings or groups with shared cultural identities), whether the interests are immediate or longer term, and whether the issues are about survival or transformation.

12.4 Bringing about change

The preceding section implies that this question can only be asked and answered meaningfully in relation to particular groups of people in particular contexts. In what contexts, under what conditions – and with whom – might the 'weak' actually effect change? To return to the original questions of this chapter, do actions from below represent a way forward? are they an alternative? what can be learnt from them?

'Rural development' is an uneven process and has largely involved the creation of wealth in certain branches and types of farming at the expense of others. Broadly speaking, policies have been directed to two main objectives: the development of capitalist agriculture, and/or creating a viable class of commercial small farmers (petty commodity producers). There are many policy variants, and many stories of relative success and failure in terms of policy objectives and who benefits.

A first response to the question above, then, is to consider individuals' possibilities of improving their livelihoods: how and under what conditions does that happen? Individual accumulation by small producers clearly can take place. We have witnessed it in conditions of technical change, especially the application of Green Revolution technologies (Chapters 3, 7); where there have been possibilities for acquiring land (Chapters 2, 6, 9); and where producers have been able to command the labour of others (also Chapters 2, 6 and 9). Often these conditions presume an inherent 'head start' in the accumulation process, an existing base of relative wealth and differentiation which has enabled certain farmers to take advantage of credit schemes, changes in land tenure laws, new technical openings, and the availability of a ready labour force where extra labour inputs are needed. Often such producers are labelled 'progressive farmers': they take risks and try out new things, and have the resources to do so.

But many, and in some places most, rural people are not in this situation. If 'actions from above' have generally had limited access in creating possibilities of improved livelihoods (Chapters 7 and 11), what alternative proposals might rural people put forward themselves?

There are many types of organization of peasants and rural workers, ranging from informal solidarity networks based on kinship and custom to peasant associations, workers' unions, federations of co-operatives, and so on. In many parts of the world, the rural poor have launched their own struggles to defend existing access to resources, jobs and wages. But the 'organized poor' also engage in struggles over obtaining resources (which almost always implies redistribution of land, as well as access to credit and other resources) and have often put forward alternative forms of organization of labour and distribution of income. For example, in many Latin American countries, and in parts of India and Africa, peasant organizations have campaigned to obtain land, involving 'recuperations' (where land is regarded as having been taken from those who used to farm it), squatting and demonstrations, as well as court claims or negotiations with land reform bodies or representatives from agricultural ministries (Figure 12.6).

In Latin America, peasant groups have often tried to establish collective forms of land tenure and land use rather than individual 'household' farms. For example, in the Honduran land reform programme, which was initiated in the 1960s and continued in different degrees during the 1970s and 80s, land was redistributed to organized groups of landless peasants. (Groups often had to occupy the land and confront reprisals from former landowners before gaining legal access.) The groups often allocated a communal plot for marketed grain and emergency reserves, while each household would have a family plot for its own subsistence. This form of organization worked with varying success (depending on the type of land the group was allocated, whether the group had access to credit and technical assistance, and the degree of the group's social cohesion). Some groups set up large-scale commercial co-operatives or other types of enterprise producing export crops. Others established projects in the production of staple foods, but many groups have had great difficulty in surviving, and some have dispersed or divided their land among themselves to farm individually.

Such initiatives on the part of peasants can rarely succeed without external support of some kind – financial and technical, as well as training in administration, book-keeping, marketing and so on. This usually means that peasant groups have to engage the help of either the state or NGOs (local or foreign), which in turn can result in conflicts of interests or the diversion of goals and benefits from what was originally intended. In addition, the heterogeneity of interests among rural people can make common programmes for rural development from below difficult to achieve. It is salutary to look at some of the problems that can arise.

Peasant organizations and the state

The following example illustrates some of the dilemmas facing peasant organizations and their attempts to establish alternatives for rural development. It is taken from a case study by León Zamosc who followed the progress of a Colombian organization called ANUC in the 1970s (Zamosc, 1989). ANUC is the *Asociación Nacional de Usuarios Campesinos,* Colombia, or the National Peasant Association of Colombia. ANUC is unusual in the history of peasant organizations because it was actually launched with the support of the *state*, not 'from below', and was later appropriated for peasant ends in opposition to the state.

In the 1960s, Colombia was trying to expand its industrial base after a period of import-substituting industrialization (see *Allen & Thomas, 1992*, Box 11.2). Industrial expansion required an increasing supply of raw materials and food for urban consumption, which in turn placed demands on Colombia's agriculture, to supply raw materials and earn extra foreign exchange from exports to pay for industrial inputs. There had been growth of capitalist agriculture in the valley areas but *haciendas* (Chapter 2) persisted, while peasant farming was relegated primarily to the mountainous zones. Growing pressure on land available to peasants was also leading to colonization of forest areas and increasing rural–urban migration. The peasant population was heterogeneous: some were seasonal wage workers on capitalist farms, others share-cropped or had service-based tenancies with the *haciendas*, those who had colonized new lands were trying to consolidate their holdings, and others had farmed in the mountains over a long period of time (Table 12.3).

Figure12.6 Different types of organized rural action: (above) direct action, blocking train lines during a rural workers' strike, Ayaviri, Peru; (right) peasants embarking on a land seizure, Ayaviri, Peru; (below) Quechua women working an idle plot of land in an occupied hacienda, San Juan, Ecuador.

Table 12.3 Colombian agrarian structures, main class sectors and their demands (1960s)

Geotopographical settings	Prevailing agrarian structure	Main class sectors	Aspiration in relation to an independent peasant economy	Necessary condition	Main demands	Directed towards/against
Andean ranges	peasant economy	→ stable peasants	→ consolidate	→ improvement of existing reproduction conditions	→ improvement of services and credit	→ state
Piedmont forest frontiers in the Amazon and Orinoco basins, Pacific littoral, and other marginal areas	colonization →	precarious peasants	→ stabilize	→ obtaining the basic reproduction conditions	→ access to services and credit	→ state
Atlantic coast, eastern llanos, marginal areas in the Andes and the inner valleys	traditional latifundios	→ landless peasants	→ establish	→ gaining access to the means of production	→ land	→ state, landowners
Magdalena and Cauca inner valleys	agrarian capitalism	→ agricultural workers	→ –	–	→ employment, wages and labour conditions	→ agricultural entrepreneurs

Source: Zamosc, L. (1989) 'Peasant Struggles of the 1970s in Colombia', in Eckstein, S. (ed.) *Power and Popular Protest: Latin American social movements,* University of California Press, Berkeley, Fig. 3.1.

This heterogeneity, and hence the existence of different needs and demands, was important in the history of the peasant movement (Figure 12.7).

Another set of critical factors affecting the countryside arose from a long-running national conflict, called *La Violencia* (1948–58) in which 200 000 people were killed (Zamosc, 1989, p.106). It began with traditional political party rivalries which spread to the countryside, and near its end the conflict took on a class character. The political coalition which ended *La Violencia*, the National Front, had a strong interest in pacifying any incipient rural unrest or discontent, as well as continuing the development of capitalist agriculture. The creation of ANUC in 1967 as a Ministry of Agriculture initiative was an attempt to carry out both objectives. The then president wanted to create a climate of pressure for agrarian

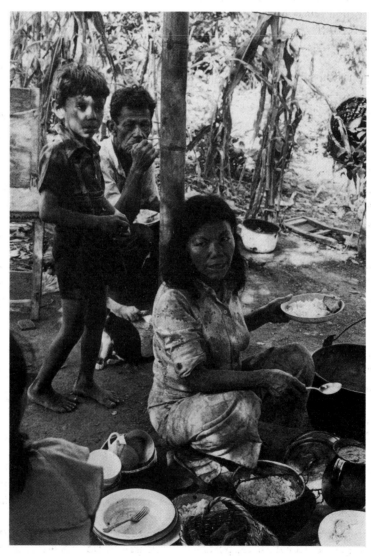

Figure 12.7 The rural poor of Colombia had different sources of livelihood, different needs and interests: this poor family is from Neiva, which lies in the mountainous regions of Colombia.

reform which would encourage agricultural modernization. ANUC had state support for land appropriations and for technical assistance for peasant farming.

Contradictions arose when ANUC took on a more 'authentic' peasant existence of its own and carried out programmes and initiatives independently of government policy. ANUC's membership comprised mainly landless and tenant farmers working for the *latifundios* (Chapter 2), who wanted land of their own, and peasant colonizers of new lands who needed basic services to build up their holdings. There were spontaneous land seizures and demonstrations, which increased in intensity under a conservative change of government (1970). After further land invasions in 1971, the government froze ANUC's funds and sacked agrarian reform officials working in alliance with the organization. Land invasions continued and were repressed. Finally, the state encouraged part of the ANUC membership to set up a parallel organization which worked predominantly with small peasants or *minifundistas* in the mountains and areas of stable colonization. The original (or 'radical') ANUC concentrated on land struggles in the valleys and new colonization areas. It also went further and began to create an 'alternative strategy' for Colombian development which was based on Maoist ideas of peasant leadership and was in alliance with urban workers and intellectuals.

This 'alternative path' for Colombian development foundered for a number of reasons: many producers had benefited from early land redistributions and wanted to consolidate their gains rather than enter the arena of political struggle; there was considerable factionalism and fragmentation within ANUC resulting from debates about strategy between the 'Maoist cadres' who had entered the struggle and ANUC's leadership (ibid., p.119). The Maoists were expelled while ANUC tried to form its own political party. There were also important changes in the economic and political context, as the government began to set up Integrated Rural Development Projects (Box 12.1). Rural employment structures were also changing with increased urban migration and greater absorption of rural labour in

> ### Box 12.1 Integrated Rural Development Projects
>
> Such projects are promoted by government ministries, and may include rural credit, technical assistance, crop diversification, income generation projects for women, consciousness-raising about 'social development' and, occasionally, land purchasing and rural housing schemes.

capitalist agriculture, as well as the incorporation of many peasant producers in the 'marijuana economy' (ibid., p.121). ANUC itself became more and more bureaucratic, distanced from its membership and divided within itself, and ultimately merged again with the state-sponsored parallel organization in 1981.

This abbreviated version of ANUC's history shows some of the difficulties of establishing a 'peasant alternative' to the development of a capitalist agriculture in which poor farmers grow basic foods and provide ready wage labour for enterprises in the countryside and towns which offer precarious and poorly paid employment. It was hard to establish (and sustain) a commonality of interests among the different groups, and 'radical' ANUC did not succeed in working with urban political interests on a broader strategy for achieving national development in Colombia.

Although the 'peasant alternative' foundered, some peasants gained land through the reforms, but even these gains were qualified by the low level of support they obtained from the state, and peasants involved in these co-operatives became semi-proletarianized. Zamosc concludes: 'ANUC's struggles failed to modify the patterns of agrarian development to the advantage of the peasantry as a whole' (ibid., p.124). The state did not in the end support peasant initiatives, and the attempts by ANUC to form urban political alliances led to fragmentation rather than a broad-based movement. In part, this resulted from a conflict of interests between those who were concerned with immediate economic gains and survival and those who sought longer term and more transformative goals.

This short study shows that effecting change at a national level from below can have many difficulties and contradictions. There are of course examples of revolutionary movements combining peasants, workers, and intellectuals in alliances that have established a 'new order', but in practice the representation of the various interests of the rural poor in the state is extremely difficult to achieve. While peasant movements sometimes succeed in exerting considerable pressure on the state to redistribute land and provide credit, technical assistance and marketing facilities, these achievements tend to benefit a small percentage of the rural poor. In fact, such programmes largely benefit those who are already relatively better off, as suggested above.

There are other issues which the case of ANUC raises. In general, it is difficult for a single organization to represent the interests of all rural people, especially where they are differentiated by access to resources, types of employment and geographical location. Secondly, as such organizations grow, they are almost inevitably subject to internal differences about strategy and tactics, especially how to combine 'productivist' goals such as obtaining land and credit and organizing cooperatives with more 'political' goals such as the representation of the interests of the rural poor in the state. Many peasant organizations depend on outside funding for survival and this too can have divisive effects over the control and use of resources. Finally, the question of alliances with other groups or organizations to work towards longer term transformation is always fraught with problems in negotiating a common set of demands that satisfies all groups in such a coalition.

Peasants and NGOs

Are there 'alliances' that organizations from below can make to achieve more *limited* goals, and to suggest or bring about reforms to existing paths of development? The question of who works with the rural poor in some kind of collaborative effort (if not political alliance, as in the Colombian case) raises the issue of the roles played by non-governmental organizations. In Chapter 11, such organizations were seen as initiators of development alternatives from above, although at their best they were shown as aspiring to transcend the divisions of 'above' and 'below', of the 'outside' and 'inside' of rural communities. Can NGOs form effective alliances with the poor? (Note that I am differentiating between NGOs and actual organizations of the poor, as in Chapter 11. A rather different approach suggests that organizations of the poor may themselves be regarded as a type of NGO; see, for example, *Wuyts et al., 1992*, ch.5.)

There are several issues about the role of NGOs. NGOs are highly heterogeneous and are often beset with ambiguity, not least when they seek to 'side with the poor'. NGOs are usually established by groups of people who have a common 'cause' – such as fighting poverty and powerlessness – which they might try to achieve through providing finance, technical or advisory services, training and so on. Some NGOs set up and run their own projects on behalf of rural people or try to integrate them into the NGOs' work. Others may support existing organizations of the rural poor through the services mentioned above. NGOs' non-profit-making character and their 'moral imperatives' do not, however, mean that they do not have their own interests, which may differ from those they are purporting to help. Moreover, the activities of NGOs are rarely analysed and evaluated from the point of view of the poor themselves. We tend to know what NGOs think they are doing, but not what the poor think of NGO efforts and their effects.

A second issue is that of the 'outsider/insider' relationship between NGOs and the rural poor, their communities and organizations. This relationship has several aspects: whether the NGO is international or local, whether it is urban or rural based, what kinds of services and assistance it offers, whether it sees itself as initiating programmes and projects or as facilitating initiatives by the poor, and so on. In addition, the outsider/insider dimension affects how needs and interests are seen, understood and interpreted in practice – the case of Médecins Sans Frontières programmes in Laropi discussed in Chapter 10 is an example.

A third issue is whether NGOs are concerned with immediate, material improvements or are engaged in a longer term project of change, either locally in particular rural areas or at a national level. Some international NGOs can also have 'hidden agendas', as with some evangelizing religious organizations or those with underlying political goals. Finally, NGOs might well have strong links to states or international institutions through funding and potential policy involvement. Collaboration from and with NGOs can thus be a complex – and often problematic – option for the poor.

However, NGOs may often provide the only way that small, locally based groups of the rural poor can obtain support for their projects and improvements to their livelihoods – whether through material resources, skills training or other 'enabling' inputs. But this too can have problems, such as the 'dependence trap' where certain rural groups

end up relying on financial aid for their survival and social cohesion. In addition, the injection of funds into low-income contexts can be extremely divisive where there are different local interests involved. The control over such funds can reinforce existing power relations or establish new ones. Some questions which arise for NGOs are: which groups/strata/interests should NGOs support in a particular instance? What happens if the local balance of power changes over time? What if those who were the 'targets' of support have not been helped? (For further discussion of these issues, see *Wuyts et al., 1992*, ch.5.)

On the other hand, the activities of NGOs can also be extremely positive. An example is the Projet Agro-Forestier (PAF) scheme in Burkina Faso, cited in Chapter 8, and supported by the international NGO, Oxfam (Figure 12.8). Inevitably there were some ambiguous effects. For example, it was difficult to reach the poorest producers in

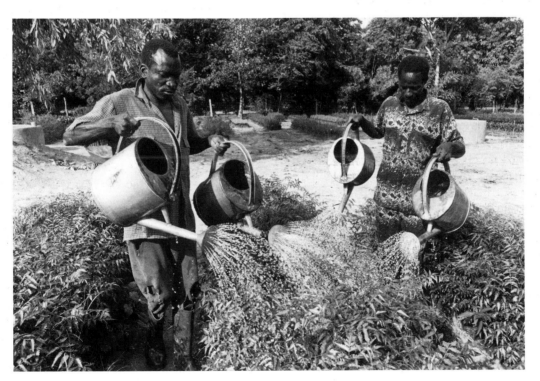

Figure 12.8   Partnership between state, NGOs and rural people can work: tree nursery of Burkina Faso's Ministry of Environment and Tourism, used by Projet Agro-Forestier and other NGOs to distribute trees for planting in villages throughout the country.

the PAF scheme. There was also a need for a much wider scale village initiative in order to achieve more generalized land conservation in the area.

Another Oxfam-supported project in West Africa has helped women increase their commercial processing and marketing of sheanuts through financing a loan scheme. Sheanut trees grow wild but the nuts can be either parboiled and dried or made into butter and sold. It is instructive to see how Oxfam discusses this project:

> "Loans of various sizes were distributed to existing women's groups. The philosophy of the scheme was subsumed under the overall philosophy of OXFAM which is to work with the underprivileged with a view to raising their quality of life and promoting greater social justice. The motivation for the scheme was a strong desire to encourage women to stand up for themselves in the face of what looked like an organized and determined attempt by local businessmen to dislodge them. The rural women required credit to enable them to assert their traditional role as pickers, processors and users of sheanuts and shea-products for the overall family benefit. It was anticipated that increased off-farm income would lead to increased consumption of goods and services, better nutrition and consequently overall improvement in family well being."

> (Pugansoa & Amuah, 1990, p.1)

This statement tells us about Oxfam's 'cause' or 'moral imperative'. It also reveals that the project had to confront local power relations (or help overcome the powerlessness of the women sheanut pickers in relation to local, male, traders) to improve low incomes and poor nutrition. A possible 'solution' to these different problems was to give women's groups credit for the commercialization of the sheanut crop and its products. How does Oxfam evaluate the outcomes?

> "The loan scheme has been beneficial to precisely those women who needed support most. The institution of the loan scheme has contributed significantly...to the needs of participating groups for productive credit which was otherwise not available to them. Loan repayments have been extremely high, and group turnovers healthy. The scheme has provided women's groups with the opportunity to build 'buffer food stocks' which are essential to food security; this has extended the benefits to the whole community...

> Sheanuts have become a cash crop. Women believe that it is no longer possible for men to overlook sheanuts' economic importance; but because the sheanut industry has become politicized, women need a louder voice and more political patronage in order to assert their traditional role in the industry. The current low levels of organization and literacy make this difficult...

> The sheanut is a wild crop and, therefore, subject to unpredictable yields. Sometimes, therefore, women's groups had deviated from the stated application of the loan, in response to other business opportunities...This had no adverse effect on the scheme.

> The loan scheme guidelines should be amended to reflect this and the need to apply credit generally in support of women's off-farm income generating activities."

> (ibid., p.8)

This evaluation shows that the project 'succeeded' within its original terms of reference, as well as helping to stimulate a politicization of the sheanut industry. In other words, recognizing sheanuts as a valuable commodity, and safeguarding part of its value for women, made crop picking, processing and marketing a more prominent area of competitive social relations. Women were still disadvantaged because of their illiteracy and lack of organization, but access to credit helped women undertake other income-generating activities as well as the commercialization of sheanuts. Oxfam felt these initiatives should be supported by an extension and broadening of the existing scheme.

This little 'cameo' shows a number of the possibilities and contradictions in NGO collaboration with rural poor and peasant farmers, both for the NGO and for the local groups, because the project's 'success' generated new (and potentially conflictive)

Figure 12.9   The 'insider/outsider' relationship between an NGO (with its professional workers) and local people can be difficult to negotiate: here an extension worker tries to discuss problems with women workers at an agricultural co-operative in Zimbabwe.

social relations as well as offering new possibilities for improving livelihoods. Thus, while the 'success' of such projects lies in defined goals established by community groups, their wider (and longer term) implications can be more complex. Projects which lead to increased commercialization, like the sheanut scheme, are bound to result in changes in social relations, which in turn have their own dynamic irrespective of the presence of the NGO. However, anything other than a welfare approach to poverty will inevitably involve changes in social relations in some direction.

Many international NGOs are sensitive to these issues. Do national NGOs have an easier task? This question is hard to answer in the abstract: much depends on the perspectives and aims of the NGOs as well as the context in which they work – the 'insider/outsider' relationship is a relative one and all organizations have their own interests and agendas (Figure 12.9). With any organization purporting to help or represent the interests of the poor, the questions: what kind of change? for whom? who benefits? always have to be asked.

Effecting change from below

The examples in this section illustrate how complex the process of effecting change from below can be. The case of ANUC shows how some gains can be made by national organizations of the rural poor, while longer term transformation in the interests of the poor remains very difficult. ANUC's history raises questions about the relationship between organizations of the poor and other political and economic interests, as well as the state (and different governments). It also raises questions about the different interests among many rural people – peasants, small farmers, rural workers. While it is salutary to realize that there is no easy 'solution', it is also instructive to see possible links (and forms of collaboration, difficult though these might be to achieve) between different groups and interests, between 'above' and 'below', 'outside' and 'inside', the local and the national, the short term and the longer term.

NGOs can try to build up such links in order to help the rural poor fight poverty and powerlessness, but their role is also a very delicate one. Through the resources NGOs provide and the interventions they make, they also help to challenge, change or reinforce inequalities in local communities.

12.5   Conclusions

"The political task as I see it is…to work to eliminate the conditions of possibility for *all* such forms of humiliation and degradation. This amounts to a political choice in favor of focusing broadly on empowerment, not narrowly on poverty; freeing the slaves, not feeding them better."

(Ferguson, 1990, p.288)

The issue of empowerment, its relationship to improving rural livelihoods, and how 'outsiders' can 'help' are ongoing themes of debate. So is the question of 'whose empowerment?' As we have seen, rural differentiation is combined with a wide range of types of action and outcomes, as well as types of collaboration with the rural poor (whether by NGOs, researchers, or progressive governments).

The different types of action offer different possibilities. While, as Scott suggests, actions by the 'weak' and poor are frequently individualistic or apparently irrational, attempts to transform and empower as well as to resist do involve collective and coherent action, even though establishing the basis of such action may be fraught with problems and contradictions. We are then left with a set of questions to reflect on about different types of action:

- Are they about survival or do they go beyond it?

- Do they have short-term effects or are they about longer term change?

- Are they individually or collectively based?

- Do they just relate to local conditions or are they directed to larger scale or even national change?

- Who can effectively collaborate, and in what way? Is there room for alliances with NGOs, political organizations or the state?

- If empowerment is key to improving rural livelihoods, whose empowerment? and how can it be achieved?

Some of these issues and how they are linked through different kinds of public action are explored further in the fourth volume in this series (*Wuyts et al., 1992*).

Summary

1   Poverty is multidimensional and culturally complex: powerlessness is a particularly important dimension. Struggles against poverty may be individual or collective, and may be directed to immediate or longer term survival, short term or longer term change.

2   The poor have to cope with persisting poverty as well as the effects of change or sudden crisis. The borderlines between these conditions are fine because the margins for manoeuvre are small.

3   Individually or collectively, rural people may resist poverty, powerlessness and indignity by many means, such as sabotage, folk tales and religious movements. Such actions have been termed 'weapons of the weak' or 'everyday forms of resistance'. The dynamics and effects of such processes are often contradictory, and whether such actions reinforce poverty and powerlessness or whether they can bring about change needs to be investigated.

4   Collective organizations of the rural poor may be able to bring longer term changes in rural livelihoods as well as some degree of empowerment. Longer term change may bring 'actions from below' into conflict with the state and local interests. The empowerment of the rural poor often involves forming alliances or requires the support of external resources, but such involvement may have unintended effects.

References

Agarwal, B. (1985) 'Women and technological change in agriculture: the Asian and African experience', in Ahmed, I. (ed.) *Technology and Rural Women: conceptual and empirical issues*, Allen & Unwin, London.

Agarwal, B. (1986) 'Women, poverty and agricultural growth in India', *Journal of Peasant Studies*, 13(4), July, pp.167–80.

Agarwal, B. (1990) 'Social security and the family in rural India coping with seasonality and calamity', *Journal of Peasant Studies*, 17(3), pp.341–412.

Allen, T. & Thomas, A. (eds) (1992) *Poverty and Development in the 1990s*, Oxford University Press/ The Open University, Oxford (Book 1 of this series).

Amin, S. (1972) 'Underdevelopment and dependence in Black Africa', *Journal of Modern African Studies*, 10(4).

Amsden, A. (1979) 'Taiwan's economic history. A case of Etatisme and a challenge to dependency theory?', *Modern China*, 5(3).

Amsden, A. (1989) *Asia's Next Giant: South Korea and late industrialization*, Oxford University Press, Oxford.

Annis, S. (1987) 'Can small scale development be large scale policy? The case of Latin America', *World Development*, 15 (supp.) August, pp.129–34.

Archetti, E. P., Cammack, P. & Roberts, B. (eds) (1987) *Sociology of Developing Societies: Latin America*, Macmillan, London.

Bandhyopadhyay, D. (1985) 'An evaluation of policies and programmes for the alleviation of rural poverty in India', in Islam, R. (ed.) *Strategies for Alleviating Poverty in Rural Asia*, pp.99–151, International Labour Organization, Bangkok.

Bandhyopadhyay, D. (1986) *A Study on Poverty Alleviation in Rural India Through Special Employment Creation Programmes*, Asian Employment Programme Working Paper, ILO-ARTEP, New Delhi.

Bandhyopadhyay, D. (1988) 'Direct intervention programmes for poverty alleviation: an appraisal', *Economic and Political Weekly*, 23(26) June 25, pp.A-77 to 95.

Bannerman, R., Burton, J. & Wen-Chieh, C. (eds) (1983) *Traditional Medicine and Health Care Coverage*, World Health Organization, Geneva.

Barker, J. (1984) 'Politics and production', in Barker J. (eds) *The Politics of Agriculture in Tropical Africa*, Sage, Beverly Hills.

Barker, J. (1989) *Rural Communities Under Stress*, Cambridge University Press, Cambridge.

Barrett, M. & McIntosh, M. (1982) *The Anti-Social Family*, Verso, London.

Basant, R. & Kumar, B. L. (1988) *Rural Non-agricultural Activities in India: a review of the available evidence*, Gujarat Institute of Area Planning Working Paper no. 20, Gandhinagar, Gujarat.

Bates, R. H. (1981) *Markets and States in Tropical Africa*, University of California Press, Los Angeles.

Bauer, A. J. (1979) 'Rural workers in Spanish America: problems of peonage and oppression', *Hispanic American Historical Review*, 59(1).

Beinart, W. (1990) 'Transkeian migrant workers and youth labour on the Natal sugar estates 1918–1940', unpublished seminar paper, Institute of Commonwealth Studies, University of London.

Bell, C. (1990) 'Reforming property rights in land and tenancy', *World Bank Research Observer*, 5(2) July, pp.143–66.

Bernstein, H. (1990) 'Agricultural 'modernisation' and the era of structural adjustment: observations on sub-Saharan Africa', *Journal of Peasant Studies*, 18(1).

Bernstein, H. & Crow, B. (1988) 'The expansion of Europe', in Crow, B., Thorpe, M. *et al.* (eds), *Survival and Change in the Third World*, Polity, Cambridge.

Berry, S. (1980) 'Rural class formation in West Africa', in Bates, R. H. & Lofchie, M. F. (eds) *Agricultural Development in Africa*, Praeger, New York.

Berry, S. (1984) 'The food crisis and agrarian change in Africa: a review essay', *African Studies Review*, 27(2).

Bhaduri, A. (1983) *The Economics of Backward Agriculture*, Academic Press, London.

Bhalla, G. S. & Tyagi, D. S. (1989) *Patterns in Indian Agricultural Development: a district level study*, Institute for Studies in Industrial Development, New Delhi.

Bhalla, S. (1987) 'Trends in employment in Indian agriculture, land and asset distribution', *Indian Journal of Agricultural Economics*, 42(4), Oct–Dec.

Bharadwaj, K. (1985) 'A view on commercialisation in Indian agriculture and the development of capitalism', *Journal of Peasant Studies*, 12(4).

Bharadwaj, K. (1989) *The Formation of Rural Labour Markets: an analysis with special reference to Asia*, World Employment Programme Working Paper, International Labour Organization, Geneva.

Boateng, M., Ratchford, C. & Blase, M. (1987) 'Profitability analysis of a farming system in Africa,' *Agricultural Systems*, 24(2), pp.81–93.

Bonnevie, H. (1987) 'Migration and malformation: case studies from Zimbabwe', CDR Project paper A87.1, Centre for Development Research, Copenhagen.

Brass, T. (1986) 'The elementary strictures of kinship', *Social Analysis*, 20, pp.56–68.

Bundy, C. (1979) *The Rise and Fall of the South African Peasantry*, Heinemann, London.

Byres, T. J. (1991) 'The agrarian question and differing forms of capitalist agrarian transition: an essay with reference to Asia', in Breman, J. C. & Mundle, S. (eds) *Rural Transformation in Asia*, Oxford University Press, Delhi.

Byres, T. & Crow, B. (1988) 'New technology and new masters for the Indian countryside', in Crow, B., Thorpe, M. *et al.* (eds) *Survival and Change in the Third World*, Polity, Cambridge.

Cernea, M. (1988) 'Nongovernment organizations and rural development', *World Bank Discussion Paper* no.40, The World Bank, Washington DC.

Chambers, R. (1983) *Rural Development: putting the last first*, Longman, London.

Chambers, R. (1988a) *Poverty in India: concepts, research and reality*, Discussion Paper 241, Institute of Development Studies, University of Sussex.

Chambers, R. (1988b) 'Bureaucratic reversals and local diversity', *IDS Bulletin*, 19(4), pp.50–6.

Chatterji, L. (1984) *Marginalization or the Induction of Women into Wage Labour*, World Employment Programme Research Paper, International Labour Organization, Geneva.

Clairmonte, F. (1960) *Economic Liberalism and Underdevelopment*, Asia Publishing House, Bombay.

Colburn, F. D. (1989) 'Foot dragging and other peasant responses to the Nicaraguan Revolution', in Colburn, F. D. (ed.) *Everyday Forms of Peasant Resistance*, M. E. Sharpe, Inc., New York.

Cooper, F. (1981) 'Africa and the world economy', *African Studies Review*, 24(2/3).

Copestake, J. G. (1988) 'Government sponsored credit schemes in India: proposals for reform', *Agric. Admin. and Extension*, 28, pp.265–82.

Crehan, K. (1985) 'Production and gender in north-western Zambia', in Pottier, J. (ed.) *Food Systems in Central and Southern Africa*, School of Oriental and African Studies, London.

Critchley, W. (1991) *Looking After Our Land*, Oxfam Publications, Oxford.

Crow, B. & Thomas, A. (1983) *Third World Atlas*, Open University Press, Milton Keynes.

Dandekar, V. M. (1988) 'Agriculture, employment and poverty', in Lucas, R. E. B. & Papanek, G. F. (eds) *The Indian Economy: recent development and future prospects*, Oxford University Press, Delhi.

Dankelman, I. & Davidson, J. (1988) *Women and Environment in the Third World*, Alliance for the Future, Earthscan, London.

Dantwala, M. L. (1987) 'Rural assets distribution and composition of the labour force', *Indian Journal of Agricultural Economics*, 42(3), July–Sept.

de Janvry, A. (1981) *The Agrarian Question and Reformism in Latin America*, Johns Hopkins University Press, Baltimore.

de Janvry, A. (1987) 'Peasants, capitalism and the state in Latin American culture', in Shanin, T. (ed.) *Peasants and Peasant Societies* (second edition), Blackwell, Oxford.

de Janvry, A., Sadoulet, E. & Wilcox Young, E. (1989) 'Land and labour in Latin American agriculture from the 1950s to the 1980s', *Journal of Peasant Studies*, 16(3).

Djukanovic, V. & Mach, E. (1975) *Alternative Approaches to Meeting Basic Needs in Developing Countries*, World Health Organization, Geneva.

Douglass, M. (1983) 'The Korean *Saemaul Undong*: accelerated rural development in an open economy', in Lea, D. A. M. & Chaudhuri, D. P. (eds) *Rural Development and the State*, Methuen, London and New York.

Drèze, J. & Sen, A. K. (1989) *Hunger and Public Action*, Clarendon Press, Oxford.

Ferguson, J. (1990) *The Anti-Politics Machine: 'development', depoliticization and bureaucratic state power in Lesotho*, Cambridge University Press.

Ghai, D. *et al.* (1979) *Agrarian Systems and Rural Development*, Macmillan for ILO World Employment Program, London.

Gibbon, P. (1975) 'Colonialism and the great Irish famine of 1845–49', *Race and Class*, 17(2).

Gittinger, J. P. (1982) *The Economic Analysis of Agricultural Projects*, Johns Hopkins University, Baltimore.

Goodman, D. & Redclift, M. (1981) *From Peasant to Proletarian*, Blackwell, Oxford.

Guyer, J. (1983) 'Women's work and production systems: a review of two reports on the agricultural crisis', *Review of African Political Economy*, nos 27/8.

Guyer, J. (1987) 'Introduction', in Guyer, J. (ed.) *Feeding Africa's Cities*, Manchester University Press, Manchester.

Harrison, P. (1987) *The Greening of Africa*, Paladin, London.

Harriss, B. & Harriss, J. (1984) ''Generative' or 'parasitic' urbanism? Some observations from the recent history of a south Indian market town', in Harriss, J. B. & Moore, M. (eds), *Development and the Rural–Urban Divide*, Frank Cass, London.

Harriss, J. (1987) 'Capitalism and peasant production: the Green Revolution in India', in Shanin, T. (ed.) *Peasants and Peasant Societies* (second edition), Blackwell, Oxford.

Hartmann, B. & Boyce, J. K. (1983) *A Quiet Violence: view from a Bangladesh village*, Zed, London.

Hewitt, T., Johnson, H. & Wield, D. (eds) (1992) *Industrialization and Development*, Oxford University Press/The Open University, Oxford (Book 2 of this series).

Hill, P. (1963) *The Migrant Cocoa Farmers of Southern Ghana*, Cambridge University Press, Cambridge.

Hirway, I. (1986a) *Abolition of Poverty in India, with Special Reference to Target Group Approach in Gujarat*, Vikas Publishing House, New Delhi.

Hirway, I. (1986b) *Wage Employment Programmes in Rural Development*, Oxford University Press/IBH, New Delhi.

Hobsbawm, E. J. (1969) *Industry and Empire*, Penguin, Harmondsworth.

Hobsbawm, E. J. & Ranger, T. (eds) (1983) *The Invention of Tradition*, Cambridge University Press, Cambridge.

Holt , S. & Ribe, H. (1991) 'Developing financial institutions for the poor and reducing barriers to access for women', *World Bank Discussion Paper* no.117, The World Bank, Washington DC.

Hopkins, A. (1978) 'Innovation in a colonial context', in Dewey, C. & Hopkins, A. (eds) *The Imperial Impact*, Athlone Press for the Institute of Commonwealth Studies, London.

Hopkins, A. G. (1973) *An Economic History of West Africa*, Longman, London.

Hyden, G. (1980) *Beyond Ujamaa in Tanzania: underdevelopment and an uncaptured peasantry*, Heinemann, London.

Iliffe, J. (1987) *The African Poor: a history*, Cambridge University Press, Cambridge.

Jagannathan, N. V. (1987) *Informal Markets in Developing Countries*, Oxford University Press, Oxford.

Jodha, N. S. (1986) 'Common property resources and rural poor in dry regions of India', *Economic and Political Weekly*, July.

Johnson, H. & Bernstein, H. (1987) *Third World Lives of Struggle*, Heinemann Educational Books, London.

Johnson, T. & Sargent, C. (1990) *Medical Anthropology: contemporary theory and method*, Praeger, London.

Jose, A. V. (1987) *Agricultural Wages in India*, Asian Employment Working Paper, ILO-ARTEP, New Delhi.

Kabeer, N. (1985) 'Do women gain from high fertility', in Afshar, H. (ed.) *Women, Work and Ideology in the Third World*, Tavistock Publications, London.

Kabeer, N. (1990) 'Poverty, purdah and women's survival strategies in rural Bangladesh' in Bernstein, H. *et al.* (eds) *The Food Question*, Earthscan, London.

Kabeer, N. (1991) 'Gender dimensions of rural poverty: analysis from Bangladesh', *Journal of Peasant Studies*, 18(2).

Kandiyoti, D. (1985) *Women in Rural Production Systems*, UNESCO, Paris.

Kay, C. (1974) 'Comparative development of the European manorial system and the Latin American hacienda system', *Journal of Peasant Studies*, 2(1).

Kimmage, K. (1991) 'Small-scale irrgation initiatives in Nigeria: the problems of equity and sustainability', Applied Geography, 11, pp.5–20.

Kjekshus, H. (1977) *Ecology Control and Economic Development in East African History*, Heinemann, London.

Kuznets, S. (1965) *Economic Growth and Structure*, W.W. Norton, New York.

Last, M. & Chavunduka, G. L. (eds) (1986) *The Professionalisation of African Medicine*, Manchester University Press, Manchester.

Lewis, J. (1988) 'Strengthening the poor: some lessons for the international community' in Lewis, J. (ed.) *Strengthening the Poor: what have we learned?* Overseas Development Council, Washington DC.

Lewis, O. (1963) *The Children of Sánchez: autobiography of a Mexican family*, Vintage Books, New York.

Lipton, M. (1984) 'Farm technology, urban institutions and the rural poorest: some lessons of tractorization in Sri Lanka', in Farrington, J. *et al.* (eds) *Farm Power and Employment in Asia: performance and prospects*, ARTI and ADC, Colombo and Bangkok.

Low, D. A. (1973) *Lion Rampant*, Frank Cass, London.

Mackintosh, M. (1989) *Gender, Class and Rural Transition: agribusiness and the food crisis in Senegal*, Zed Books, London.

Mahendradev, S. (1986) 'Growth of labour productivity in Indian agriculture: regional dimensions', *Economic and Political Weekly*, June 21–8.

Mamdani, M. (1972) *The Myth of Population Control*, Monthly Review Press, New York.

Mamdani, M. (1982) 'Karamoja: colonial roots of famine in north-east Uganda', *Review of African Political Economy*, 25.

Mamdani, M. (1987) 'Extreme but not exceptional: towards an analysis of the agrarian question in Uganda', *Journal of Peasant Studies*, 14(2).

Manners, R. A. (1962) 'Land use, labour and the growth of market economy in Kipsigis Country', in Bohannan, F. & Dalton, G. (eds) *Markets in Africa*, Northwestern University Press, Evanston.

Martin, M. R. (1990) 'Bias and inequity in rural incomes in post-reform China', *Journal of Peasant Studies*, 17(2).

Mbilinyi, M. (1990) ''Structural adjustment', agribusiness and rural women in Tanzania', in Bernstein, H. *et al.* (eds) *The Food Question*, Earthscan, London.

Minhas, B. S. (1990) *A Note on the Measurement of Poverty*, Indian Statistical Institute Working Paper, Delhi.

Mitchell, J. (1975) *Psychoanalysis and Feminism*, Penguin, Harmondsworth.

Moore, B. (1966) *Social Origins of Dictatorship and Democracy*, Beacon Press, Boston.

Murray, C. (1981) *Families Divided*, Cambridge University Press, Cambridge.

Parthasarathy, G. (1987) 'Changes in the incidence of rural poverty and recent trends in some aspects of agrarian economy', *Indian Journal of Agricultural Economics*, 42(1), Jan–Mar.

Patnaik, U. (1990a) 'Some economics and political consequences of the Green Revolution in India', in Bernstein, H. *et al.* (eds) *The Food Question*, Earthscan, London.

Patnaik, U. (1990b) 'Introduction', in Patnaik, U. (ed.) *Agrarian Relations and Accumulation*, Oxford University Press, Delhi.

Pearse, A. (1975) *The Latin American Peasant*, Frank Cass, London.

Planning Commission (1986) *Evaluation Report on IRDP*, New Delhi.

Planning Commission (1987) *Evaluation Report on NREP*, New Delhi.

Planning Commission (1990) *Employment: past trends and prospects for 1990s*, Planning Commission Working Paper, New Delhi.

Prasad, P. H., Rodgers, E. B., Gupta, S. I., Sharma, A. N. & Sharma, B. (1981) *The Pattern of Poverty in Bihar*, World Employment Programme Working Paper No. 152, International Labour Organization, Geneva.

Pugansoa, B. & Amuah, D. (1990) *The Oxfam Sheanut Loan Scheme*, GADU Pack No.11, Oxfam, Oxford.

Pulley, R. V. (1989) 'Making the poor creditworthy: a case study of the integrated rural development project in India', *World Bank Discussion Paper* no.58, The World Bank, Washington DC.

Raikes, P. (1988) *Modernising Hunger*, James Currey, London.

Ranger, T. (1985) *Peasant Consciousness and Guerrilla Warfare in Zimbabwe*, James Currey, London.

Rao, C. H. H., Ray, S. K. & Subbarao, K. (1988) *Unstable Agriculture and Droughts*, Vakas Publishing House, New Delhi.

Rath, N. (1985) ''Garibi Hatao': can IRDP do it?' *Economic and Political Weekly*, 20(6) February 9, 238–46.

Richards, P. (1983) 'Ecological change and the politics of African land use', *African Studies Review*, 26(2), 1–72.

Richards, P. (1985) *Indigenous Agricultural Revolution*, Hutchinson, London.

Rohrback, D. (1987) 'A preliminary assessment of factors underlying the growth of communal maize production in Zimbabwe', in Rukuni, M. & Eicher, C. (eds) *Food Security for Southern Africa*, UZ/MSU Food Security Project, Department of Agricultural Economics and Extension, University of Zimbabwe, Harare.

Saith, A. (1990) 'Development strategies and the rural poor', *Journal of Peasant Studies*, 17(2).

Sarvekshana (1990) *Results of the Fourth Quinquennial Survey on Employment and Unemployment, National Sample Survey 43 Round*, Government of India, Ministry of Planning, New Delhi.

Saul, S. B. (1960) *Studies in British Overseas Trade 1870–1914*, Liverpool University Press, Liverpool.

Scott, J. C. (1985) *Weapons of the Weak: everyday forms of peasant resistance*, Yale University Press, New Haven.

Scott, J. C. (1989) 'Everyday forms of peasant resistance' in Colburn, F. D. (ed.) *Everyday Forms of Peasant Resistance*, M. E. Sharpe, Inc., New York and London.

Seabright, P. (1990) 'Quality of livestock assets under poverty alleviation schemes: evidence from South Indian data', Paper to ESRC Development Economics Study Group.

Sen, A. K. (1981) *Poverty and Famines*, Clarendon Press, Oxford (1982 revised edition, Oxford University Press).

Shanin, T. (1987) 'Introduction: peasantry as a concept', in Shanin, T. (ed.) *Peasants and Peasant Societies* (second edition), Blackwell, Oxford.

Sharma, U. (1980) *Women, Work and Property in North-west India*, Tavistock Publications, London and New York.

Siddle, G. & Swindell, K. (1990) *Rural Change in Tropical Africa*, Blackwell for the Institute of British Geographers, Oxford.

Skocpol, T. (1979) *States and Social Revolutions*, Cambridge University Press, Cambridge.

Slikkerveer, L. J. (1990) *Plural Medical Systems in the Horn of Africa*, Kegan Paul, London.

Sontag, S. (1979) *Illness as Metaphore*, Allen Lane, London.

Stamp, P. (1986) 'Kikuyu women's self help groups', in Robertson, C. & Berger, I. (eds) *Women and Class*, Holmer and Meier, New York.

Standing, H. (1992) 'Gender and employment an Indian case study' in Ostergaard, L. (ed.) *Gender and Development: a practical guide*, Routledge, London.

Stolcke, V. & Hall, M. M. (1983) 'The introduction of free labour on São Paulo coffee plantations', in Byres, T. J. (ed.) *Sharecropping and Sharecroppers*, special issue of *Journal of Peasant Studies*, 10(2/3).

Thorner, D. (1956) *The Agrarian Prospect in India*, Delhi School of Economics, Delhi.

Thorner, D. & Thorner, A. (1962) *Land and Labour in India*, Asia Publishing House, Bombay.

Timberlake, H. (1988) *Africa in Crisis*, Earthscan, London.

Townsend, P. (1979) *Poverty in the United Kingdom*, Penguin, Harmondsworth.

UNDP (1991) *Human Development Report 1991*, Oxford University Press, Oxford and New York.

UNICEF (1990) *State of the World's Children*, Oxford University Press, Oxford.

Vaidyanathan, A. (1988) 'Agricultural development and rural poverty', in Lucas, R. E. B. & Papanek, G. F. (eds) *The Indian Economy: recent development and future prospects*, Oxford University Press, New Delhi.

van Arkadie, B. (1989) *The Role of Institutions in Development*, Proceedings of the World Bank Annual Conference on Development Economics, The World Bank, Washington DC.

Van Onselen, C. (1990) 'Oral histories: reconstructing rural lives', unpublished seminar paper, Institute for Commonwealth Studies, University of London.

Vaughan, M. (1983) 'Which family?: problems in the reconstruction of the history of the family as an economic and cultural unit', *Journal of African History*, 24, pp.275–83.

Visaria, P. (1980) *Poverty and Unemployment in India: an analysis of recent evidence*, World Bank Staff Working Paper, no. 417, Washington, DC.

Wade, R. (1990) *Governing the Market: economic theory and the role of government in East Asian industrialization*, Princeton University Press, Princeton.

Warriner, D. (1969) *Land Reform in Principle and Practice*, Clarendon Press, Oxford.

Watts, M. (1983) *Silent Violence: food, famine and peasantry in northern Nigeria*, University of California Press, Berkeley.

Whitcombe, E. (1980) 'Whatever happened to the Zamindars?', in Hobsbawm, E. J. *et al*. (eds) *Peasants in History*, Oxford University Press, Delhi.

Whitehead, A. (1981) ''I'm Hungry Mum': the politics of domestic budgeting', in Young, K., Wolkowitz, C. & McCullagh, R. (eds) *Of Marriage and the Market*, CSE Books, London.

Whitehead, A. (1990) 'Food crisis and gender conflict in the African countryside', in Bernstein, H. *et al*. (eds) *The Food Question*, Earthscan, London.

Wolf, E. (1959) *Sons of the Shaking Earth*, University of Chicago Press, Chicago.

Wood, G. (1984) 'Provision of irrigation services by the landless – an approach to agrarian reform in Bangladesh', *Agricultural Administration*, 17, pp.55–80.

Wood, G. & Palmer-Jones, R. (1990) *The Water Sellers: a cooperative venture by the rural poor*, Kumarian Press, West Hartford, Conn.

World Bank (1988) *Rural Development: World Bank experience 1965–86*, The World Bank, Washington DC.

World Bank (1989a) *Sub-Saharan Africa: from crisis to sustainable growth*, IRBD/The World Bank, Washington DC.

World Bank (1989b) *India: poverty, employment and social services*, The World Bank, Washington DC.

World Bank (1990) *World Development Report 1990*, Oxford University Press, New York.

World Bank (1991a) *World Development Report 1991*, Oxford University Press, New York.

World Bank (1991b) *Poverty and Gender in India*, World Bank Country Study, The World Bank, Washington DC.

Wuyts, M., Mackintosh, M. & Hewitt, T. (eds) (1992) *Development Policy and Public Action*, Oxford University Press/The Open University, Oxford (Book 4 of this series).

Zamosc, L. (1989) 'Peasant Struggles of the 1970s in Colombia', in Eckstein, S. (ed.) *Power and Popular Protest: Latin American social movements*, University of California Press, Berkeley.

Acknowledgements

Grateful acknowledgement is made to the following sources for permission to reproduce material in this book:

Text

Chapter 1: Hartmann, B. & Boyce, J. K. (1983) *A Quiet Violence: view from a Bangladeshi village,* Zed Press, London; *Chapter 3:* Harriss, J. (1987) 'Capitalism and peasant production: the green revolution in India', *Peasants and Peasant Societies,* 1st edition, Penguin Books Ltd; *Chapter 12:* Bandypadhyay, D. P. 'Indra Lohar and the due process of law', Manohar Publications, Delhi; Manhhezi, A. 'Interviews with Mozambican peasant women', Centre of African Studies, Eduardo Mondlane University, Maputo; Omvedt, G. 'Women have to do double work', Zed Press, London.

Tables

Tables 1.1, 1.2: The World Bank (1990) *World Development Report,* © The World Bank; *Table 2.1:* from Kandiyoti, D. (1985) *Women In Rural Production Systems: problems and policies,* © UNESCO 1985; *Tables 2.2, 2.3, 2.4, 2.5:* Janvry, A. de, Sadoulet, E. & Wilcox Young, E. (1989) 'Land and labour in Latin American agriculture from the 1950s to the 1960s, *Journal of Peasant Studies,* April 1989, vol. 16, no. 3, reproduced by permission of Frank Cass & Co Ltd, London; *Tables 4.1, 4.2, 4.3:* The World Bank (1989) *Sub-Saharan Africa From Crisis to Sustainable Growth,* © The World Bank; *Table 12.1:* Scott, J. C. (1989) 'Everyday forms of peasant resistance' in Colburn, F. D. (ed.) Everyday Forms of Peasant Resistance, M. E. Sharpe Inc., New York; *Table 12.3:* Zamosc, L. (1989) 'Power struggles of the 1970s in Colombia', in Eckstein, S. (ed.) *Power and Popular Protest: Latin American social movements,* copyright © 1989 The Regents of the University of California.

Diagrams

Figure 8.1(a): from *Enhancing Agriculture in Africa: a role for U.S. Development Assistance,* 1st edition, September 1988, National Technical Information Service, Springfield, VA, © Congress of the U.S. Office of Technological Assessment; *Figure 8.1(b):* Farmer, G. & Wigley, T. M. L. (1985) *Climatic Trends for Tropical Africa,* report for the U.K. Overseas Development Administration, 136 pp, (limited distribution), Climatic Research Unit, University of East Anglia, Norwich.

Photographs and cartoons

Cover (background): Ed Parker/Still Pictures; *cover (inset):* Mark Edwards/Still Pictures; *Section title pages:* p.11: Oxfam; p.85: Mark Edwards/Still Pictures; p.137: Ron Giling/Still Pictures; p.249: Julio Etchart; *Figure 1.1:* Sharma Studios, New Delhi; *Figure 1.2:* Camera Press, photo by Alfred Gregory; *Figure 1.3:* Mark Edwards/Still Pictures; *Figure 2.1:* Sharma Studios, New Delhi; *Figure 2.2(both):* reproduced with permission from Rugendas, J. M., *Viagem pittoresca através do Brasil,* Livraria Martins Editora S.A., São Paulo; *Figure 2.3:* copyright © Romano Cagnoni; *Figure 2.4:* copyright © Janet Marsh-Penney; *Figure 2.5:* Judy Blankenship and *New Internationalist; Figure 2.6:* Hutchison Library © Michael MacIntyre; *Figure 2.7:* Oxfam; *Figure 2.8:* Paul Harrison/Panos Pictures; *Figure 3.1:* The Hulton-Deutsch Collection; *Figure 3.2:* Topham Picture Library; *Figure 3.3:* Maggie Murray/Format; *Figures 3.4, 3.6:* Sharma Studios, New Delhi; *Figures 3.5, 3.7:* John and Penny Hubley; *Figure 4.1:* Mary Evans Picture Library; *Figures 4.2, 4.3, 4.5:* Cartoons by Tony Namate from Zipperer, S. (1987) *Food Security: agricultural policy and hunger,* c/o ZIMFEP, Harare; *Figure 4.4(top):* UNICEF photo by Hewett; *Figures 4.4(right), 4.7(left and right):* Maggie Murray/Format; *Figures 4.4(bottom), 4.6, 4.7(far left):* Mark Edwards/Still Pictures; *Figures 5.1(top), 5.10(d):* Maggie Murray/Format; *Figure 5.1(bottom):* Tom Hanley; *Figure 5.2(top):* Ron Giling/Panos Pictures; *Figure 5.2(bottom):* Olivia Graham/Oxfam; *Figures 5.3, 5.8:* Oxfam; *Figures 5.4, 5.6:* Jeremy Hartley/Panos Pictures; *Figure 5.5(both):* Bernard Taylor/Oxfam; *Figure 5.7:* Sean Sprague/Panos Pictures; *Figure 5.9:* F. Mattidi/Centre for World Development Education/World Wide Fund for

Nature; *Figure 5.10(a):* Sharma Studios, New Delhi; *Figure 5.10(b):* Sue Darlow/Format; *Figure 5.10(c):* John and Penny Hubley; *Figure 6.1:* Anwar Hossain/Mark Edwards/Still Pictures; *Figure 6.2:* International Planned Parenthood Federation; *Figures 6.3, 6.6(both), 6.7:* Ron Giling/Still Pictures; *Figure 6.4(top):* Bernard Taylor/Oxfam; *Figure 6.4(bottom):* B. Storey/Oxfam; *Figure 6.5(top):* Carlos Reyes-Manzo/Andes Press Agency; *Figure 6.5(bottom):* Julio Etchart; *Figure 7.2(both):* Mark Edwards/Still Pictures; *Figures 7.4, 7.5, 7.12:* Oxfam; *Figure 7.6:* Sharma Studios, New Delhi; *Figures 7.8, 7.9:* Mark Edwards/Still Pictures; *Figure 7.10:* Cartoon by Laxman, R. K. (1979) *You Said It*, no.6, The Times of India; *Figure 7.11:* Laxman, R. K. (1968) *You Said It*, no.3, The Times of India, © R. K. Laxman 1968; *Figures 8.3(a), 8.4:* Mark Edwards/Still Pictures; *Figure 8.3(b):* Ron Giling/Still Pictures; *Figure 8.3(c):* Julia Rowley/Oxfam; *Figures 8.5, 8.9(both):* Maggie Murray/Format; *Figure 8.6:* Oxfam; *Figure 8.7:* FAO; *Figure 8.10:* Jeremy Hartley/Oxfam; *Figure 8.11:* Neil Cooper/Panos Pictures; *Figures 9.2 to 9.7:* Tim Allen; *Figure 9.8:* Maggie Murray/Format; *Figures 10.1 to 10.15:* photographs by Tim Allen; *Figures 11.1, 11.2, 11.10:* Mark Edwards/Still Pictures; *Figure 11.3:* Topham Picture Library; *Figure 11.4:* Tom Hanley; *Figure 11.5:* Cartoon by Laxman, R. K. (1971), *You Said It*, no.2, The Times of India, © R.K.Laxman 1971; *Figures 11.6, 11.7:* Cartoons by Laxman, R. K. (1979), *You Said It*, no.6, The Times Of India, © R. K. Laxman 1979; *Figure 11.8:* Cartoon by Laxman, R. K. (1968), *You Said It*, no.3, The Times of India, © R. K. Laxman 1968; *Figure 11.9(both):* Intermediate Technology/John Young; *Figure 11.11:* R.O. Cole/Oxfam; *Figures 12.1, 12.5, 12.6(bottom):* Julio Etchart; *Figure 12.2:* Melissa Parker; *Figure 12.3:* Tim Allen; *Figure 12.4:* Popperfoto; *Figure 12.6(top):* Melchor Lima (Ayaviri)/Tafos/Panos Pictures; *Figure 12.6(right):* Sebastián Turpo (Ayaviri)/Tafos/Panos Pictures; *Figure 12.7:* Sean Sprague/Panos Pictures; *Figure 12.8:* Jeremy Hartley/Panos Pictures; *Figure 12.9:* Oxfam.

Editors' and authors' acknowledgements

The editors are pleased that colleagues from Third World countries – Dr Jayati Ghosh and Professor Krishna Bharadwaj (India) and Professor Mahmood Mamdani (Uganda) – were able to collaborate in preparing this book. They are also grateful to Dr Hilary Standing of the University of Sussex for many useful comments on draft chapters.

Tim Allen would like to thank Paul Baxter, Melissa Parker and Helen Lambert for detailed comments on drafts of Chapter 10.

List of acronyms and abbreviations

AIDS	acquired immune deficiency syndrome
ANUC	Asociación Nacional de Usuarios Campesinos, Colombia
CPR	common property resources
EAP	economically active population
FAO	Food and Agriculture Organization (United Nations)
FRELIMO	Frente de Libertação de Mozambique (Mozambique Liberation Front)
GDP	gross domestic product
GNP	gross national product
ha	hectare
HIV	human immunodeficiency virus
HYV	High Yield Variety
IMF	International Monetary Fund
IRDP	Integrated Rural Development Programme (India)
MSF	Médecins Sans Frontières
NGOs	non-governmental organizations
NRA	National Resistance Army (Uganda)
OECD	Organization for Economic Co-operation and Development
PAF	Projet Agro-Forestier (Oxfam)
UK	United Kingdom
UN	United Nations
UNDP	United Nations Development Programme
UNHCR	United Nations High Commissioner for Refugees
UNICEF	United Nations (International) Children's (Emergency) Fund
USA	United States of America
USSR	Union of Soviet Socialist Republics
WHO	World Health Organization (United Nations)

Index

cultural differences
    and development agencies 274
    in health care 246
currency
    convertible 187
    exchange rates 73
    foreign exchange 156, 210
cycles of prosperity 123

D

dance as resistance 284
death *see* mortality
debt
    India 141
    and poverty 19
    sub-Saharan Africa 74, 75, 76
    West Africa 180, 181
    see also credit; loans
debt bondage
    and fictive kinship 127–8
    Latin America 38
    Peru 126
decentralized planning 162
defensive resistance of Ugandan
        peasants 212–15
deforestation
    Southern Rhodesia 178
    sub-Saharan Africa 73
    Uganda 211, 283
demography *see* population
dependency of poor people 18, 20, 297
deprivation *see* poverty
devaluation in sub-Saharan Africa
        188
development
    and action from below 286
    alternative 5–7
    appropriate 5–7
    centralization 266
    and gender 269
    effect on poverty 5
    skewed 144
    and structural change 286
    uneven and combined 29–30
development agencies
    and cultural differences 274
    and social relations 274
development projects
    and class 72, 269
    and conflict of interests 72, 265
    cost–benefit analysis 71, 72
    gender differences 72, 79
    and peasant households 72, 73
    sub-Saharan Africa 73, 78
    targeting women 80
developmental state in sub-Saharan
        Africa 74, 76

diamond mining 123
differentiation 278
    and households 122–8
    India 156
    social 30, 73, 78, 122–3, 192–3
    Uganda 199–202, 207, 213
disasters *see* crises
disease 218, **219**
    alcohol-related 228
    cattle, sub-Saharan Africa 177,
        193
    epidemics in Latin America 34
    resistance of children 228
    specific diseases: in Uganda 219,
        228; bilharzia 219, 228; cancer
        219, 245; hepatitis 228, 234;
        jaundice 228, 234, 235;
        trypanosomiasis (sleeping
        sickness) 193, 229; ulcers 228,
        235; worm infestation 228, 235
    see also health; illness
disguised resistance 284, 285, 286,
        288, 289
dispensaries, Uganda 228, 229,
        232–3, 234
dispersed cropping in sub-Saharan
        Africa 170, 173
distress migration 5, 284
distribution
    of income *see* income
    of land *see* land
    of power *see* power
    of resources *see* resources
diversification
    agriculture, sub-Saharan Africa
        170, 173
    livelihood strategies 23–4, 81–2,
        84, 124, 280; India 129, 140,
        153–4
diversity of societies, sub-Saharan
        Africa 65, 69, 214–15
divination *see* spirit mediums
division of labour
    households 93–9, 202–6
    international 30
    racial, sub-Saharan Africa 69
    sexual 4, 33, 46, 49, 79, 97, 100,
        126, 155, 184, 186 ; Bangladesh
        110; estate work 102, 103;India
        61, 62, 129–31, 150, 151, 152,
        257; Latin America 45, 49;
        Senegal 94–5, 104, 108;
        sub-Saharan Africa 78–82, 173,
        178; Tanzania 281–2; Uganda
        203, 204, 205
    social 25, 30
    spatial 30
domestic hierarchies in India 61
domestic industries in India 61

domestic work 4, 97, 99
    Bangladesh 110
    children 118, 119
    Senegal 97, 99, 104, 108
    Uganda 203
    see also reproduction
domination
    and resistance 285
    see also power
drama as resistance 284
drinking
    alcohol-related diseases 228
    by men 204
    as resistance 285
drinking groups in Uganda 200
drought 131, 156, 169, 199
drug therapy 233, 234
dry period crop loss 169
dry season work in sub-Saharan
        Africa 101, 180, 184
dryland farming areas 61, 266

E

East Asia
    agrarian reform 259–62
    industrialization 261
    poverty incidence 13–14
East India Company 51, 52
Eastern Europe, poverty incidence
        13–14
eating arrangements 108, 110
ecological crises in Uganda 210–12
ecological management in sub-
        Saharan Africa 165–6, 170,
        189–90, 297
ecological zones of sub-Saharan Africa
        166–73
economic disenfranchisement, India
        157–8
economically active population (EAP)
        of Latin America 43, 44, 45
economies of scale 57
Ecuador, income of households 47
education
    and socialization 118–20
    women 147
effective demand in India 158
Egypt, poverty incidence 15
El Salvador
    farm size 46
    income of households 47
elasticity of labour demand 150
employment
    generation 7, 270; India 139–64;
        through industrialization 40,
        56